Extractives Industry Law in Africa

Damilola S. Olawuyi

Extractives Industry Law in Africa

Springer

Damilola S. Olawuyi
Hamad bin Khalifa University
Doha, Qatar

ISBN 978-3-030-07386-2 ISBN 978-3-319-97664-8 (eBook)
https://doi.org/10.1007/978-3-319-97664-8

This Springer imprint is published by the registered company Springer Nature Switzerland AG
The registered company address is: Gewerbestrasse 11, 6330 Cham, Switzerland

To Oluwabunmi, Titilayo, and Oluwatoni Olawuyi, for their inspiration and support.

Preface

The African continent has for several decades accounted for a significant portion of the global production and supply of oil, gas and solid minerals. For example, Nigeria, Libya, Algeria, Egypt and Angola are historical giants in oil production in Africa, while Kenya, Zimbabwe, South Africa, Ghana, South Sudan, Namibia, Gabon, Cameroon, Tunisia, Equatorial Guinea, the Democratic Republic of the Congo and Côte d'Ivoire have attracted significant investment and income from mineral resources, ranging from gold, bauxite, cobalt, industrial diamond, phosphate rock, platinum-group metals, vermiculite and zirconium.

Despite the geological affluence of the continent, resource extraction and production in Africa has not fostered corresponding economic, social and environmental development. Many resource-rich African countries have fallen prey to the resource curse conundrum, which refers to the dichotomy between Africa's abundant resource wealth and the inability of its people to benefit therefrom. Poor resource management, lack of transparency and accountability, resource theft, governmental corruption, environmental pollution, land grabs and forced evictions from ancestral lands, poor labour conditions and lack of stakeholder engagement, among others, have exacerbated the cycles of poverty and conflict across the African continent.

To address these questions, several road maps, action plans, declarations, laws and frameworks have been adopted across Africa, at regional, national, provincial and municipal levels, to regulate petroleum and solid mineral production, marketing and distribution. The result has been a proliferation of extractive laws and regulations, to such an extent that one needs a "road map" to work through the legal labyrinth, when investing in extraction projects across Africa. The plenitude of sources makes the task of understanding the contents, scope and the follow-up policy measures prescribed by these legal instruments daunting and nearly insurmountable, even to the most zealous investor, student, lawyer, administrator or policymaker.

Furthermore, several African countries have developed separate, and at times overlapping, legal frameworks for petroleum and solid mineral sectors, making it difficult to gain a systemic and holistic understanding of the key intersections, opportunities, volatilities and challenges in the extractive industry value chain, covering both subsectors. A comparative analysis and evaluation of the African extractive legal and fiscal regime, in the light of emerging international law norms on business and human rights, has yet to receive a book-length, exhaustive and rigorous exposition and analysis. While earlier scholarships have focused on applicable legal regimes on petroleum, mining or environment, little or no attention has been paid to the wider extractive industry, specifically the need to reflect emerging global norms on systemic integration and regulatory coherence in the extractive sector value chain. This book is an attempt to fill that gap.

This book provides an authoritative exposition of the legal, fiscal and institutional frameworks applicable to the commercial development of petroleum and mineral resources in Africa. It explores the values, assumptions and guiding principles that underpin the extractive industries and explores how questions of resource ownership and control, human rights risks mitigation, accountability, corporate social responsibility, environmental justice and participatory development have been, and could be better, addressed.

Overall, the book aims to enhance an understanding of the nature and scope of extractive industries in Africa; appraise the general legal and fiscal framework governing African extractive industries; explore the laws and institutions that address specific sustainability issues such as gas flaring, spillage, pollution, water pollution, waste management and environmental health in African extractive industries; foster an understanding of the unique legal, policy and governance risks and challenges faced in African extractive industries; provide practical discussions on how corporations can identify, mitigate and prevent human rights risks when investing in African extractive industries; and unpack the legal reform processes required across the African continent to achieve a sustainable and rights-based extractive industry. Also provided is a discussion of the diverse institutions and structures that influence and govern the extractive industry.

The book is prepared in a user-friendly style to enhance its utility among its primary audience, namely students, corporations, investors, environmental and human rights administrators, advocates, policymakers, judges, international negotiators and government officials. It interweaves descriptive materials on the state of extant laws and institutions in the African extractive industries, with analytical evaluations of how international law could influence and inform thoughts on how best to revitalize the underpinning legal and fiscal regime in African extractive industries to make them more accountable, sustainable and rights-based.

Doha, Qatar
30 June 2018

Damilola S. Olawuyi

Acknowledgements

It remains for me to register my profound gratitude to many helping hands, without whom the publication of this book would not have been possible. First and foremost, my thanks and appreciation go to God the Almighty for the successful publication of this book. Secondly, I appreciate the kind motivation and support of the President and Founder of Afe Babalola University—Aare Afe Babalola, OFR, CON, SAN, FNIALS, FCIArb., LL.D—for his unflinching support to me and my family. Aare Afe Babalola's journey and global achievements provide reinforcing pillars and instructive pathways that one could only aspire to follow.

Furthermore, I am especially indebted to everyone at Qatar Foundation and Hamad Bin Khalifa University, Doha, Qatar, for their exceptional love and support over the years. I am deeply grateful for research funding and support from the Qatar National Research Fund, which significantly facilitated the initial framing of this book. I would also like to extend a word of special gratitude to Umair Dogar, Bolaji Akintoye and Irene Ekweozoh for providing remarkable and thoughtful research assistance; and to my assistants Curtis Snow, Moronkeji Kolawole and Tosin Adewunmi who have all worked tirelessly for the success of this book.

Finally, I would like to thank my dear wife Oluwabunmi and our amazing twin girls, Titilayo and Oluwatoni, for their love, support and inspiration. Thanks are also due to my exceptional students whose class contributions and ideas served as timely seeds that blossomed into this book. I thank you all for your support and kindness, and I commit this book to your scholastic minds.

I have endeavoured to state the position of the law as of 30 June 2018, although I have been able to take into account subsequent developments in one or two instances.

Contents

List of Abbreviations

ACHR	American Convention on Human Rights
AfCHPR	African Charter on Human and Peoples Rights
ATCA	United States Aliens Torts Act
ATI	Access to Information
BITs	Bilateral Investment Treaties
CEDAW	Convention on the Elimination of All Forms of Discrimination against Women
CER	Certified Emissions Reductions
CERD	Convention on Elimination of Racial Discrimination
CESCR	Committee on Economic, Social and Cultural Rights
CDM	Clean Development Mechanism
CFCs	Chlorofluorocarbons
CFU	Carbon Finance Unit
CO_2	Carbon dioxide
COP	Conference of the Parties
ECs	Extractive Companies
ECHR	European Convention on Human Rights
EIA	Environmental Impact Assessment
ERU	Emission Reduction Unit
ET	Emissions Trading
EU	European Union
FAO	Food and Agriculture Organization of the United Nations
FPIC	Free Prior Informed Consent
GEF	Global Environment Facility
GHGs	Greenhouse Gases
HRBA	Human Rights-Based Approach
HRIA	Human Rights Impact Assessment
ICCPR	International Covenant on Civil and Political Rights
ICESCR	International Covenant on Economic, Social and Cultural Rights
ICERD	International Convention on the Elimination of All Forms of Racial Discrimination
ICJ	International Court of Justice

IFC	International Finance Corporation
ILC	International Law Commission
ILO	International Labour Organization
IMF	International Monetary Fund
IOCs	International Oil Companies
IPCC	Intergovernmental Panel on Climate Change
IUCN	International Union for Conservation of Nature
JI	Joint Implementation
KPCS	Kimberley Process Certification Scheme
LDC	Least Developed Countries
LDCF	Least Developed Country Fund
LRTAP	Convention on Long-Range Transboundary Air Pollution
MARPOL	International Convention on the Prevention of Pollution by Ships
MEA	Multilateral Environmental Agreement
MIGA	Multilateral Investment Guarantee Agency
MOU	Memorandum of Understanding
MOE	Ministry of Environment
MOP	Meeting of the Parties
MPR	Ministry of Petroleum Resources
NGOs	Non Governmental Organizations
NIMBY	Not-in-My-Backyard
OAU	Organization of African Unity
OECD	Organisation of Economic Co-operation and Development
OHCHR	Office of the High Commissioner for Human Rights
PCBs	Polychlorinated biphenyls
PDD	Project Design Document
PDT	Participation Development Theory
PIC	Prior informed consent
POPs	Persistent organic pollutants
PPMs	Process and Production methods
SO_2	Sulphur dioxide
SALs	Structural adjustment loans
SEA	Strategic Environmental Assessment
UN	United Nations
UNCLOS	United Nations Convention on the Law of the Sea
UNCTAD	United Nations Conference on Trade and Development
UNECE	United Nations Economic Commission for Europe
UNEP	United Nations Environment Programme
UNESCO	United Nations Educational, Scientific and Cultural Organization
UNFCCC	United Nations Framework Convention on Climate Change
UNECE	United Nations Economic Commission for Europe
UNESCO	United Nations Educational, Scientific and Cultural Organization
UNDRIP	United Nations Declaration on the Rights of Indigenous Peoples
UNDG	United Nations Development Group
UNHRC	United Nations Human Rights Council

UDHR	Universal Declaration on Human Rights
VCLT	Vienna Convention on the Law of Treaties
VOCs	Volatile Organic Compounds
WBIP	World Bank Inspection Panel
WHO	World Health Organization
WMO	World Meteorological Organization
WSSD	World Summit on Sustainable Development
WTO	World Trade Organization
WWF	Worldwide Fund for Nature

Part I
Introductory Context and Principles

Chapter 1
Introduction

Abstract This chapter provides a conceptual overview of the extractives industry in Africa. It provides foundational definitions and information on the key processes and actors in the extractives industry, and then identifies many of the essential challenges that require dynamic legal solutions in the African extractives industry. These include legal challenges relating to infrastructure deficits, business and human rights, and environmental sustainability. The chapter also outlines the overall aim and structure of the book.

1.1 Background

This book provides an overview of the body of legislation, regulatory framework, institutions, industry guidelines, contractual arrangements and practices that govern the commercial development of oil, gas and solid minerals in Africa. It provides a multijurisdictional and systemic survey of the legal and institutional frameworks established to govern the commercial development of oil, gas and mineral resources across the African continent.

Although Africa is not one homogeneous geographical unit, the history, processes, volatilities, and governance challenges associated with the commercial development of oil, gas and minerals in the continent have been largely consistent and similar.[1] This book examines how these key processes and stages have been regulated across Africa; unpacks regulatory gaps and challenges in extractive sector investment and activities in Africa; and discusses contemporary instruments and norms aimed at addressing those challenges and gaps.

The extractives industry consists of people, companies, financial institutions, trade organizations and national authorities involved in removing raw materials, metals, coal, oil, mineral and natural resources from the earth for ultimate use by

[1] With 44 countries, Africa is the world's second-largest and second-most-populous continent covering about over 30 million square kilometres, which is more than three times the size of the United States, and was home to about 1.1 billion people in 2013. Africa accounts for about 15% of the world's human population. See National Geographic (2013).

D. S. Olawuyi, *Extractives Industry Law in Africa*,
https://doi.org/10.1007/978-3-319-97664-8_1

consumers.[2] The extractives industry comprises two main subsectors: petroleum and solid minerals.[3]

Examples of extractive processes include oil and gas exploration, solid minerals mining, dredging and quarrying.[4] With respect to the oil and gas subsector, it consists of operations that remove crude oil (or any related hydrocarbon) or natural gas, as it exists in its natural state in strata, from the earth. In the solid mineral sector, extractives operations include activities that remove naturally occurring metallic and non-metallic minerals, including coal, salt, quarry and pit material, and all rare and precious minerals and metals from the earth. The entire extractives industry value chain involves diverse complementary processes, activities and stakeholders that facilitate the extraction of natural resources from the earth (oil, metals, mineral and aggregates), and ultimately, processing and transporting them to end users.[5]

Extractive industry law therefore is the network of laws, institutions and fiscal regimes established to govern the commercial development of oil, gas and mineral resources. This includes rules and procedures relating to: the exploration or extraction of oil, gas or minerals; the acquisition or holding of a permit, licence, lease or any other authorization to carry out oil, gas or mineral exploration or extraction; and the institutions established at national, regional and supra national levels to oversee commercial exploration of oil, gas and minerals.[6]

This chapter provides a conceptual overview of the extractives industry in Africa. It provides foundational information and definitions, which would help readers to firmly grasp the underpinning concepts and terminologies relevant to the extractives industry. Section 1.2 of this chapter identifies many of the essential challenges that require dynamic legal solutions in the African extractives industry, most especially legal challenges relating to business, human rights, and environmental sustainability. It lays the groundwork for the detailed discussions on the extractives industry, which follows in this book. Section 1.3 outlines the overall aim and structure of the book.

---

[2] See Sigam and Garcia (2012), pp. 3–4. See also s. 21 of *The Nigeria Extractive Industries Transparency Initiative (NEITI) Act 2007* which defines an extractive industry company as "any company in Nigeria that is engaged in the business of prospecting, mining, extracting, processing and distributing minerals and gas, including oil, gold, tin, bitumen, diamonds, precious stones and such like; and includes any agency or body responsible for the payment of extractive industry proceeds to the federal government or its statutory recipient." See also Kenya (2016) which defines extractive industry as "any process that involves the extraction of raw materials from the earth to be used by consumers. It consists of any operations that remove metals, mineral and aggregates from the earth and includes oil and gas extraction, mining, dredging and quarrying." Examples of extractive industries in Kenya are oil and gas extraction, mining, dredging and quarrying.

[3] Sigam and Garcia (2012), pp. 3–4. Also Extractive Industries Transparency Initiative, EITI Glossary.

[4] Sigam and Garcia (2012), pp. 3–4.

[5] Ibid. See also Alba (2009), pp. 1–2.

[6] The entire value chain, *exploitation commerciale de pétrole, de gaz ou de minéraux*, is described in Canada's Extractive Sector Transparency Measures Act (S.C. 2014, c. 39, s. 376).

1.2 The African Extractives Industry: Legal Context and Overview

The extractives industry is the chief economic mainstay of many African countries.[7] For many African countries, a significant part of the gross domestic product (GDP) is derived from the extractives sector. For example, Nigeria, Libya, Algeria, Egypt, and Angola are historical giants in oil production in Africa,[8] while South Africa, Zimbabwe, Botswana, Democratic Republic of Congo, Ghana, Sierra Leone, Liberia, Namibia, Angola, Cameroon, Tunisia, Equatorial Guinea, Morocco, Kenya and Côte d'Ivoire have attracted significant investment and income from mineral resources, ranging from gold, lithium, rhodium, diamond, soda ash, bauxite, iron ore, cobalt, industrial diamond, phosphate rock, platinum-group metals, vermiculite, and zirconium.[9] Notably, the African continent has perennially accounted for a significant portion of global resource production and supply, establishing great potential for the extractive industry to catalyse economic development in many low-to-middle income African countries.[10] The revenue generated from the extractive industry has been a significant source of finance for public infrastructure development, human capital investments in education and health and the development of other economic sectors across the continent.

Despite these statistics however, resource extraction and production in many parts of Africa has not fostered corresponding economic, social, and environmental development and growth for citizens.[11] In African countries, resource extraction has been closely associated with resource mismanagement, conflicts and complex environmental and human rights challenges, arguably more so than in any other region.[12] For example, estimates indicate that Nigeria has had at least $400 billion of its oil revenue stolen or misspent since independence in 1960.[13] In addition to concerns relating to lack of transparency in the management of resource wealth, many African countries have failed to establish robust law and governance frameworks that can

[7] See United States Energy Information Administration (2013). See also KPMG (2007); National Geographic (2015).

[8] According to the 2017 BP Statistical Energy Survey, Africa had proven oil reserves of 128.438 billion barrels at the end of 2016, equivalent to 41.2 years of current production and 8.01% of the world's reserves. Similarly, Africa produced an average of 7892 thousand barrels of crude oil per day in 2016 or 8.6% of the world production. See British Petroleum (2017), also see African Development Bank (2017). See also UN Economic Commission for Africa (UNECA) (2011), pp. 10–12.

[9] UNECA (2011), pp. 10–12.

[10] In 2008, Africa produced about 483 tons of gold, or 22% of the world's total production. See UNECA (2011), pp. 10–12.

[11] See Sigam and Garcia (2012), p. 14, stating that in spite of their potential positive economic impacts, extractive activities tend to leave a strong environmental footprint that must be addressed.

[12] See for example wars and conflicts relating to illicit diamond mining in Sierra Leone, Liberia and many parts of Africa. Le Billon (2008). See also Elba (2017), pp. 18–25; Sigam and Garcia (2012), p. vii.

[13] Annan (2012).

address negative environmental impacts of resource extraction projects. For example, gold mining in the Migori Gold Belt, in western Kenya, has resulted in significant negative impacts on environmental and human health in that region. Studies show that the concentrations of heavy metals, mainly Hg, Pb and As, from gold sites are above acceptable levels.[14] Similarly, salt mining in Magarini District, Kilifi County of Kenya has elicited a myriad of human rights issues including: environmental pollution and degradation, forced evictions from ancestral lands, health hazards, unacceptable labor conditions, police intimidations and lack of stakeholder engagement among others.[15] These problems have reduced the benefits of extractive resources to the overall quality of lives of citizens.

These concerns are not peculiar to Kenya. According to a 2017 resource governance index, virtually all African countries score very low in terms of law and governance structure to promote transparency, accountability and resource maximization in the extractive sector.[16] According to the index, frontier energy jurisdictions such as Norway, United Kingdom, Canada and the United States rank on top because they have established laws and practices which ensure that extractive resource wealth benefits citizen. On the other hand, several African countries rank in the bottom ladder. Ghana is the strongest ranked African country and it only appears in the satisfactory category that requires reforms and improvements in some key areas. Other African countries such as South Africa, Burkina Faso, Cameroon, Mali, Morocco, Tanzania, Zambia and Mozambique feature in the weak category. While Angola, Algeria, Madagascar, Nigeria, Ethiopia, Egypt, Gabon, Guinea are all ranked in the poor category. In the failing category are Democratic Republic of Congo, Eritrea, Libya, Sudan, Equatorial Guinea and Zimbabwe categorized as having almost no governance framework to ensure resource extraction proceeds in a sustainable, transparent and equitable manner. This index is an accurate depiction of how resource governance across Africa remains weak, with very little potential to maximize and equitably allocate benefits of resource wealth to citizens.

Nigeria offers a particularly worrisome example of how resource extraction, if not properly regulated, could engender complex human rights concerns. Decades of oil spillage, gas flaring, and effluent discharge in Nigeria's Niger Delta have severely damaged the ecosystem and environment of that region.[17] Oil production in the Niger Delta has also resulted in significant environmental degradation of the region, arbitrary confiscation of traditional and indigenous lands without compensation, siting of resource processing projects in poor and vulnerable communities, lack of opportunities for participation by citizens in decision-making processes, criminalization and persecution of rights advocates and opposition movements (often labelled as economic saboteurs), governmental corruption and lack of accountability

[14] See Ogola et al. (2002), p. 141.

[15] Ocholla et al. (2013), p. 213.

[16] Natural Resource Governance Institute (2017).

[17] According to a 2011 report of the United Nations (UN), reversing many of the environmental and social consequences of oil spillage in the Niger Delta is possible but may take 25–30 years. UN Environment Programme (2011).

on resource use, and the absence of judicial and quasi-judicial remedies for victims of the above-mentioned problems.[18] With a life expectancy of less than 47 years; a modest household income of $1 a day, and the lack of access to clean water due to oil spills,[19] Nigeria has been rightly cited in several studies as a classic example of a resource cursed nation.[20] Generally, people living in Africa's resource- rich countries are 3% less literate, have shorter life expectancy by 4.5 years, and have higher rates of malnutrition among women and children, compared to other countries in the region.[21] The resource curse conundrum is therefore a complex and recurring phenomenon across resource rich African countries.[22]

Furthermore, extractive industries across Africa suffer from huge infrastructural deficits and decay.[23] Due to many years of corruption, sabotage, neglect and lack of proper maintenance, some of the essential infrastructure—such as pipelines, barge, cargo ship, refineries and road networks—that are essential for transporting, refining, storing and processing extracted resources are either non-existent, most especially across the West African region; or are preforming below optimal levels.[24] In Nigeria for example, an investor in the extractive sector faces logistical challenges relating to inadequate electricity supply; ageing refineries; timeworn pipeline infrastructure; and lack of motorable roads to sites of mineral deposits.[25] With the exception of South Africa,[26] these problems are noticeable across Africa, especially in Congo DR, Kenya, Ghana, Tanzania, Zimbabwe.[27] As capacity increases with new investments in the generation, transmission and distribution sectors, these challenges can themselves provide opportunities for investors to spearhead infrastructural rehabilitation and expansion projects that are required to connect extracted resources with refineries,

[18] See generally Oluduro (2014), p. 165 (providing an account of the nature and scope of the problems discussed); see also Human Rights Watch (1999), p. 47, discussing the impact of governmental corruption in the forms of systematic kickbacks for the award of contracts, special bank accounts in the control of the presidency, and fraudulent allocation of oil or refined products to political favourites, on sustainable resource governance in Nigeria.

[19] See UNICEF (2013).

[20] Sala-i-Martin and Subramanian (2003), p. 570; Mahler (2010), p. 25; van der Ploeg (2011), p. 366.

[21] Chuhan-Pole et al. (2017), p. xvii.

[22] Former UN Secretary General, Mr. Kofi Annan, rightly summarized the resource curse problem in Africa when he noted as follows: "Used wisely, these natural resource revenues could lead to sustainable economic growth, new jobs and investments in health, education and infrastructure. But sadly, history teaches us that a more destructive path is likely — conflict, spiraling inequality, corruption and environmental disasters are far more common consequences of resource bonanzas. The cliché remains true: striking oil is as much a curse as a blessing." Annan (2012). Also van der Ploeg (2011), pp. 366–420.

[23] Infrastructure covers physical infrastructure (rail, roads, ports, bridges, energy, pipelines, refineries) and soft infrastructure (telecommunications, electricity, logistics, and the institutions required to maintain physical infrastructures in place). See Ramdoo (2015).

[24] Collier (2011); also, Escribano et al. (2008).

[25] See KPMG (2017).

[26] According to the International Energy Agency (2014).

[27] See Ramdoo (2015), pp. 1–2. Also, Fister and Briceno-Garmendia (2010).

processing plants, and ultimately to end users.[28] The expansion potential of the African extractive industry, especially the midstream sector, can be enhanced if supported by robust and dynamic legal and governance frameworks that simplify and incentivize investment opportunities for potential investors.

Another key problem is several African countries is the perennial over dependence on oil and gas resources, despite overwhelming evidence of significant solid minerals deposits across the continent.[29] While the petroleum sector has received significant economic and regulatory attention for several decades, many African countries have failed to optimize the economic, social and sustainability potential of the solid minerals sector.[30] With the notable exceptions of South Africa, Zimbabwe, Kenya and Ghana that have robust and matured solid mineral industries and regulations, several African countries are only now waking up to the potential economic importance of solid minerals in the quest to combat poverty and under development in the continent.[31] For example, Nigeria loses about $40 billion annually in unexploited gold alone.[32] The result is the rise in illegal mining activities, under development of the continent's geological affluence; and perpetuation of poverty and employment in countries such as Nigeria with robust but untapped solid mineral resources.[33] For example, as much as 75% of the mining in Nigeria is estimated to be taking place illegally.[34] With good governance; coherent and transparent development of the entire extractive sector; and full respect for human rights, environment and traditional knowledge of resource-based communities, these revenues from extractive industries can have a significant impact on reducing poverty and boosting economic diversification and prosperity in the African continent.[35]

Furthermore, lack of technological and financial resources have hindered the rapid and commercial development of natural resources that have been found to have commercial prospects and viability.[36] In South Africa for example, exploration studies have found significant quantities of natural gas in shale formations in Western Karoo, Central Karoo and Eastern Karoo basins.[37] However, due to lack of experience and technological capacity required for leveraging hydraulic fracturing technologies in the country, shale gas development projects have not proceeded as hitherto expected. This same challenge has hindered the development of shale gas projects in Nigeria despite different exploratory studies that show the economic potential of shale

[28] OECD (2012).

[29] Sigam and Garcia (2012), pp. 10–11; also Banerjee et al. (2014).

[30] Ayodele et al. (2013) and David et al. (2016).

[31] Fayemi (2016).

[32] See Nigerian Investment Promotion Council, Solid Minerals.

[33] Ibid, stating that the Nigerian mining industry is underdeveloped, currently accounting for only 0.5% of Nigeria's GDP, and leading to the importation of commodities that can be produced locally.

[34] Lwanda (2017).

[35] Sigam and Garcia (2012), pp. 1–10.

[36] For excellent and detailed analysis of this, see van den Berg and Moster (2018), pp. 257–260.

[37] See PASA Reference No 12/3/219; PASA Reference No 12/3/220; PASA Reference No 12/3/221.

projects in Nigeria.[38] For African economies to diversify their energy mixes and tap into unexplored extractive resources, technological expansion and capacity development must be understood in Africa as urgent priorities. The quest for economic diversification and low carbon energy transition in Africa must be backed by dynamic legal innovations that can support technology absorption and deployment.[39]

Similarly, a legal hindrance to the full-scale development of the extractive industry across Africa is the artificial bifurcation of the regulatory frameworks, institutions and legal instruments applicable to the industry.[40] In virtually all surveyed African countries, separate legislation, institutions and knowledge resources exist respectively for the petroleum and solid minerals sectors. Across Africa, there are at least over hundred distinct committees, departments, and agencies established across different levels of government with mandates to respectively supervise and coordinate petroleum operations; oversee solid minerals development; and protect the environment.[41] With several agencies mandated to tackle various aspects of the extractive sector value chain, the perennial result is a one-pipe-at-a-time approach to extractive sector development.[42] Lack of coherent and holistic approach, coupled with frequent changes in governance arrangements, often mean that different agencies pursue parallel and at times conflicting programs without fully delineating the entire scope of the human rights, environmental, accountability and transparency challenges in the extractive industry.[43] This concern is further exacerbated by the fact that petroleum and mining law research and scholarship, including the teaching of related law courses, have equally progressed as separate and unrelated fields in the continent.[44]

---

[38] See Olawuyi (2014).

[39] See van den Berg and Moster (2018); also Olawuyi (2017).

[40] See Alba (2009), p. 2, calling for an integrated and comprehensive approach to managing the full extractive industry (EI) value chain, including all of the steps of EI development and impact in oil, gas, and mining industries.

[41] There are several examples across Africa. See for example Kenya's Ministry of Energy; The National Oil Corporation of Kenya (NOCK); Kenya's National Environment Council (NEC) and the National Environment Management Authority (NEMA) established under Kenya's Environmental Management and Coordination Act no. 8 of 1999. Also South Africa's Petroleum Agency; PetroSA (National Oil Company of South Africa); African Exploration Mining and Finance Corporation; and South Africa's Department of Environmental Affairs. In Nigeria, there is the Nigerian Ministry of Petroleum Resources; Ministry for Solid Minerals Development; Nigerian National Petroleum Corporation; National Environmental Standards and Regulations Enforcement Agency (NESREA) established under the NESREA Act; the Niger Delta Development Commission (NDDC) of Nigeria established under the NDDC (Establishment etc) (Amendment) Act of 2017, the Energy Commission of Nigeria, the National Oil Spill Detection and Response Agency (NOSDRA),National Water Resources Institute (NWRI) and the Nigerian Tourism Development Corporation (NTDC). Recently the Government of Nigeria established the Hydrocarbon Pollution Restoration Project (HYPREP) to investigate, evaluate and make recommendations on the clean-up of hydrocarbon impacted sites in the Niger Delta region of Nigeria.

[42] See Murombo (2015), p. 323; also Roberts (2014), pp. 1–4.

[43] See van den Berg and Moster (2018), pp. 254–256; Drohan (2012); Halvor et al. (2006), pp. 1–20.

[44] On the need for cohesiveness and integration in energy law research and teaching, see Heffron et al. (2016), pp. 346–356.

Over the last decade, international law has increasingly recognized the need for a coherent, coordinated and sustainable utilization and development of natural resources, most especially the need to integrate transparency, accountability and respect for human rights into policy frameworks on resource utilization to avoid unintended negative consequences of resource exploitation.[45] This is part of broader theoretical debates and efforts aimed at coordinating the systemic integration of human rights and sustainable development goals in international law to ensure regime interaction and policy harmonization.[46] International law has increasingly promoted a rights-based understanding of the concept of corporate responsibility (CSR) and risk management.[47] This requires private actors and business enterprises to have a degree of responsibility not only for the commercial or economic consequences of their activities, but also for the social and environmental implications.[48] International law underscores the importance of mainstreaming human rights norms into the design, approval, finance and implementation of resource extraction projects to avoid negative human rights impacts.[49] Intricately intertwined with the concepts of CSR and risk management is the idea of sustainable development which emphasizes the need for investments or resource utilization that meet human needs, while preserving the abilities of future generations to meet their own needs.[50]

Despite increased global attention on environmental and human rights impacts of resource utilization, especially in mining and oil and gas production activities, it was only recently that an explicit human rights-based approach began to emerge in legal regimes on resource management and governance in Africa. Since 2006, African policy leaders have increasingly sought to introduce rights-based governance models and language into oil production and mineral extraction policies and regimes.[51] The most recent of these is the *Resolution on a Human Rights-Based*

---

[45] See Alba (2009), p. 2, calling for an integrated and comprehensive approach to managing the full extractive industry (EI) value chain, including all the steps of EI development and impact in oil, gas, and mining industries; also Ruggie (2011); United Nations Common Understanding, HRBA Portal; also Dashwood (2012).

[46] Mining Working Group at the United Nations (2014); Macias (2010), pp. 10–12; MacKay (2004), p. 43; Petersmann (2002), p. 626. See also Petersmann (2004); International Law Commission (2006), pp. 11–17; Mclachlan (2005); Tzevelekos (2010), p. 631.

[47] For a summary of these efforts, see Lindsay (2013); United Nations Conference on Trade and Development (2009); also Frey (1997).

[48] Saether and Aguilera (2008).

[49] See Olawuyi (2016), pp. 1–25.

[50] The Brundtland Report (1987).

[51] In 2006, the Southern African Development Community (SADC) adopted a "Framework for the Harmonization of Mining Policies Standards and Regulatory Frameworks" to create policy guidelines in key areas such as governance, environmental management and public participation. Similarly, in 2007, the United Nations Economic Commission for Africa (ECA) convened the *Big Table* meeting on "Managing Africa's Natural Resources for Growth". The Big Table meeting triggered initiatives, such as the Extractive Industries Transparency Initiative (EITI++), and the International Study Group to Review Africa's Mining Regimes (ISG), aimed at establishing human rights safeguards in resource extraction projects. See United Nations Economic Commission for Africa, Africa Review: Report on Mining (Executive Summary) at 7–12.

Approach to Natural Resources Governance adopted in 2012 by the African Commission on Human and Peoples' Rights (the "2012 African Resolution").[52]

As evidenced by the 2012 African Resolution and other recent efforts in the continent on CSR and human rights in the extractive sector, the obligations of IOCs to consider and prevent environmental, human rights and social impacts of resource extraction in the African continent is increasingly taking an important dimension.[53] In response to concerted international call for action, over the last decade, African countries have progressively enacted laws aimed at enhancing accessibility, transparency, accountability, environmental protection and respect for human rights in the extractive sector.[54] Some of these requirements are scattered in a wide range of instruments which makes it difficult for even the most diligent investor to clearly identify the key requirements across petroleum and mining sectors. There is therefore a strong business case, in terms of cost, reputation and effectiveness for prospective participants in African extractive markets to gain a detailed knowledge of the range of participation and compliance requirements across these two sectors in order to identify areas of convergence, divergence and possible risk points.

This book provides an in-depth exploration of the legal, institutional and fiscal framework and regimes that promote accountability, sustainability and efficiency in the extractives sector. Taking the extractive industries in Africa from a systemic and multi-disciplinary perspective, the book highlights the need for dynamic legal innovations to address business, human rights and environmental risks associated with oil, gas and mineral extraction in the African continent.

1.3 Aim, Scope and Structure of the Book

As noted earlier, most of the existing legal scholarship on the extractive industry in Africa have focused on specific and sectorial issues in the industry (for example petroleum, mining, human rights, environment or infrastructure) and/or specific country, and have therefore not provided comparative and multijurisdictional perspectives and evaluation of crosscutting themes, laws, institutions, and practical requirements on conducting extractive investments across Africa.[55] This analytical bifurcation and the ensuing regulatory disconnect have not allowed for systemic and multisector perspectives on the key regional, national and local actors and institutions who play related and pivotal roles in promoting transparency, accountability and efficiency across the African extractive industry as a whole.[56] Given the overlapping legal, governance, accountability and human rights challenges facing the

[52] Resolution on a Human Rights-Based Approach to Natural Resources Governance (2012).

[53] See Olawuyi (2015); also Ogola et al. (2002), p. 141; UN Environment Programme (2011), p. 12.

[54] Olawuyi (2015).

[55] See Elbra (2017), Zillman et al. (2002), Franklin and Kaplan (1982), Atsegbua (2004), Dale et al. (2005), Badenhorst and Mostert (2005), Mostert (2012) and Usa (2009).

[56] On the importance of systemic innovation, see Godden et al. (2018), pp. 13–15.

petroleum and solid mineral sectors across the continent, a systemic and coherent examination of governance and institutional settings, including opportunities for multisector collaborations is warranted. This book fills this gap. It offers an exhaustive and book length exposition of the legal regimes and risks in African extractives investments from a holistic perspective that has not been offered in existing literature. The book takes a systemic and multisector view of the African extractive industry; and discusses the opportunities, legal requirements, guidelines and risks in executing petroleum and solid mineral projects and investments in the African continent.

While this book offers a scan of the sources and underpinning principles of extractive industry law in the African continent, it is clearly acknowledged that the substantive chapters cannot unpack and analyse every applicable legislation and instruments in all African countries. The central organizing principle and aim of this book therefore is to enhance an understanding of the guiding principles of laws, applicable instruments, fiscal framework and institutions that govern African extractive industries; analyse the nature and scope of extractive industries in Africa; explore specific sustainability issues such as gas flaring, spillage, pollution, water pollution, waste management, extractive sector transparency, and infrastructural development in African extractive industries; unpack the unique legal, policy and governance risks and challenges faced in African extractive industries; provide practical viewpoints on how corporations can identify, mitigate and prevent human rights risks when investing in African extractive industries; and provide perspectives on the nature of legal protection required to achieve a sustainable and rights-based extractives industry in the continent.

1.4 Structure

The book is organized into three parts to reflect a transition from conceptual and theoretical issues to practical and substantive issues. Part I, consisting three chapters, provides introductions on the meaning, nature, scope and participation requirements in African extractives industries. Chapter 2 examines the meaning, nature and key activities in the petroleum and solid mineral sectors of the African extractive industry. Chapter 3 examines legal frameworks applicable to the transportation of petroleum and mineral resources. The key legal requirements for the processing and transportation of petroleum and mineral resources will be discussed under two headings: physical and associated infrastructure.

Apart from ownership regime and structure, it is pertinent for investors and stakeholders to have a proper grasp of laws applicable to extractive investments in target African countries to avoid running into legal risks and hurdles. It is particularly important to understand specific laws that stipulate participation requirements for extractives operations to avoid any default. This is the focus of Part II. Part II, comprising six chapters, provides detailed analysis and evaluation of laws, regulations and jurisprudence on obtaining authorizations and permits and interests in

African extractive industries. Chapter 4 surveys the applicable ownership theories and regimes underpinning the African extractive industry.

Chapter 5 introduces the sources of law, regulations and governance mechanisms relating to African extractives investments. It discusses the various genres and bodies of law that contribute to extractive sector activities and investments across the African continent. Chapter 6 surveys and evaluates governance and institutional arrangements across Africa for the regulation and supervision of the extractive sector. It outlines the key institutions, ministries, agencies and government entities that supervise, oversee and approve participation licenses and contracts in frontier extractive regimes across Africa.

Chapter 7 examines the key distinctions between the types of interests granted by African regimes for petroleum, mining and metallurgical operations. To be examined are the nature, scope and fundamentals of freehold, leasehold, mineral, surface and royalty interests. Chapter 8 discusses the fundamentals of key contractual and fiscal arrangements utilized in African regimes to convey titles and interests. This chapter will examine the nature and scope of concessions, joint venture contracts, production sharing contract, service contracts, farm out contracts, pooling agreements, unitization agreements, oil sale contracts and mining development agreements. Chapter 9 evaluates legal requirements on local content, procurements, and product mandating in African regimes. It surveys the key drivers of, and approaches to, local content policies in extractive jurisdictions in Africa.

A theoretical thread that runs through this proposed book is the need to anticipate and avoid environmental, social and human rights risks in extractive investments in Africa. Given the pivotal nature of human rights risks identification and mitigation in the extractives industry, it is pertinent to analyze and discuss nature of such risks and how best to mitigate and manage them. This is the focus of Part III.

Part III, comprising two chapters, discusses environmental, social and business risks associated with extractives investments in Africa. Chapter 10 unpacks key environmental challenges and risks associated with extractive investments in Africa. The impacts of gas flaring, oil spillage, effluent discharge, strip mining, underground mining and coal mining on biodiversity, air quality, water resources and human rights are examined in this chapter. Chapter 11 discusses commonly encountered legal, business and human rights risks in African extractive industries. It discusses how business enterprises can avoid regulatory fines and reputational backlash associated with extractive projects, by incorporating human rights due diligence into corporate risk management processes. There is a strong business case, in terms of cost, reputation and effectiveness for doing so. This chapter examines contract negotiation, human rights due diligence and general corporate risk management processes steps through which investors can identify, mitigate and address their risk exposure in the implementation of extractives projects in Africa.

1.5 Conclusion

There is abundant opportunity for cross fertilization and exchange of best practices between the key actors, institutions and stakeholders operating in the petroleum and solid mineral sectors of African extractive industries. For example, in countries such as Nigeria, Libya and Algeria that have robust and mature petroleum sectors, but emerging solid minerals sectors, key lessons on how governance and regulatory arrangements may be improved in solid minerals development to avoid the perennial pitfalls in petroleum sector management will be pivotal. The same applies to countries such as South Africa, Congo DR, Ghana, Kenya, Liberia and Zimbabwe that have significant and mature solid mineral sectors and increasingly growing portfolio of petroleum sector developments. A systemic and multisector linkage of the extractives industry can ensure that regulation, governance and capacity development in the sector are transformed, holistic, coordinated, well informed, effective, and underpinned by lessons from extant governance arrangements.

The next chapter provides an overview of the key activities, actors, and opportunities in the African extractives industry.

References

African Commission on Human and Peoples' Rights (2012) Resolution on a human rights-based approach to natural resources governance. 2012. 51st Sess. http://www.achpr.org/sessions/51st/resolutions/224. Accessed 9 June 2018

African Development Bank (2017) African Economic Outlook 2017. https://www.afdb.org/fileadmin/uploads/afdb/Documents/Publications/AEO_2017_Report_Full_English.pdf. Accessed 13 June 2018

Alba E (2009) Extractives industry value chain: a comprehensive integrated approach to developing extractive industries. World Bank: Extractives Industries for Development Series #3

Annan K (2012) Momentum rises to lift Africa's resource curse. https://www.nytimes.com/2012/09/14/opinion/kofi-annan-momentum-rises-to-lift-africas-resource-curse.html. 4 September 2012

Atsegbua L (2004) Oil and gas law in Nigeria, theory and practice, 2nd edn. New Era Publishers, Lagos

Ayodele O, Akongwale S, Nnadozie U (2013) Economic diversification in Nigeria: any role for solid mineral. Mediterranean J Soc Sci 4(6):691

Badenhorst PJ, Mostert H (2005) Mineral and petroleum law of South Africa. Juta and Company (Pty) Press, Sandton

Banerjee S, Romo Z, McMahon G, Toledano P, Robinson P, Arroyo I (2014) The power of the mine: a transformative opportunity for Sub-Saharan Africa. World Bank, Washington, DC

BP Statistical Review of World Energy (2017) https://www.bp.com/content/dam/bp/en/corporate/pdf/energy-economics/statistical-review-2017/bp-statisticalreview-of-world-energy-2017-full-report.pdf. Accessed 13 June 2018

Canada's Extractive Sector Transparency Measures Act (S.C. 2014, c. 39, s. 376)

Chuhan-Pole P, Dabalen A, Christopher Land B (2017) Mining in Africa: are local communities better off. World Bank, Washington, DC

Collier P (2011) Building an African infrastructure. Finance Dev 48(4):18–21

Dale M, Bekker MO, Bashall LFJ, Chaskalson M, Dixon C, Grobler GL, Loxton CDA (2005) South African mineral and petroleum law. LexisNexis Butterworths, Durban
Dashwood H (2012) The rise of global corporate social responsibility: mining and the spread of global norms. Cambridge University Press, Cambridge
David O, Noah O, Agbalajobi S (2016) An empirical analysis of the contribution of mining sector to economic development in Nigeria. Khazar J Humanit Soc Sci 19(1):88–104
Drohan M (2012) The 9 habits of highly effective resource economies: lessons for Canada. Canadian International Council, Toronto
Elbra PA (2017) Governing African gold mining: private governance and the resource curse. Springer, Berlin
Escribano A, Guasch J, Pena J (2008) Assessing the impact of infrastructure quality on firm productivity in Africa: cross country comparisons based on investment climate surveys from 1999 to 2005. Policy Research Working Paper 5191. World Bank, Washington, DC
Extractive Industries Transparency Initiative, EITI Glossary. https://eiti.org/glossary. Accessed 9 June 2018
Fayemi K (2016) Nigeria's solid minerals sector: alternative investment opportunities. Chatham House, London. https://www.chathamhouse.org/sites/files/chathamhouse/events/2016-05-19-Nigeria-solid-minerals-appg-transcript.pdf. Accessed 9 June 2018
Fister V, Briceno-Garmendia (2010) Africa's infrastructure: a time for transformation. World Bank, Washington, DC
Franklin BLS, Kaplan M (1982) The mining and mineral laws of South Africa. Butterworths, Durban
Frey B (1997) The legal and ethical responsibilities of transnational corporations in the protection of international human rights. Minnesota J Glob Trade 6:153
Godden L, Paddock L, Roggenkamp M (2018) Introduction. In: Zillman D, Godden L, Paddock LR, Roggenkamp M (eds) Innovation in energy law and technology: dynamic solutions for energy transitions. Oxford University Press, Oxford
Halvor M, Moene K, Torvik R (2006) Institutions and the resource curse. Econ J 116(508):1–20
Heffron RJ, Roberts P, Cameron P, Johnston A (2016) A review of energy law education in the UK. J World Energy Law Bus 9(5):346–356
Human Rights Watch (1999) The price of oil: corporate responsibility and human rights violations in Nigeria's oil producing communities. Human Rights Watch, Washington, DC http://hrw.org/reports/1999/nigeria/nigeria0199.pdf. Accessed 9 June 2018
International Energy Agency (IEA) (2014) World Energy Outlook 2014 special report: African Energy Outlook
International Law Commission (2006) Fragmentation of international law: difficulties arising from the diversification and expansion of international law, 58th General Assembly Session. UN Doc A/CN.4/L.682
Kenya's Environmental Management and Coordination Act no. 8 of 1999
Kenya (2016) Ministry of Mining, Draft National Extractive Policy. http://www.ke.undp.org/content/dam/kenya/docs/IEG/National%20Extractives%20Policy%20(draft).pdf. Accessed 12 Aug 2018
KPMG (2007) Nigerian mining sector brief. https://assets.kpmg.com/content/dam/kpmg/ng/pdf/advisory/ng-Nigerian-Mining-Sector.pdf. Accessed 9 June 2018
KPMG (2017) Oil and gas in Africa: Africa's reserves, potentials and prospects. https://www.resourcedata.org/hr/document/rgi-oil-and-gas-in-africa-africa-s-reserves-potential-and-prospects. Accessed 23 Sept 2017
Le Billon P (2008) Diamond wars? Conflict diamonds and geographies of resource wars. Ann Assoc Am Geogr 98(2):345–372
Lindsay R et al (2013) Human rights responsibilities in the oil and gas sector: applying the UN guiding principles. J World Energy Law Bus 6(1):13
Lwanda GC (2017) Can the Nigeria solid minerals development fund deliver? World Policy
Macias R (2010) Public participation in energy and natural resources development: a theory and criteria for evaluation. Canadian Institute of Resources Law Occasional Paper No. 34

MacKay F (2004) Indigenous peoples' right to free, prior and informed consent and the World Bank's extractive industries review. Sustain Dev Law Policy 4(2):43
Mahler A (2010) Nigeria: a prime example of the resource curse? Revisiting the oil-violence link in the Niger Delta. German Institute of Global and Area Studies 25
Mclachlan C (2005) The principle of systemic integration and Article 31(3)(C) of the Vienna Convention. Int Comp Law Q 54:279
Mining Working Group at the United Nations (2014) A rights-based approach to resource extraction in the pursuit of sustainable development. Advocacy brief. https://miningwg.com. Accessed 9 June 2018
Mostert H (2012) Mineral law: principles and policies in perspective. Juta and Company (Pty) Press, Sandton
Murombo T (2015) Regulating energy in South Africa: enabling sustainable energy by integrating energy and environmental legislation. J Energy Nat Resour Law 33(4):320–348
National Environmental Standards and Regulations Enforcement Agency Act (NESREA) Act
National Geographic (2013) Getting to know Africa: 50 interesting facts. http://voices.nationalgeographic.com/2013/10/31/getting-to-know-africa-50-facts/. Accessed 9 June 2018
National Geographic (2015) Africa: resources. https://www.nationalgeographic.org/encyclopedia/africaresources/. Accessed 23 Sept 2017
Natural Resource Governance Institute (NRGI) (2017) 2017 Resource governance index. https://resourcegovernance.org/sites/default/files/documents/2017-resource-governance-index.pdf. Accessed 9 June 2018
Nigeria Extractive Industries Transparency Initiative (NEITI) Act 2007
Nigeria's Niger Delta Development Commission (Amendment) Act of 2017
Nigerian Investment Promotion Council. Solid Minerals. https://www.nipc.gov.ng/?wpfb_dl=11. Accessed 4 June 2018
Ocholla G, Bunyas M, Asoka G, Pacha O, Mbugua H, Mbuthi P, Mbiti S, Wendo H, Kamau P (2013) Environmental issues and socio-economic problems emanating from salt Mining in Kenya; a case study of Magarini District. Int J Humanit Soc Sci 3(3):213–223
Ogola J, Mitullah W, Omulo M (2002) Impact of gold mining on the environment and human health: a case study in the Migori Gold Belt, Kenya. Environ Geochem Health 24:141–158
Olawuyi D (2014) Hydraulic fracturing technology and shale gas production in Nigeria: legal and sustainability assessment. Petr Technol Dev J 4(2):38–40
Olawuyi D (2015) Increasing relevance of right-based approaches to resource governance in Africa: shifting from regional aspiration to local realization. McGill Int J Sustain Dev Law Policy 11(2):113–158
Olawuyi D (2016) The human rights based approach to carbon finance. Cambridge University Press, Cambridge
Olawuyi D (2017) From technology transfer to technology absorption: addressing climate technology gaps in Africa. J Energy Nat Resour Law 36:1
Oluduro O (2014) Oil exploitation and human rights violations in Nigeria's oil producing communities. Afrika Focus 25(2):160
Organisation for Economic Co-operation and Development (2012) Mapping support for Africa's infrastructure investment. OECD, Paris
Petersmann E (2002) Time for a United Nations 'Global Compact' for integrating human rights into the law of worldwide organizations: lessons from European integration. Eur J Int Law 13(3):621
Petersmann E (2004) The 'human rights approach advocated by the UN high commissioner for human rights and by the International Labour Organization: is it relevant for WTO law and policy? J Int Econ Law 7(3):605
Ramdoo I (2015) Synergising and optimising mineral infrastructure in regional development strategies. E15 Expert group on trade and investment in extractive industries
Roberts CWJ (2014) Extractive sector regulations and policy in Africa: old practices and new models for change. North-South Institute, Davie

Ruggie J (2011) United Nations guiding principles on business and human rights: implementing the United Nations 'Protect, Respect and Remedy' framework. UN Document A/HRC/17/31

Saether K, Aguilera R (2008) Corporate social responsibility in a comparative perspective. In: Crane A (ed) The Oxford handbook of corporate social responsibility. Oxford University Press, Oxford

Sala-i-Martin X, Subramanian A (2003) Addressing the natural resource curse: an illustration from Nigeria. J Afr Econ 22(4):570

Sigam C, Garcia L (2012) Extractive industries: optimizing value retention in host countries. United Nations Conference on Trade and Development, Geneva

Tzevelekos V (2010) The use of article 31(3)(c) of the VCLT in the case law of the ECtHR: an effective anti-fragmentation tool or a selective loophole for the reinforcement of human rights teleology? Between evolution and systemic integration. Mich J Int Law 31:621

UN Economic Commission for Africa (2011) Minerals and Africa's development: the international study group report on Africa's mineral regimes. UN Economic Commission for Africa, Addis Ababa

UN Environment Programme (2011) Environmental assessment of Ogoniland. UN Environment Programme, Nairobi https://postconflict.unep.ch/publications/OEA/UNEP_OEA.pdf. Accessed 9 June 2018

United Nations. The human rights based approach to development cooperation: towards a common understanding among UN agencies. UN practitioners' portal on human rights based approaches to programming. hrbaportal.org/the-human-rights-based-approach-to-development-cooperation-towards-a-common-understanding-among-un-agencies. Accessed 9 June 2018

United Nations Children's Fund (UNICEF) (2013) At a Glance: Nigeria. Statistics report. https://www.unicef.org/infobycountry/nigeria.html. Accessed 9 June 2018

United Nations Conference on Trade and Development (UNCTAD) (2009) The social responsibility of transnational corporations. http://unctad.org/en/Docs/poiteiitm21_en.pdf. Accessed 2 Dec 2015

United Nations Economic Commission for Africa. Africa review: report on mining (Executive summary). https://www.uneca.org/sites/default/files/PublicationFiles/aficanreviewreport-on-miningsummary2008.pdf. Accessed 9 June 2018

United Nations World Commission on Environment and Development Report (The Brundtland Report) (1987) Our common future. Oxford University Press, Oxford

United States Energy Information Administration (2013) Oil and natural gas in Sub-Saharan Africa. https://www.eia.gov/pressroom/presentations/howard_08012013.pdf. Accessed 9 June 2018

Usa I (2009) Kenya mining laws and regulations handbook. International Business Publications

Van den Berg H, Moster H (2018) Challenges to regulating hydraulic fracturing in South Africa: technological innovation and law making for climate change at the cross roads. In: Zillman D, Roggenkamp M, Paddock L, Godde L (eds) Innovation in energy law and technology: dynamic solutions for energy transitions. Oxford University Press, Oxford

Van der Ploeg F (2011) Natural resources: curse or blessing? J Econ Lit 49(2):366

Zillman DM, Lucas A, Pring G (eds) (2002) Human rights in natural resource development: public participation in the sustainable development of mining and energy resources. Oxford University Press, Oxford

Chapter 2
Nature and Scope of the African Extractives Industry

Abstract This chapter provides an overview of the origin, nature, scope and attributes of the two main subsectors of the extractives industry: petroleum and solid minerals. After providing conceptual clarifications on the meaning and attributes of petroleum and solid minerals, the chapter unpacks the origin of petroleum and minerals in general, as well as their early discoveries in the African continent; outlines the key stages, processes and investment opportunities involved in the discovery and exploration of petroleum and mineral resources in Africa; and discusses the role of law in the regulation and administration of petroleum and solid minerals production in Africa.

2.1 Introduction

Despite some similarities in the nature, operation and evaluation methods in petroleum and solid mineral sectors of the extractives industry, there are several distinguishing characteristics and features in the attributes, exploration methods, production, transportation, commercialization and decommissioning strategies respectively applicable to each. After providing detailed conceptual clarifications on the meaning and attributes of petroleum and solid minerals, the chapter will unpack the key stages, processes and investment opportunities involved in the discovery and exploration of petroleum and mineral resources in Africa.

At the end of this chapter, you should be able to: provide basic definitions of petroleum and minerals; understand the nature and origin of petroleum and minerals in general, as well as their early discovery in the African continent; understand the physical properties of petroleum and minerals found in your country; appreciate the importance of petroleum as well as the solid minerals sector to the African continent; analyse the stages and processes associated with petroleum, natural gas and solid minerals development, as well as the major sectors in these industries; and appreciate the role of law in the regulation and administration of these industries.

D. S. Olawuyi, *Extractives Industry Law in Africa*,
https://doi.org/10.1007/978-3-319-97664-8_2

2.2 Petroleum

Petroleum is a naturally occurring mixture of hydrocarbon compounds, trapped in underground rock, oily and inflammable in character.[1] It is a thick, flammable, yellow to black mixture of gaseous, liquid and solid hydrocarbons, that occurs naturally beneath the earth's surface. The word 'petroleum', itself comes from the Latin words '*petra*', which means rock, and '*oleum*', meaning oil.[2] Petroleum is mainly fossil fuel that has been formed from long-buried plants and microorganisms.[3] Fossil fuels are formed from prehistoric plants and animals that lived hundreds of millions of years ago, died and fell to the bottom of the sea where they remained buried under layers of accumulating sediment. As additional sediment layers built up over these organic deposits, the material was subjected to increasing temperatures and pressures. The plant and animal remains were 'cooked' by this process, and slowly changed into organic compounds. Over millions of years, these physical conditions chemically transformed the organic material into hydrocarbons, *i.e* organic compounds consisting of hydrogen and carbon atoms. In its compound form, petroleum is comprised primarily of hydrogen and carbon, which is where it derives its other nomenclature of 'hydrocarbon'.[4] Consequently, oil and gas have been formed as a result of the continuous processes of organic materials under the soil over a long period of time.

In its solid form, petroleum could take the form of shale, tar sands or bitumen.[5] In liquid form, it is referred to as oil, while in its gaseous form it is known as natural gas.[6] Oil and gas come in different grades or qualities, which are often classified by the location or origin, as well as by sweetness and heaviness. The sweetness of oil

---

[1] Martin and Kramer (2009), p. 704.

[2] Ibid.

[3] Bazhenova (2009).

[4] Ibid, see also American Association of Petroleum Geologists.

[5] South Africa's law excludes shale from the definition of petroleum. It notes that Petroleum means "any liquid, solid hydrocarbon or combustible gas existing in a natural condition in the earth's crust and includes any such liquid or solid hydrocarbon or combustible gas, which gas has in any manner been returned to 35 such natural condition, *but does not include coal, bituminous shale or other stratified deposits from* which oil can be obtained by destructive distillation or gas arising from a marsh or other surface deposit" See South Africa's Mineral and Petroleum Resources Development Act (MPRDA), Act 28 of 2002.

[6] Section 1 of Kenya's *Petroleum (Exploration and Production) Act* defines petroleum as "mineral oil and includes crude oil, natural gas and hydrocarbons produced or capable of being produced from oil shales or tar sands." While the definition in the Kenyan law recognizes oil shale as petroleum, the Nigerian law excludes shale oil. Section 15 of the Nigerian Petroleum Act, defines petroleum as "mineral oil (or any related hydrocarbon) or natural gas as it exists in its natural state in strata, and does not include coal or bituminous shales or other stratified deposits from which oil can be extracted by destructive distillation." The Nigerian approach is consistent with the definition in the Petroleum (Production) Act 1934 of the United Kingdom which defines petroleum as including "any mineral oil or relative hydrocarbon and natural gas existing in its natural condition in strata, but does not include coal or bituminous shales or other stratified deposits from which oil can be extracted by destructive distillation."

refers to the amount of sulphur in the oil. Oil with less sulphur is referred to as 'sweet', which is containing little sulphur, and requires less processing before use. Oil with more sulphur is referred to as 'sour' that is containing a substantial amount of sulphur.[7] The heaviness of oil refers to its density. Lighter crude can be refined into higher value products, such as the gasoline (or petrol) used by car owners. Heavier crude flows more slowly and has more unwanted chemicals that must be refined out. According to a density matrix created by the American Petroleum Institute (API), light crude measures above 31.1° API, while heavy crude measures below 22.3° API.[8] Natural gas is also described as either sweet or sour depending on the amount of hydrogen sulphide in the gas reservoir. When gas is refined, leaving mostly methane, it is called dry gas. Often natural gas is condensed into natural gas liquids, such as propane and butane. The British thermal unit (BTU) is the measure for the energy output of gas.[9]

2.2.1 History of the Petroleum Industry

The petroleum industry has a long history that dates back to the fourth century when the world's first oil wells were drilled in China in 347 AD. The Chinese utilized simple bamboo poles to drill these wells, to the depths of up to about 800 ft (240 m), and the oil that was extracted was used primarily as a source of fuel.[10] In later centuries, oil was found across Asia and Europe. The modern oil industry began in the mid-nineteenth century, following the 1859 discovery of the first underground oil reservoir near Titusville, Pennsylvania, in the United States of America ("USA") by Colonel Edwin Drake.[11] Colonel Drake's well initially produced 30 barrels of oil per day ("b/d").[12] Its success marked the beginning of global interests in the commercial viability and economic relevance of oil and gas resources. Following its commercial discovery, a variety of products were eventually developed from crude oil. These products include kerosene, gasoline and diesel amongst others.[13]

Africa has been a major global source of petroleum resources since the early 1900s. However, knowledge and history of the quantities and exploration of these resources is limited due to lack of comprehensive historical information and

[7] See generally, American Association of Petroleum Geologists.

[8] Reserves of oil are generally measured in tons or barrels of oil. Production quantities are abbreviated using "bbl" (or barrels of oil per day, bbl/d or bpd). One ton is approximately between six and eight barrels of oil. Reserves and production quantities of gas are measured in cubic meters (m^3) or standard cubic feet (scf). Ibid. see also Natural Resource Governance Institute (2015a, b).

[9] Natural Resource Governance Institute (2015a, b).

[10] Gao (1998), pp. 4–10.

[11] Llewellyn et al. (2013), pp. 1–5.

[12] One barrel of oil is equal to 159 L or 42 US gallons. See generally, OPEC (2013), p. 20.

[13] Ibid.

records.[14] For example, oil production in Egypt is considered to have begun as early as in 1886 when the Egyptian Government discovered oil in Ras Gemsah.[15] In 1907, the Egyptian Oil Trust Ltd. was registered, with the objective of receiving concessions, developing, drilling, purifying, supplying, reserving, distributing and handling petroleum products.[16] A year later, the Egyptian Oil Trust Ltd. started the drilling process. Similarly, in 1909, the same company was able to produce commercial amounts of oil from Ras Gemsah at depth of 1287 ft. The well was reportedly producing two barrels of oil per minute.[17] In 1911, more commercial discoveries were made in Ras Gemsah and Egypt's first refinery was established. In the same year, another important discovery was made in the Hurghada field by Anglo Egyptian Oilfields Ltd., a joint venture between Shell and BP (50% Shell and 50% BP). Commercial discovery of oil was made in 1938 in Egypt in Ras Gharib, located between Hurghada and the Gulf of Suez. At a production rate of 150 tons of oil per day, production rate reached its peak of 5.1 million barrels in 1939.[18] Similarly, around this same period, Gabon's oil industry started gaining attention when several oil deposits were discovered in Libreville.

The 1950s however proved pivotal to the African continent, due to commercial discoveries of oil in Angola (1955), Algeria (1956), Nigeria (1956), and Libya (1959). More commercial discoveries were to later follow in Cameroun (1972), Democratic Republic of Congo (1975), and Equatorial Guinea (1996).[19] Except for a slowdown in the early 1980s, owing to the collapse in oil prices, oil production in Africa has been increasing gradually ever since the 1960s with more recent commercial discoveries in Kenya, Ghana, Congo and Tanzania.[20]

As a whole, sub-Saharan Africa holds one-tenth of the world's crude oil, and has been one of the fastest growing oil regions in the world over the past 5 years, accounting for 30% of all new discoveries.[21]

As Table 2.1 shows, majority of African oil reserves (and production) are located in Libya, Egypt, Nigeria, Algeria, Angola, and South Sudan, which together account for more than 90% of the continent's reserves. The largest reserves are in Libya and

[14] See Magloire (1970), pp. 489–501; Traut et al. (1998).

[15] Ibid. also Foster et al. (1991); also Egypt Oil and Gas Web Portal.

[16] Grayson (1981).

[17] Egypt Oil and Gas Web Portal.

[18] Ibid.

[19] In Nigeria, oil was first discovered in commercial quantities in Oloibiri, Bayelsa State, in 1956. Libya's first productive oil well was struck in 1959 at Amal and Zelten, now known as Nasser. The country began exporting oil in 196. Algeria's first commercial oil discovery was Edjelleh in 1956, followed immediately by the Hassi Messaoud oil field the same year. Production began in 1958. Angola first produced oil from the Benfica oilfield in the Cuanza Basin in 1955. See OPEC, Profile of Member Countries.

[20] Since the discovery of the offshore Jubilee field in Ghana in 2007, with an estimated 700 million barrels of oil (MMbo) and 800 billion cubic feet (Bcf) of gas, Ghana has emerged as a commercial oil and gas producer. See Tullow Oil (2014); also Control Risks (2012).

[21] See British Petroleum (2017), p. 12. See also International Energy Agency (2015), p. 48.

Table 2.1 History and volume of petroleum discovery in African Countries[a]

Country	Commercial discovery	Proven reserves (Billion)/barrels (2016)
Egypt	1938	3.5
Algeria	1956	12.2
Angola	1955	11.6
Nigeria	1956	37.1
Libya	1959	48.4
Ghana	1970	6.6
Cameroon	1972	2.0
Congo (Brazzaville)	1975	1.6
Cote d'Ivoire	1977	1.0
Sudan	1979	1.5
South Sudan	1979	3.5
Gabon	1996	2.0
Mauritania	2001	0.2
Uganda	2009	2.5

[a]See 2017 BP Statistical Energy Survey and U.S. Energy Information Administration, International Energy Statistics

Nigeria, which account for 3.4 and 3.0%, respectively, of world reserves.[22] In terms of production, the four major producers in Africa (Nigeria, Algeria, Libya, and Angola) together account for 77% of the continent's production and contribute 9.2% to world oil production.[23] For example, it is estimated that Nigeria has about 37.2 billion barrels of proven oil reserves, the second largest in Africa after Libya. Furthermore, with an estimated oil production capacity of 2.5 million barrels per day, Nigeria ranks as Africa's largest producer of oil and the sixth largest oil producing country in the world.[24] The majority of reserves are found along Nigeria's Niger River Delta and offshore in the Bight of Benin, the Gulf of Guinea, and the Bight of Bonny. Current exploration activities are mostly focused in the deep and ultra-deep offshore. Nigeria produces mostly light, sweet (low sulphur) crude oil of which the vast majority is exported to global markets. In addition, with an estimated 188 trillion cubic feet of gas, Nigeria has the largest proven natural gas reserves in Africa and the ninth-largest in the world. Nigeria produced 1.35 trillion cubic feet ("Tcf") of dry natural gas in 2013, ranking among the world's top 30 largest natural gas producers. Most natural gas reserves are located in Nigeria's Niger Delta.[25]

Since their respective discoveries, Africa has remained an important player in world oil production, with a total share of 12.1% in 2006. According to the 2017 BP Statistical Energy Survey, Africa had proven oil reserves of 128.438 billion barrels at the end of 2016, equivalent to 41.2 years of current production and 8.01% of the world's reserves. Similarly, Africa produced an average of 7892 thousand barrels of

[22] See U.S. Energy Information Administration (2015).

[23] Ibid.

[24] Ibid. Also, Yager et al. (2012).

[25] U.S. Energy Information Administration (2015). Also, Yager et al. (2012).

crude oil per day in 2016 or 8.6% of the world production.[26] Consequently, African countries continue to provide international oil and gas companies with viable opportunities to spearhead new oil and gas discoveries and to consolidate on oil and gas exploration and production in previously explored fields, especially marginal fields. Much of Africa's crude oil production is primarily for the export market, as well as for domestic consumption of refined fuels, and feedstock for the production of petrochemicals and related products. Similarly, the primary market for Africa's natural gas was historically the export market. However, there is now increased local demand for natural gas. Domestic consumption of natural gas is mainly for, power generation, fertilizer production, methanol production, aluminium smelting, cement production, steel manufacturing, and residential consumption. It is estimated that about 80% of natural gas consumed in Nigeria is used for power generation. The remaining 20% is mostly used in the cement, fertilizer, methanol, aluminium and steel industries, while a negligible proportion is consumed as Liquefied Propane Gas ("LPG"). LNG is rarely used in Nigeria as it is exclusively for export. Domestic demand for gas was projected to reach five billion standard cubic feet a day by 2013 with the majority of the demand coming from the growing power sector. In addition, gas production is expected to rise to 80 billion cubic metres by the end of the 2020, with demand rising by 246% between 2010 and 2020, with export potential at about 35 billion cubic metres, largely in the form of LNG, leaving about 45 billion cubic metres for local consumption.[27]

2.2.2 *Importance of Oil and Gas to African Economies*

Many oil-producing countries in Africa operate largely as monocultures with near total dependence on oil and gas as the chief source of revenue. In Libya for example, oil resources, account for approximately 95% of export earnings, 75% of government receipts, and over 50% of GDP.[28] Similarly in Nigeria, about 80% of government income, 90–95% of the export earnings and more than 90% of the foreign exchange revenues in Nigeria are derived from the oil sector.[29] Ever since the first drop of oil was struck in the Eastern Niger Delta, now Bayelsa State in 1956, oil and gas have been the dominant source of Nigeria's wealth and income. In the period of 2014–2017, oil business generated revenues of approximately US $200 billion, and an equivalent of nearly $1.2 trillion from over the last 40 years.[30] Although many years of ineffective resource management, governmental corruption and lack of accountability, has seen several African countries wallow in poverty despite the

[26] See BP Statistical Review of World Energy (2017).

[27] Ibid.

[28] World Bank, Libya's Economic Outlook (2018a).

[29] World Bank, Nigeria's Economic Report (2018b).

[30] See US Energy Information Administration (2016a, b).

significant oil wealth, petroleum production has, and continues to play pivotal roles in economic development and recovery in Africa.

In addition to its economic significance, petroleum is a commodity with very many industrial uses and applications. Its importance for energy generation and supply in Africa cannot be overemphasized. Transportation is the greatest single use of petroleum, accounting for over 75% of all African petroleum consumed in 2015.[31] Given Africa's fast rising population, projected growth in urbanization, and a significant reliance on automobile transportation especially by middle class consumers, Africa has witnessed a significant increase in transportation energy demand in the last 5 years. According to the IEA, the consumption of light-duty vehicle fuels has increased by 3.9%/year in Africa, accounting for 52% of total transportation energy use in the region. Petroleum products, especially motor gasoline, diesel, and jet fuel, provide virtually all the energy consumed in the transportation sector.[32]

The industrial sector is the second largest petroleum-consuming sector and accounts for about 20% of all petroleum consumption in Africa.[33] Energy consumption in this sector includes use of fossil fuel fired manufacturing plants, process and assembly, steam and cogeneration, process heating and cooling, and lighting, heating, and air conditioning for buildings.[34] The commercial and electric utility sectors account for the remaining 5% of petroleum consumption. In its refined form, petroleum products are the pivot of everyday life and commerce, motor gasoline/petrol are used for fuel for cars and trucks, jet fuel powers aircrafts. Distillate fuel oil, which includes diesel oil, heating oils, and industrial oils, is used to power diesel engines in buses, trucks, trains, automobiles, and other machinery. It is also used to heat residential and commercial buildings and to fire industrial and electric utility boilers. Furthermore, there is asphalt or bitumen, which is used to pave roads; and naphtha, which is the main feedstock for petrochemicals. Liquefied petroleum gas also known as butane and propane are an automotive fuel and are used for household heating and cooking. Kerosene is used for residential cooking in many parts of Africa and commercial production processes.

The oil industry is a major, and one of the largest, employers of labour across Africa. In 2010, 6 out of the 10 largest companies in terms of revenues were from the energy and mining sectors.[35] In Nigeria for example, the oil industry alone creates over 65,000 direct jobs, and more than 250,000 jobs in non-direct employment.[36] Similarly in Libya, Congo and Angola, the oil sector remains a primary source of employment generating significant household income and opportunities

[31] Ibid, 129–130.

[32] Ibid. Also US Energy Information Administration (2018).

[33] US Energy Information Administration (2016a, b).

[34] Ibid, 113.

[35] Sigam and Garcia (2012), p. vii.

[36] US Energy Information Administration (2016a, b), pp. 114–120.

for citizens. The oil sector has also accounted for a rapid growth and boost in tourist activities in several oil producing oil communities across Africa.

Given the importance of the oil industry to governments and households across Africa the importance of effective regulation to ensure maximum and sustainable management of petroleum resources cannot be overemphasized.

2.3 Key Stages in Oil and Gas Development

The key stages in the oil and gas development process are: exploration, prospecting, production, refining, transportation and distribution.[37] Prior to these key stages however, there are some preliminary operations and planning that take place. Before commencing any search for oil in an area, oil companies must be awarded the license and right by relevant authorities. As discussed in Chap. 4, an oil exploration license allows the holder to make preliminary search by surface geological and physical methods, for hydrocarbon reservoirs in a closely defined area; sometimes even the depth is specified. In awarding exploration licenses resource based countries are often in search of oil companies that have the best technical track record and that can offer the best exploration work program commitments and returns. That is why oil companies most often have to compete in international licensing or bidding rounds to be allowed to explore and search for oil in sedimentary basins. During the bid round, companies are invited to submit detailed proposals and plans (bids) that set out the total amount they intend to invest to explore an area, the amount of work they intend to carry out, and evidence of the company's technical and financial track record. The company typically bears the entire cost of exploration work and will only be reimbursed if a field is brought on stream.

Following a successful licence round/bid, seismic surveys are carried out on the prospective oil field to develop a picture of the geological structures below the ground. It is at this stage, that the oil and gas company embarks on site survey to acquire more information about the area and coverage of a potential prospect. This includes acquiring data on the environment around the drilling location. After the site survey, oil production activities will proceed under the various the key stages of oil development. These stages are briefly discussed below.[38]

The activities in petroleum sector supply chain can be divided into three key chronological sectors: upstream, midstream and downstream.

---

[37] See Wolf (2009).

[38] See generally, Cairn Energy (2010) and Oil and Gas Project Life Cycle.

2.3.1 *Upstream*

The upstream oil sector is commonly known as the exploration & production (E&P) sector.[39] It comprises activities related to searching for, recovering, and assessing the quantity of crude oil and/or natural gas from underground or underwater fields, and subsequently operating the wells that recover and bring the crude oil and/or raw natural gas to the surface. It includes different operations such as the searching for prospective underwater oil and gas fields through aerial or satellite photography and magnetic surveys, seismic surveys (2D/3D), drilling of exploratory wells and also making requisite operations on the well so as to bring the crude oil or natural gas to the ground surface.[40]

2.3.1.1 Exploration

Upstream activities start with exploration, which comprises seismic surveys (activities of finding oil and gas and assessing its quantity for possible exploitation) and then exploratory well drilling.[41] The exploration and appraisal process may take 3–10 years to complete. During this stage, one or more exploratory wells are drilled to determine if a prospect exists and also to gain further information on the subsurface conditions, which may exist in that field. This is also known as exploratory drilling. If a hydrocarbon reservoir is discovered, the well is surveyed and tested to see if the reservoir is viable for production.[42] Well testing at this stage will provide hydrocarbon samples as well as information on flow rate, temperature and pressure. If on the other hand, no viable reserves are encountered, an operator may carry out further drilling if initial tests show that hydrocarbons may still exist in that license area. Wells that are found to be unsuitable for further development are sealed, tested to ensure that they are fully secured, and then abandoned.

2.3.1.2 Oil Prospecting

This is the reservoir appraisal stage. Oil prospecting is the stage in which petroleum geologists and geophysicists appraise and determine the size of the reservoir and the best production methods for achieving commercial recovery.[43] Following the

[39] Sigam and Garcia (2012).

[40] Sigam and Garcia (2012).

[41] Ibid, also Olawuyi (2018).

[42] Proven reserves are oil finds that are considered commercially viable—that is, the company is at least 90% certain that it would make money getting petroleum out of the ground and taking it to market. If oil is held in a *probable reserve*, then the company thinks there is at least a 50% chance of recovering the oil. Unprovable or possible reserves are those with a 10–50% probability of profitable extraction. Ibid.

[43] Conoway (1999), pp. 10–25; Economides and Oligney (2002), pp. 1–20.

exploration stage, if a promising amount of oil and gas are confirmed to be present in the location, the right to prospect (oil prospecting license) must be secured to test if the reservoir is viable for production. Field appraisal is used to establish the size of the field as well as the most appropriate production method to assess if the oil field has commercial quantities. If an appraisal confirms a commercial reserve, then the operator may proceed to production. Once a prospect has been shown to be technically and commercially viable, a development plan is submitted to the relevant authorities for approval. The results of the seismic survey may show that there are potential good reservoir rocks or not (a reservoir is a formation that contains hydrocarbons).

2.3.1.3 Drilling and Production

Once a commercially viable reservoir has been found, the right to drill must be secured before drilling can begin. At this stage of the project, test wells are drilled using various drilling methods. The most common method of drilling in use today is rotary drilling, *i.e* boring a hole by continuous turning of a bit- an important drilling tool. Once the development drilling is completed, and a stable production level is reached, the facility comes on stream and formally starts production.

Production is the process of getting the oil and gas out of the ground. Production can be accomplished through primary, secondary or tertiary recovery methods. Primary recovery efforts allow the oil to flow to the surface under its own pressure. After natural reservoir drive diminishes, water or gas may have to be injected into the reservoir to increase the reservoir's pressure and lift the oil, in which case recovery is considered secondary. Tertiary, or enhanced, removal (EOR) involves the use of various techniques for increasing the amount of crude oil that can be extracted from an oil field. There are three primary techniques EOR: thermal recovery, gas injection, and chemical injection. Using EOR, 30–60%, or more, of the reservoir's original oil can be extracted.[44]

Once the prospect is online, production is gradually increased to maximum or peak production. Peak production is maintained until it starts to decline, at which time it may be feasible to consider additional recovery methods. These may include injecting water or gas into the reservoir to maintain production levels or to drill new wells in nearby reservoirs and connect them.[45]

2.3.1.4 Decommissioning

Being the final stage of oil exploration activities, decommissioning happens mostly after a long period of oil production when platforms have reached the end of their productive lives. Studies show that on average, offshore fields mature for

---

[44] Electric Power Research Institute, "Enhanced Oil Recovery Scoping Study" Final Report, No. TR-113836. Palo Alto, CA (1999).

[45] See generally, Cairn Energy, Oil and Gas Project, (n. 36).

decommissioning between 20 and 50 years after they are opened for exploration activities.[46] After this period of production, the limits of economic recovery of the hydrocarbons are generally reached and the production structure is dismantled to rehabilitate the area. When the commercial exploitation ends, the decommissioning of production installations starts. This involves the removal of buildings and equipment, the restoration of the site to environmentally-sound conditions, the implementation of measures to encourage site re-vegetation, and the continued monitoring of the site after closure.[47] While decommissioning is an inevitable end process of oil exploration and production, it becomes a source of complex and far-reaching environ- mental problems when offshore oil exploration facilities are disposed in an environmentally unsound manner.[48]

2.3.2 *Midstream*

The second sector in the petroleum industry value chain is the midstream sector. The key business in this sector is the transportation and storage of oil and natural gas. It provides a central link between oil production and processing facilities; and final use by customers. This involves moving the oil and natural gas, retrieved in the upstream sector, from production sites to storage facilities, and then to downstream processing facilities, such as refineries, so that the recovered products can be turned into various finished products for consumers' day to day use.[49]

2.3.2.1 Storage

Oil and gas products are stored mainly for future refining or use. However, oil and gas can be stored for use in production line fills, or to provide inventories for price hedging and speculation purposes.[50] For example, oil-storage trade, also referred to as contango, has since early 1990s emerged as a market strategy in which large oil

[46] See Ayoade (2002), pp. 1–15.

[47] See generally, Sigam and Garcia (2012), pp. 3–4.

[48] See *The Petroleum (Drilling and Production) Regulations to the Petroleum Act 1969*. Regulation 36 establishes procedures for safe abandonment and decommissioning of wells. Under Regulation 37 operators must maintain all production apparatus and appliances in good repair and condition, and must carry out all operations in a proper and workmanlike manner to prevent the escape of petroleum into any water, well, spring, stream, river, lake, reservoir, estuary or harbour; and to cause as little damage as possible to trees, crops, buildings, structures and other property on land.

[49] See Olawuyi (2015a, b), pp. 190–192. See also Sigam and Garcia (2012).

[50] Oil speculation involves storing, trading and making profits from oil sale by anticipating changes in prices and taking appropriate positions. See Yglesias (2014).

companies purchase oil for immediate delivery and storage when the price of oil is low, and hold it in storage until the price of oil increases.[51]

Similarly, countries store and maintain strategic emergency oil stockpile to guarantee security of supply for domestic use. For example, The United States' Strategic Petroleum Reserve (SPR), with a capacity to hold 727 million barrels of oil, is the world's largest storage of emergency crude oil and stores unrefined oil in huge underground salt caverns along the coastline of the Gulf of Mexico.[52] The US established the reserve in 1975 to mitigate future oil supply disruptions in the aftermath of the 1973–1974 oil embargo.[53] The mission of the OPR is to protect the United States economy from severe petroleum supply interruptions through the acquisition, storage, distribution and management of emergency petroleum stocks.[54]

Oil storage facilities, such as fixed bulk tanks, terminal oil storage facilities, consumer storage, pumps and drainage systems used in the storage of oil, as well as conditioned gas tanks and reservoirs, are constructed to connect the produced petroleum products with refineries that will separate and distill them into end products. While some storage facilities are constructed to receive unrefined crude oil, others are specially conditioned to store and receive processed and refined petroleum products.

2.3.2.2 Transportation

Transportation and distribution in the midstream oil and gas sector relates to moving crude oil from the oil fields, where oil has been discovered, to petroleum refineries, where the oil is refined and processed, and to storage areas after it has been refined. It also involves getting the end products from the refineries/storage terminals to the end users. The logistics network that is required to supply petroleum products from refineries to end users is often complex and requires a complex system of ships, railways, trucks, trains and pipelines. The various modes of transportation are needed to move petroleum products from the refineries, ports and large terminals to markets across the country, internationally and to domestic end users.

In its crude state, oil tankers and pipelines primarily transport crude oil. Pipelines are by far one of the safest, most reliable and cost-effective means of transporting petroleum products worldwide. Vast networks of fuel oil storage and pipeline systems are constructed and maintained by midstream operators. For example, getting oil and gas to travel the long distances to the downstream refinery locations requires extreme amounts of compression and inline filling systems. After the initial pipeline

---

[51] Contango is a situation where the futures price of a commodity is above the expected spot price. See Commission of the European Communities, *Is there a speculative bubble in commodity markets?* (Report of the Task force on the role of speculation in agricultural commodities price movement, 2008 SEC (2008)2971).

[52] US Department of Energy, Office of Fossil Energy.

[53] Ibid.

[54] Ibid.

transfer, additional transportation and storage may be required before the oil is processed and distributed in the downstream sector. To accomplish this, oil may be transported, by trucking over highways and roads; by barges or other seafaring vessels; or by rail, to various downstream facilities and processing plants.

2.3.3 Downstream

The downstream oil sector is the closest to the consumers. It covers activities ranging from the refining/processing of hydrocarbons, to marketing and selling to final consumers.[55] The sector encompasses the processing, refining, marketing, selling and distributing petroleum products, such as LPG, gasoline or petrol, jet fuel, diesel oil, other fuel oils, asphalt and petroleum coke, to the final end users.

2.3.3.1 Refining

Crude oil, to be useful to end users, must be converted or refined into petroleum products. Petroleum refining is the physical, thermal and chemical separation of crude oil into major distillation components or fractions.[56] These distillation fractions can be further processed through separation and conversion into finished or end use petroleum products. Petroleum refining activities take place in a petroleum refinery, and involve processes where petroleum is distilled and separated to make the different end use products that are utilized in vehicles, homes and industries. The refining process involves a number of steps, which include distillation and separation, conversion or upgrading, desulphurization or sweetening and in some cases, the inclusion of additives to improve and enhance the fuel's performance.[57]

In distillation, separation and conversion, high levels of heat and pressure, along with catalysts are used to break down the crude oil into component parts or to clean it by removing sulphur or other impurities. The first step is atmospheric distillation, where the crude oil vapours rise in the column, while the heaviest molecules remain at the bottom. After separation of the molecules, the next step is conversion, which involves a "cracking" process, which cracks the molecules that are still too heavy, then producing gas, gasoline and diesel. Furthermore, molecules that are corrosive or cause air pollution such as sulphur are removed.

---

[55] Olawuyi (2018), pp. 1–2.

[56] Ibid.

[57] See generally, Environment Canada, Petroleum Refining.

2.3.3.2 Marketing and Retailing

After refining and processing, oil products marketing includes the activities necessary to sell and deliver refined products to end consumers through wholesale, direct industrial sales and retail in petrol stations. The downstream industry touches consumers through thousands of products such as petrol, diesel, jet fuel, heating oil, asphalt, lubricants, synthetic rubber, plastics, fertilizers, antifreeze, pesticides, pharmaceuticals, natural gas and propane.

Key players in the downstream sector include oil refineries, petrochemical plants operators, petroleum product distributors, retail outlets and natural gas distribution companies. It also includes blending of products through petrochemical processes, packaging and trading in lubricants, as well as distributing and selling petroleum products to the various consumer markets.[58]

2.4 The Solid Mineral Sector

In line with our discussions of foundational concepts in the extractive sector, this section examines the meaning, nature and scope of solid minerals, the nature and origin of solid minerals. It will also provide a brief overview of the solid minerals sector in the African continent, as well as examine the discovery and development of solid minerals in Nigeria with emphasis on a few identified minerals. Finally, this section examines the importance of solid minerals as well as the stages of mineral exploration and development.

2.4.1 Meaning, Nature and Scope of Solid Minerals

A mineral is generally a naturally occurring metallic or non-metallic substance, formed through biogeochemical processes. This includes coal, salt, quarry and pit material, and all rare and precious minerals and metals.[59] Depending on the jurisdic-

---

[58] See generally, Bhardwaj (2013).

[59] Canada's *Extractive Sector Transparency Measures Act* (S.C. 2014, c. 39, s. 376) s. 2. See also Section 1 of the Alberta Mines and Minerals Act Chapter M-17 which defines "mineral" to mean "all naturally occurring minerals, and without restricting the generality of the foregoing, includes (1) gold, silver, uranium, platinum, pitchblende, radium, precious stones, copper, iron, tin, zinc, asbestos, salts, sulphur, petroleum, oil, asphalt, bituminous sands, oil sands, natural gas, coal, anhydrite, barite, bauxite, bentonite, diatomite, dolomite, epsomite, granite, gypsum, limestone, marble, mica, mirabilite, potash, quartz rock, rock phosphate, sandstone, serpentine, shale, slate, talc, thenardite, trona, volcanic ash, sand, gravel, clay and marl, but (2) does not include (A) sand and gravel that belong to the owner of the surface of land under section 58 of the Law of Property Act, (B) clay and marl that belong to the owner of the surface of land under section 57 of the Law of Property Act, or (C) peat on the surface of land and peat obtained by stripping off the overburden, excavating from the surface, or otherwise recovered by surface operations."

tion, the exact legal connotation of a mineral can be enlarged to cover all forms of natural resources, be it solid, liquid or gaseous. For example, a mineral is defined in the *Nigerian Minerals and Mining Act*, the governing legislation for the exploration and exploitation of solid materials in Nigeria, as:

> any substance whether in solid, liquid or gaseous form occurring in or on the earth, formed by or subjected to geological processes including occurrences or deposits of rocks, coal, coal bed gases, bituminous shales, tar sands, any substances that may be extracted from coal, shale or tar sands, mineral water, and mineral components in tailings and waste piles, but with the exclusion of petroleum and waters without mineral content.[60]

This expansive definition includes all forms of natural resources except for petroleum and water. Similarly, the South African *Mineral and Petroleum Resources Development Act* defines mineral as:

> any substance, whether in solid, liquid or gaseous form, occurring naturally in or on the earth or in or under water and which was formed by or subjected to a geological process, and includes sand, stone, rock, gravel, clay, soil and any mineral occurring in residue stockpiles or in residue deposits, but excludes —(a) water, other than water taken from land or sea for the extraction of any mineral from such water; (b) petroleum; or (c) peat.[61]

Furthermore, *Kenya's Mining Act* defines minerals to include:

> all minerals and mineral substances, other than mineral oil…and may be precious metals, precious stones or non-precious minerals, but…does not include clay, murram, limestone, sandstone or other stone or such other common mineral substances as the Minister may by notice in the *Gazette* declare not to be minerals for the purposes of this Act, always provided these do not contain any precious metal or precious stone in economically workable quantities.[62]

The above definitions mirror the categorizations of minerals in the laws of several other African countries, which generally identify minerals to be natural substances (other than petroleum, water or peat) that exist in solid, liquid or gaseous forms under the ground or on land which may be mined or exploited.[63] A mineral is usually solid and inorganic, and has a crystal structure. It is different from a rock, which can be an aggregate of minerals or non-minerals and does not have a specific chemical composition. A rock is a solid composed of different minerals and other substances.[64]

---

[60] Nigerian Minerals and Mining Act No. 20 of 2007.

[61] South African Mineral and Petroleum Resources Development Act No. 28 of 2002.

[62] The Republic of Kenya, Mining Act, Chapter 306.

[63] Peat is a highly organic mixture of more or less decomposed plant (humus) material found in marshy or damp regions. It is cut and dried for use as fuel. See International Peatland Society (2015).

[64] A rock that contains a commercially profitable quantity of one or more minerals (or metals in particular) is called an *ore*. The left-over, worthless rock material from the ore is called gangue. The content of the ore, the cost per unit weight of the extracted mineral, is measured on a scale called the grade. See Natural Resource Governance Institute (2015a, b).

In broad and general terms, a mineral is a naturally occurring solid substance; stable at room temperature; usually abiogenic; and has an ordered atomic arrangement. The first criterion means that a mineral has to be formed by a natural process, which excludes anthropogenic compounds. Stability at room temperature, in the simplest sense, is synonymous to the mineral being solid. The requirement that a valid mineral must be abiogenic means it is formed by inorganic means (not produced by the decomposition of living organisms). Finally, the requirement of an ordered atomic arrangement means a mineral is crystalline (i.e. having the structure and form of a crystal, strikingly clear or sparkling).[65]

There are over 5300 known mineral species.[66] Minerals are distinguished by various chemical and physical properties. Differences in chemical composition and crystal structure distinguish various species, and these properties in turn are influenced by the mineral's geological environment of formation. Changes in the temperature, pressure, or bulk composition of a rock mass cause changes in its minerals. Minerals can also be described by various physical properties, which relate to their chemical structure and composition.

The quality of a mineral is often graded using a set of criteria known as the four Cs, namely color, cut, clarity and carat weight.[67] Other common distinguishing characteristics include crystal structure and habit, hardness, lustre, diaphaneity, streak, tenacity, cleavage, fracture, parting, and specific gravity. More specific tests for minerals include magnetism, taste or smell, radioactivity and reaction to acid. Solid minerals are also usually classified according to their key chemical constituents.[68]

Minerals are often categorized based on their monetary value and worth.[69] Base metals are non-precious metals such as lead, zinc, copper, nickel, and tungsten amongst others, while precious metals include gold, silver and platinum group metals (PGMs).[70]

---

[65] See generally, CM Enterprises, Minerals.

[66] Ibid.

[67] See American Gem Society, 4 Cs of Diamonds; also Nassau (1978).

[68] See generally, The Visual MD, Minerals.

[69] For example in Nigeria, the Ministry of Mines and Steel Development categorized Nigeria's mineral resources as follows: (a) *World class minerals:* These include gold, coal, bitumen, gemstones, iron ore, cassiterite, rutile, tantalite, columbite, manganese, galena and sphalerite; (b) *Industrial minerals*: These include limestone, barites, marble, clay, mica, feldspars, silica sand, diatomite, gypsum, kaolin, talc, magnesite, kyanite, phosphate rock, bentonite and fluorite; (c) *Newly discovered minerals:* These include chromite, sillimanite, uraninite, sperrylite, vanadinite, cobaltite, chalcopyrite, lanthanum, stibnite and pentlandite; and (d) *Low world demand minerals*: These include wolframite, bismuth, argentite, molybdenite, ilmenite and lithium. See Nigerian Investment Promotion Council, Solid Minerals.

[70] The Republic of Kenya, Mining Act, Chapter 306, s 2.

Precious stones include diamond, emerald, opal, ruby, sapphire, turquoise and any other stones with significant monetary value of worth. Generally, the top 10 most valuable minerals in the world are: lithium,[71] blue garnet,[72] jadeite,[73] rubies,[74] gold,[75] platinum,[76]

---

[71] Lithium is a highly valued, and one of the world's most sought after, minerals. First discovered in 1817 in Stockholm, by Swedish chemist Johan August Arfvedson, lithium is a crucial ingredient in the production of rechargeable batteries. Lithium-ion batteries are used in portable electronic devices, electric tools and electric vehicles. Zimbabwe is one of the world's largest producers of lithium. The country's privately owned Bikita Minerals controls nearly all of the country's lithium mining and holds the world's largest-known deposit of lithium, at over 11 metric tones (MT). Given the growing importance of energy metals and lithium-ion batteries, securing a consistent supply of lithium is a top priority for technology companies around the world and could make lithium a most sought-after metal in coming years. See Zimbabwe Herald (2016).

[72] Garnet has been used as a gemstone for thousands of years. Garnets are found in various colors such as brown, green, orange, pink, purple, red and yellow. Among all these colors, red and blue garnets are generally considered the most valuable. This mineral was first discovered in the 1990s in Madagascar, and since then it has been mined in Russia, Turkey and the United States. The current value of the blue garnet is $1.5milion per carat. See Mining Greece, Ten Most Expensive Minerals in the world.

[73] Jadeite is one of the rarest and most valuable gemstones in the world, often used as jewellery, ornaments, tools, weapons, gemstone. Jadeite's color commonly ranges from white, through pale, apple green, to deep jade green, blue-green, pink, lavender and a multitude of other rare colors. Jadeite is a rare mineral that's most often harvested in places like California, US; Myanmar; New Zealand; Guatemala; Itoigawa, Japan; Kazakhstan; Russia; British Columbia, Canada; Italy and Turkestan. Jadeite is currently valued at around $3 million per carat. See Mining Greece, Ibid.

[74] Rubies are the world's most expensive coloured gemstone. The world's rarest and most expensive ruby is the Sunrise Ruby. The Sunrise Ruby is recently valued at $30m (£19.1m). The largest supply of this mineral was harvested in Burma, which is known as the Mecca for rubies. African countries where rubies have been discovered include Kenya, Madagascar, Malawi Nigeria, Rwanda, Sierra Leone, Tanzania and Mozambique. See Shor and Weldon (2009). See also BBC News (2015).

[75] Gold is the most mined mineral resource in Africa. Perhaps because of its extensive use and long history of mining-more than 6000 years—Gold is erroneously thought of as the most valuable and most expensive mineral in the world. Though not by any means the most expensive, Gold is still highly valuable with a worth of around $56 per gram. Gold mining is a global business with operations on every continent except Antarctica. Gold has been discovered in about 95 countries of the world. For many years South Africa has been one of the world's dominant gold producers. Located in South Africa, the Witwatersrand Basin represents the richest gold field ever discovered. It is estimated the 40% of all of the gold ever mined has come out of the Basin. In 2017, South Africa's output accounted for about 10% of the world's gold production with a production of 140 Metric Tonnes. Other African countries with significant Gold deposits are Ghana, Nigeria, Algeria, Benin, Botswana, Burkina Faso, Cameroon, Central African Republic, Equatorial Guinea, Eritrea, Ethiopia, Liberia, Madagascar, Mali, Morocco, Mozambique, Namibia, Rwanda, Senegal, Sierra Leone, Sudan, Tanzania, Uganda and Zimbabwe. See US Geological Survey (2017).

[76] Platinum Group Metals (PGMs-Platinum, palladium, rhodium, ruthenium, iridium, osmium) are mainly used in the automobile industry as catalytic converters to decrease harmful emissions from automobiles; in glass industry; in jewelry; in the chemical industry for catalysts in nitric acid and other bulk-chemical production, for refining petroleum, and for fabricating laboratory equipment; and in the electronics industry in computer hard disks to increase storage capacity, in multilayer ceramic capacitors, and in hybridized integrated circuits. The largest global reserves of PGMs are in the Bushveld Complex in South Africa. South Africa ranks amongst the top two largest and most important producers of platinum in the world. South Africa accounts for almost 96% of the entire

black opal,[77] diamond,[78] painite,[79] and rhodium.[80] Minerals are generally measured in ton, although gemstones are measured in carats. (One carat is equal to approximately 200 mg).

2.4.2 *History and Overview of Solid Minerals Development in Africa*

Mining is one of the oldest industries to have ever developed in Africa. Africa is richly endowed with mineral resources and has been a significant source of some of the world's most valuable minerals since the pre-colonial era.[81] The African continent produces 12 of the world's most expensive and sought after mineral resources: cobalt, gold, lithium, diamond, PGM, copper, coal, uranium, aluminium, bauxite, soda ash, and iron ore.[82] South Africa, Zimbabwe, Democratic Republic of Congo, Botswana and Namibia are some of the historical giants in the African mining industry.

platinum production in the African continent with 110,000 kg in 2016. Zimbabwe is also a leading source of PGMs. See U.S. Geological Survey (2017).

[77] Australia is the world's largest and most important producer of opal. Around 97% of the world's production come out of Australian opal fields, therefore this mineral is often considered as the national gemstone of Australia. The town of Coober Pedy in South Australia is a major source of opal. Black Opal has been also mined in Ethiopia, Brazil, Mexico and the United States. See Eckert (1997).

[78] Diamond is commercially the most popular mineral because of its eminent role in the world of jewellery trading. An estimated ten million people globally are directly or indirectly supported by the diamond industry. Diamond is the hardest known substance, the greatest conductor of heat, has the highest melting point of any substance and it has the highest refractive index of any natural mineral. Africa accounts for a significant amount of the world's diamond production, 65% of the world's diamonds come from African countries. See generally Diamond Facts, Conflict Diamonds.

[79] Painite is the "hardest to locate" and one of the rarest minerals in the world. This mineral was first discovered in Myanmar in 1950 by the British gem dealer Arthur C.D. Pain. Because of its rarity, painite is one of the world's most expensive minerals, with an estimated worth of $50,000–60,000 per carat. There are little or no occurrences of painite in Africa. See Rossman (2005).

[80] Rhodium, a PGM metal, is highly relied upon in the automotive industry for the manufacture of three-way catalysts, which reduce the nitrous oxide emissions from light-duty gasoline cars and motorcycles. Because of its rarity and industrial application, this silver-white noble metal is one of world's most expensive minerals. South Africa and Zimbabwe are some of the world's largest suppliers of Rhodium. See United States Department of the Interior (2015).

[81] International Council on Mining and Metals (2014).

[82] For example, the 2015 share of world production from Africa was bauxite 9%; aluminium 5%; chromite 38%; cobalt 60%; copper 10%; gold 23%; iron ore 2%; steel 1%; lead (Pb) 2%; manganese 38%; zinc 1%; cement 4%; diamond 56%; graphite 2%; phosphate rock 21%; coal 4%; mineral fuels (including coal) and petroleum 12%; uranium 18%. United States Department of the Interior (2015).

2.4.2.1 Gold

Gold is one of the most mined minerals in Africa. South Africa accounts for 11% of the world's gold reserves.[83] With an estimated in-situ mineral base of USS2.5trillion, South Africa has established itself as a leading figure in the gold mining industry.[84] South Africa's mineral wealth is typically found in geological formations and settings, such as the Witwatersrand Basin, which yields some 93% of South Africa's gold output, and contains considerable uranium, silver, pyrite and osmiridium resources.[85] Ghana is the ninth leading producer of gold in the world and the second in Africa despite the 2% decrease in production from 92 tonnes in 2010 to 91 tonnes in 2011.[86] Other African countries with significant gold deposits are Algeria, Benin, Botswana, Burkina Faso, Cameroon, Central African Republic, Equatorial Guinea, Eritrea, Ethiopia, Liberia, Madagascar, Mali, Morocco, Mozambique, Namibia, Nigeria, Rwanda, Senegal, Sierra Leone, Sudan, Tanzania, Uganda and Zimbabwe.[87]

2.4.2.2 Diamond

Africa accounts for 65% of the world's diamond production.[88] Approximately $8.5 billion worth of diamonds come from African countries every year.[89] Botswana is the leading diamond-producing country in the world in terms of value, and the second largest in terms of volume. In 2013, Botswana produced 23.2 million carats of diamond with a stated value of $3.63 billion.[90] Botswana has seven mines, the two important ones being Orapa and Jwaneng, two of the most prolific diamond mines

---

[83] United States Department of the Interior (2015).

[84] Ibid. Also US Energy Information Administration.

[85] South Africa Government, Mineral Resources.

[86] United States Department of the Interior (2015).

[87] Ibid. See Nigerian Investment Promotion Council. Solid Minerals stating that Nigeria loses about $40 billion annually in unexploited gold alone.

[88] Democratic Republic of Congo (DRC), Botswana, Zimbabwe, Angola, South Africa, Namibia and Sierra Leone rank amongst the top diamond producing countries in the world. Botswana is the leading diamond-producing country in the world in terms of value, and the second largest in terms of volume. In 2013, Botswana produced 23.2 million carats with a stated value of $3.63 billion. Botswana has seven mines, the two important ones being re Orapa and Jwaneng, two of the most prolific diamond mines in the world. Although the DRC is Africa's second largest diamond producer, most of the DRC's production is mined by artisanal diamond miners in the informal sector and not by mining companies. According to estimates, some 700,000 artisanal diamond miners mine the country's alluvial mines. Countries such as Angola, Sierra Leone, Liberia and Namibia rely on diamonds, as a major natural resource, to build hospitals, provide education and infrastructure. The diamond mining industry generates over 40% of Namibia's annual export earnings. See United States Department of the Interior (2015).

[89] World Diamond Council, The Diamond Industry Fact Sheet.

[90] Ibid. Also World Diamond Council, An Estimated 65% of the World's Diamonds come from African countries.

in the world. Similarly, Zimbabwe's diverse mineral output also includes about 9% of the world's diamond production (by volume), an estimated 6% of the world's platinum production, and about 4% of the world's palladium output.[91] Countries such Democratic Republic of Congo (DRC), Angola, South Africa, Namibia and Sierra Leone also rank amongst the top diamond producing countries in the world.[92] Although the DRC is Africa's largest diamond producer by volume, most of the DRC's production is mined by artisanal diamond miners in the informal sector and not by mining companies. According to estimates, some 700,000 artisanal diamond miners mine the country's alluvial mines.[93] Similarly, the diamond mining industry generates over 40% of Namibia's annual export earnings.[94]

Unfortunately, however, diamond has also been frequently used in Africa to fund violent wars and conflict, often described as conflict diamond or blood diamond. Conflict diamonds, as defined by the United Nations, are "...diamonds that originate from areas controlled by forces or factions opposed to legitimate and internationally recognized governments, and are used to fund military action in opposition to those governments, or in contravention of the decisions of the Security Council." Illicit rough diamonds have been used by rebels to fund conflicts in Sierra Leone Angola, Liberia, Ivory Coast, the Democratic Republic of Congo and the Republic of Congo.[95]

2.4.2.3 Copper

Zambia is the top producer of copper in the continent and the fifth top producer in the world.[96] Zambia produces 3.57% copper of the world. DRC is also a top producer of copper at over one million and thirty tonnes annually. Though the DRC does not have the highest production rate, mostly due to lack of appropriate technology, the second largest copper reserves in the world can be found in the DRC. Other copper producing countries in Africa are South Africa, Botswana, Morocco, Zimbabwe and Namibia. In Africa, a 480 km long belt extending from the northern part of Zambia North West across Katanga province of Congo Republic is the world's largest copper region.

Copper is mainly used electrical wiring, home roofing, plumbing and painting.

---

[91] Ibid.

[92] Ibid.

[93] Ibid.

[94] Ibid.

[95] See World Diamond Council, Conflict Diamonds.

[96] US Geological Survey (2017).

2.4.2.4 Cobalt

Democratic Republic of Congo is the world's biggest supplier of cobalt.[97] In addition to also holding the world's largest reserves of cobalt, the DRC produces more than half of the world's cobalt.[98] Despite these, Congo remains one of the poorest countries on the planet.[99] This is primarily because only 6% of the revenue generated from cobalt mining in the DRC go to the national treasury.[100]

Cobalt is in very high demand in the technology industry, with close to 50% of exported cobalt used for cell phone batteries and self-driving cars. Cobalt is also used to make alloys for jet engines and gas turbines, magnetic steels and some types of stainless steels.[101]

2.4.2.5 Platinum Group Metals (PGM)

South Africa is the top producer of PGM in the continent. The largest global reserves of PGMs are in the Bushveld Complex in South Africa-with associated copper, nickel and cobalt mineralization-, chromium and vanadium-bearing titanium-iron ore formations and industrial minerals, including fuorspar and andalusite.[102] South Africa and Zimbabwe are also some of the world's largest suppliers of Rhodium. Similarly, Zimbabwe is one of the world's largest and most important producers of lithium.[103]

2.4.2.6 Soda Ash

Kenya's best-known mineral export is soda ash, mined from Lake Magadi—a southernmost lake in the Kenyan Rift Valley. Soda ash is typically used in making glassware, paper and industrial chemicals. Kenya is also a major exporter of titanium and fluorspar.[104]

---

[97] US Geological Survey (2017).

[98] Ibid.

[99] DRC is ranked 176 out of 187 on the United Nations Human Development Index.

[100] Kazeem (2017).

[101] Ibid.

[102] Furthermore, the Transvaal Supergroup in South Africa contains enormous deposits of manganese and iron ore. The Karoo Basin extends through Mpumalanga, KwaZulu- Natal, the Free State and Limpopo, hosting considerable bituminous coal and anthracite resources. The Phalaborwa Igneous Complex hosts extensive deposits of copper, phosphate, titanium, vermiculite, feldspar and zirconium ores. Kimberlite pipes host diamonds that also occur in alluvial, fuvial and marine settings. Heavy mineral sands contain ilmenite, rutile and zircon. Significant deposits of lead-zinc ores associated with copper and silver are found near Aggeneys in the Northern Cape. See US Geological Survey (2017).

[103] Ibid.

[104] Daily Nation (2015).

2.4.2.7 Iron Ore

Iron is one of the most abundant rock-forming elements and, after aluminium, the most explored and widely utilized metal.[105] Iron ore production in Africa is dominated by South Africa (38.1 Mt in 2003 and the world's eighth largest producer), Mauritania (10 Mt per year) and Algeria (2.3 Mt per year).[106] Other African countries with significant iron ore deposits are Nigeria, Angola, Cameroon, Central African Republic, Cote D'Ivoire, Gabon, Ghana, Liberia, Libya, Mali, Mauritania, Zimbabwe, Morocco, Senegal, Sierra Leone, Uganda and Tunisia. Although iron has many specific uses (e.g. pipes, fittings, engine blocks) its main use is to make steel, the most useful metal in construction, transportation and machinery and tool manufacturing.

2.4.2.8 Coal

Africa is home to a large deposit of coal. South Africa remains one of the top ten coal producers in the world and the first in Africa, responsible for about 3.5% of the world's coal resources. With a production of 260 Mt of coal in 2015, South Africa produces about 3.3% of the world's annual total and its coal exports are 6% of global exports, making South Africa the sixth in the list of coal-exporting nations[107] The country also has the world's ninth largest proven reserves of coal at 30.2 billion tons.

Zimbabwe, Nigeria, Egypt, Mozambique, Malawi, Botswana, Niger, Swaziland, Zambia and Tanzania are other African countries with vast coal resources.[108] Although Zimbabwe holds vast coal reserves, particularly in the north western and southern regions, coal production measured just around 3.6 million tonnes (Mt) in 2013, mainly due to limited technology and focus on coal mining.[109] Nigeria is also believed to have abundant coal reserves left unexploited due mainly to Nigeria's one-way focus on the oil sector.[110] However, with falling global oil prices putting

[105] O'Brien (2009).

[106] US Geological Survey (2017).

[107] US Geological Survey (2017).

[108] Ibid.

[109] International Energy Agency (2016).

[110] It is estimated that about 396 million metric tonnes of proven reserves of coal can be found in the central, middle-east and south-east regions of Nigeria, while there are about 1, 134 million metric tonnes of unproven reserves in these areas. Coal is primarily found in the Nigerian states of Enugu, Imo, Kogi, Delta, Plateau, Anambara, Abia, Benue, Edo, Ondo, Bauchi, Adamawa and Kwara. Coal was first discovered in Enugu State in eastern Nigeria in 1906. The Ogbete Mine in Enugu was opened and began regular operations by 1916. The Nigerian Coal Corporation, which was established in 1950, was the government agency, with the sole responsibility for the coal industry in Nigeria. Prior to the oil and gas boom, coal was a major export product and was utilized domestically for a host of activities. In recent years, the Nigerian government has recognized the need to revive the county's coal mining sector, with a view to utilizing it for fuel for power genera-

Nigeria on the edge to diversify its economy, the increased global demand for coal could provide a new opportunity for Nigeria explore the untapped coal deposits and several other mineral resources that are available in the country.

Improved geological data over the years have revealed that Nigeria is endowed with numerous deposits of industrial, metallic and non-metallic minerals. There are over 40 different types of solid minerals found all across Nigeria in varying quantities. These include gold, barite, bentonite, limestone, coal, bitumen, iron ore, tantalite/columbite, lead/zinc, barites, gemstones, granite, marble, gypsum, talc, iron ore, lead, lithium and silver to mention a few. However not all these solid minerals are found in commercially viable quantities, and not all have been exploited. Only 13 solid minerals are being actually mined, processed and marketed in Nigeria. They are coal,[111] bitumen,[112] limestone,[113]

---

tion and domestic use. The most significant uses of coal are for electricity generation, steel production, cement manufacturing, as well as a liquid fuel. It is also used in the manufacturing process for paper, and in the chemical and pharmaceutical industries. Nigerian coal has been found to be suitable for boiler fuel, production of high caloric gas, domestic heating, briquettes, formed coke and in the manufacturing process as material for the production of a wide range of chemicals, which include waxes, resins, adhesives and dyes. See KPMG (2014), p. 4.

[111] KPMG (2014), p. 4.

[112] Bitumen is a semi-solid hydrocarbon product, which is heavy and an extremely viscous oil, that must be treated before it can be processed by refineries to produce usable fuels such as gasoline and diesel. Bitumen was discovered in Nigeria in 1900. Nigeria has an estimated deposit of 42.7 billion metric tonnes of bitumen, and is said to rank close to Canada and Russia, which have the largest deposits of bitumen in the world. Nigerian bitumen deposits have been found in Ondo, Edo, Lagos, and Ogun States, with the largest deposits found in Ondo State in Western Nigeria. The estimated probable reserves of bitumen, which have been found in Ondo State, in the south-west region of Nigeria, is estimated to be about 16 billion barrels, while that of tar sands and heavy oil is estimated to be about 42 billion barrels. This bitumen occurs both on the surface and the subsurface. Bitumen products used in Nigeria have been processed from imported heavy crudes, and bitumen imported to supplement domestic consumption. Nigerian bitumen has many uses and applications. For example, heavy and extra heavy crude can be extracted from Nigerian tar sands, and sulphur and phenol can be derived from them. They are also suitable for the production of lubricants for industrial use. See Alberta Energy, What is Oil Sands; also The Citizen (2013).

[113] Limestone is a sedimentary rock composed primarily of calcium and aragonite, which are crystal forms of calcium carbonate. Most limestone deposits are formed from the skeletal remains of marine organisms such as coral, foram and molluscs. Limestone deposits are found in Benue, Sokoto, Dahomey, and Borno basins, as well as in Calabar and Kogi States in Nigeria. The largest and purest deposits of limestone in Nigeria are found in the south-west and middle belt regions of the country. Limestone found in the south-west region of Nigeria is estimated to be about 31 million tonnes. Limestone can be used crushed stone in construction, for road base and railroad ballast. Limestone is utilized domestically primarily for cement production. See Fatoye and Gideon (2013).

iron ore,[114] barite,[115] lead zinc,[116] gold,[117] kaolin, dolomite, feldspar, glass sand, ganstones (haphazard), iron ore, lead-zinc, tin and its associated minerals and recently gypsum. One of the objectives of the new National Policy on Solid Minerals in Nigeria is to ensure the orderly development of the mineral resources of the country.[118]

From the foregoing, Africa's mineral industry is undoubtedly one of the most expansive mineral industries in the world. With Africa's abundant land resources comes significant potential for mineral resource exploration and development.

[114] Iron ore are rocks and inseams from which metallic iron can be extracted. Nigeria has several deposits of iron ore, and deposits in Nigeria have primarily been found in the north central, north east, and south-east regions of Nigeria. It is estimated that there are over 3 billion tons of iron ore to be found in Kogi, Enugu, Zamfara and Kaduna states in Nigeria. However, the purest deposits have been found in Kogi state of Nigeria. These deposits are usually found in the form of hematite, magnetite, metasedimentary, bands of ferruginous, quartzite, sedimentary ores, limonite, maghemite, goethite and siderite. It is estimated that there are at least over 882.5 million tons of estimated reserves of iron ore found in northern part of Nigeria. The National Iron Ore Mining Company was established in 1979 with the mandate to explore, exploit, process and supply iron ore to the Ajaokuta Steel Company. Iron ore is used for making steel, transformer and motorcars, ferrous sulphate, electro-magnetic devices amongst other things. See generally, Foraminifera Market Research, Iron Ore Mining in Nigeria: The Opportunities.

[115] Barite is the naturally occurring mineral form of barium sulphate. It is extracted using both surface and underground mining. Barites are mostly used as a weighting agent for drilling fluids in oil and gas exploration. It is used in the manufacture of drilling mud, which is sued in oil prospecting. Other uses are found in car, electronics, paper and plastics production, rubber, glass, ceramics and the paint industry. It also used in radiation shielding and has medical applications. Deposits of barites bearing veins in Nigeria are mostly found in the Middle Benue trough. The quality of Nigerian barites is moderate to high. Nigeria has an estimated reserve of over 2,000,000 (two million) metric tons of barite ore found in different parts of the country, which include Nassarawa, Plateau, Benue, Taraba and Cross River States. It is estimated that proven reserves of barites for Benue and Nassarawa State of Nigeria are at 111,000 million tons, while the estimated probable unproven reserves are at 21, 123, 913 metric tons. See Foraminifera Market Research (2013).

[116] Lead-Zinc is usually found in association with each other, as well as in association with gold and silver. The two principal minerals containing lead and zinc are galena and sphalerite. Lead is bluish white, very dense and has a low melting point. It is essentially a co-product of zinc mining, or a by-product of copper, silver or gold mining. Zinc on the other hand, is a shiny bluish-white metal. Zinc minerals are generally associated with other mineral metals such as lead, copper or silver. It is estimated that Nigeria has over 10,000,000 tonnes of zinc ore deposit, which are found in the Benue trough. Zinc is primarily used in galvanizing as anti-corrosion coating on steel, construction materials, brass production, pharmaceuticals and cosmetics, as well as in the automotive market. Lead-Zinc ores are also utilized in manufacturing to produce batteries, electrical cables, solders glass, and also protective coating for other metals. Lead-Zinc in association are found in commercial quantities in the north-east and central regions of Nigeria, primarily in Adamawa, Taraba, Bauchi, Benue and Ebonyi States. See Foraminifera Market Research (2013), also United States Department of Energy (2002).

[117] Gold production in Nigeria began in 1913 and peaked in the 1930s, and the Nigerian Mining Corporation was formed in the early 1980s to explore for gold. The preliminary exploration and identification of gold deposits in Nigeria has confirmed that there are at least ten sites, found in the north-east, central and south-west regions of Nigeria, which hold reserves in excess of over 50,000 ounces of high quality gold. Gold is found in Zamfara, Kaduna, Katsina, Kebbi, Niger, Kogi and Osun States. See Nigerian Investment Promotion Council, Solid Minerals.

[118] See generally, Nigerian Investment Promotion Council, Solid Minerals.

Mining has been, and remains, a major economic activity has had a long and varied history in many parts of Africa. Africa represents half of the top 20 countries globally with the highest mineral export contributions.[119] For many African countries, mineral exploration and production constitute significant parts of economic planning and revenue generation. The sector presents strong potential to generate substantial revenue and employment enough to provide more visible economic benefits to the continent and improved livelihood for its population.

2.4.3 Stages of Mineral Exploration and Development

This section examines the different stages of the solid mineral exploitation and development. Mineral exploitation refers to "operations and works related to the technical and economic utilization of mineral resources, including mine development, extraction, treatment, processing and beneficiation of mineral resources as well as the activities necessary or related to the marketing of such mineral resources."[120] The main activities in the mining process are: exploration, development and mining; processing or beneficiation; smelting; refining; transportation; and storage.[121]

Summarily, the lifecycle of solid mineral extraction can be broadly divided into five key stages: prospecting and exploration; discovery and development; operation and production; closure and decommissioning; and post closure monitoring and site reclamation.[122] These processes and stages of mineral exploitation bear striking similarities with petroleum exploitation and development processes and are considered in detail below.

2.4.3.1 Prospecting and Exploration

Mineral prospecting is the first stage in mining. Reconnaissance or prospecting involves the search for minerals, precious metals or mineral specimens, through digging by hand, physical observation, rock sampling, geological surface analysis, geophysical surveys, geochemical surveys, photo-geological surveys, remote sensing techniques, laboratory testing and assays, or by other non-obstructive surveys or studies of surface geology, to define the extent of a mineral deposit and to determine its economic value.[123] When prospecting efforts reveal that something valuable may exist in the acreage, further exploration studies is conducted.

---

[119] Ibid.

[120] See s. 164 of the Nigerian Mineral and Mining Act.

[121] Natural Resource Governance Institute (2015a, b).

[122] See Sigam and Garcia (2012), pp. 4–5.

[123] See s 1 of Kenya's Mining Act, 2016. See also S. 164, MMA (2007). Prospecting must be differentiated from fossiking. Fossicking, or recreational mining, generally means searching for

Exploration means operations and works carried out to determine and ascertain the existence, quality, quantity and economic value of minerals for the purpose of commercial exploitation.[124] It includes conducting geological, geophysical, geobotanical and geochemical surveys or drilling; taking samples for the purpose of chemical or other analysis; using appropriate instruments, equipment and techniques; or extracting; or re-processing of existing seismic data, acquiring and processing new seismic data or any other related activity necessary to define the quality, characteristics and economic value of a mineral discovery.

Exploration usually begins with survey and research aimed at locating commercially viable concentrations of minerals. It is a sequential process of information gathering aimed at assessing and evaluating the mineral potential, characteristics and the economic value of mineral resources in a given area.[125] It starts with identifying portions of land that are worthy of exploration.[126] Exploration encompasses information gathering, mapping and surveying, either on the ground or from a plane or helicopter, where special equipment measures the magnetic or electrical properties of rocks on the surface and underground (subsurface). To prospect means to search for minerals, and it includes such activities and studies as is reasonably necessary to enable the prospector to test the mineral-bearing qualities of the land.[127]

Exploring for minerals is a complex process. It requires significant time, financial resources and expertise in several disciplines, such as geology, geography, chemistry and engineering. The total cost of exploration stage may vary from $20 million to over $150 million, and can take 3–10 years before feasibility decisions are made. If the information collected from the surveys indicates something out of the ordinary (an anomaly) in specific location on-the-ground work is done. An exploration team will visit each anomaly to see if minerals are present on the surface. Samples are taken for analysis to identify and measure the presence of base and/or precious metals. Other sampling methods include taking rocks, soil, till or gravel and lake sediment with a sample spacing of hundreds of metres. When it can be inferred that there are minerals but there is insufficient evidence to be certain such mineral discoveries are termed as inferred mineral reserve. When there is reasonable confidence that there are minerals, it is termed as indicated, probable or measured mineral reserve. If analysis indicates with high confidence there might be something of interest, this is called a proven reserve. When a proven reserve is

minerals for a purpose other than for commercial gain often to a depth of 2 m by digging by hand; or using hand held instruments. This is mainly recognized in mining laws in Australia, New Zealand, and nearby islands, but has yet to be widely recognized in mining laws in Africa. See Hsu et al. (2016).

[124] See s 164 of the Nigerian Mineral and Mining Act 2007.

[125] See s 164 of the Nigerian Mineral and Mining Act 2007.

[126] See Government of Canada, Minerals & Petroleum Resources Directorate, Stages of Mineral Exploration and Development.

[127] See Kenya's Mining Act, No. 12 of 2016.

found, more closely spaced sampling, modelling and feasibility assessment of the location would be done.[128]

2.4.3.2 Discovery and Development

Discovery and development occurs when a mineral of value is located. It involves the planning and construction of the mining project. At this stage, higher impact activities such as further ground geophysics, channel sampling, trenching and diamond drilling, generally take place.[129] This work is usually conducted on a mineral claim or lease. It is at this stage where permits, leases and licenses are required and where the project may be referred for environmental assessment. Mining companies must obtain appropriate permits, licenses and necessary environmental impact assessment approvals before they can begin construction. This will generally take months and in some cases years, depending on factors such as country, location, regulatory requirements and ore processing needs. Construction may not be confined to the mine itself. Frequently, mining companies construct local infrastructure and amenities to support both logistical and operational needs, as well as employee and community welfare.

Very few discovered mineral deposits become producing mines. It may take between 10 and 15 years or more for a mine to be developed. In many African countries, operation challenges such as limited infrastructure, including roads and power, as well as associated costs such as fuel, transportation, materials and labour, make it much more difficult to move beyond this stage. Most projects never get past this stage.[130]

If it appears likely that exploitation is possible, based on economics, the development stage follows with the preparation of a feasibility study and the construction of the mine. The development includes the building of the mine, the processing plant, roads, rails, sewer and waterlines, and housing to support the operation. The construction phase requires the largest amounts of investment and is the most expensive phase of the mining cycle—it can range from $100 million to $3 billion—and can last as long as 3–4 years.[131]

2.4.3.3 Operation and Production

Operation and production is the actual mining, milling, and processing of minerals. Mining operations means the operations and works carried out in the course of mineral exploitation, beneficiation, processing and contract mining.[132] The mining stage

---

[128] See Sigam and Garcia (2012).

[129] Government of Canada, Minerals and Petroleum Resources Directorate, Stages of Mineral Exploration and Development.

[130] See Sigam and Garcia (2012).

[131] Ibid.

[132] See s 164 of the Nigerian Mineral and Mining Act 2007.

starts with the commercial exploitation of the mine and consists in the removal of the mineral value in ore from the host rock or matrix. There are four major types of mining operations: surface mining (open pit or open cast), underground, dredging and artisanal.[133] While the choice of method depends on the size, shape and depth of the ore body, all operations involve the basic steps of ore breaking, loading and hauling to a mill for treatment.

After mining, processing related activities start with the mineral processing or beneficiation stage, which involves the activities to separate the mineral from waste material, remove impurities, or prepare the ores for further refinement. This beneficiation process is usually completed at the mine and its purpose is to yield a product that has a much higher content of valued material (concentrates). Following beneficiation, the concentrates are shipped to smelters or refineries to further extract and/or refine the metal, thus preparing it for its final use or for incorporation into physical or chemical manufacturing. Finally, the value addition activities include the metal working industries that process pure metals, forming it into commercially traded shapes such as ingots.

The length of time a mine is in production (the mine life) depends on the amount (reserves) and quality (grade) of the mineral, metal or gems and whether the operation is still profitable and can last as long as 100 years. It is at this stage that companies start to see a return on their investment (make some money). But they still have expenses (operating costs) such as wages, fuel, transportation, equipment and maintenance in order to keep the mine running. When commercial activity reach to an end decommissioning and mine closure activities must be carried out. The time needed to shut a mine varies from 2 to 5 years; normally, the bigger and more complex the operation, the longer it takes to close.[134]

2.4.3.4 Closure and Decommissioning

Mines have a limited lifetime, which is determined by the size and quality of the mineral deposit being extracted. Once the mineral resource at a mine is exhausted, or operations are no longer economical or profitable, mine closure occurs as the inevitable end. Mine closure is the process of shutting down mining operations on a

[133] In surface mining, workers remove the ore is all removed from the ground through open pit or open cast. When there are hard rocks, such as coal and diamonds, companies often use an *open pit*, while *open cast* mining is used for soft rocks, such as limestone. With underground mining, the surface remains intact and workers remove the minerals through tunnels or shafts. Underground mining begins with a phase of *development mining* whereby rocks are extracted so that miners can get closer to the ore. *Production mining* is when the ore with the desired mineral is extracted. Dredging is the removal of rocks and sediments from the bottom of lakes, rivers, harbours, and other water bodies. The sediment and ores are sorted and the undesired minerals are returned to the water or deposited elsewhere. Artisanal mines can be either on the surface or below the ground but are mostly on a small scale, subsistence and largely informal basis (i.e there is no large company overseeing the extraction). See Natural Resource Governance Institute (2015a, b), pp. 1–2.

[134] Ibid, see also Sigam and Garcia (2012).

temporary or permanent basis. It typically takes 2–10 years to shut down a mine, but it can take longer if long term water monitoring or treatment is required.[135]

After a mine closure, work then focuses on its decommissioning and rehabilitation. As part of decommissioning, operators take apart the mining processing facilities and equipment, while all buildings, premises, erections, infrastructure, water reservoirs, tailings, ponds, waste, overburden and other dumps, and appliances above or below the ground that have been utilized for the purpose of mining, treating or preparing minerals, obtaining, extracting or dressing mineral ores are all repurposed, demolished, or recovered, while waste is disposed of.[136] Mine closure and decommissioning plans are required by most regulatory agencies worldwide before a mining permit is granted, and must demonstrate that the site will not pose a threat to the health of the environment or society in the future. Depending on the site, the mine may be repurposed for other human uses or restored to its pre-mining use following closure. Financial assurance is also increasingly required by regulators as a guarantee that the funds required for mine closure will be available in the event that the responsible company is unable to complete the closure as planned.[137]

2.4.3.5 Post Closure Monitoring and Reclamation

A mining company is often required to monitor the mine site long after the mine site has been closed in order to ensure that the rehabilitation of the land is successful. Activities in this stage include the periodic monitoring and maintenance of tailings containment structures, and ensuring proper reclamation of the land and watercourses to an acceptable standard of productive use, ensuring that any landforms and structures are stable, and any watercourses are of acceptable water quality.[138] Reclamation is the process whereby mined land is restored back to its natural state. Reclamation typically involves a number of activities such as removing any hazardous materials, reshaping the land, restoring topsoil, and planting native grasses, trees, or ground cover.[139] Most mines are required to have closure and reclamation plans, and to put some money aside towards reclamation costs. The goal is to integrate environmental protection right from the beginning; to make sure the site is kept safe for people and animals when the mine closes.[140]

---

[135] See Natural Resources Canada (2013), pp. 65–77.

[136] See Mining Facts (2012).

[137] Ibid.

[138] Ibid, also Sigam and Garcia (2012).

[139] ibid.

[140] See Sigam and Garcia (2012). See also generally, Minerals and Petroleum Resources Directorate, Indigenous and Northern Affairs Canada.

2.5 The Need for Dynamic Legal Innovation in the Extractives Industry

The solid minerals sector, much like the petroleum sector of the extractive industry, faces complex legal and regulatory challenges. There are concerns in Kenya, South Africa, Liberia, Sierra Leone amongst others on the environmental impacts of mining activities.[141] Furthermore, issues of corruption and poor governance; lack of accountability and transparency; inadequate protection of labour and human rights of mining workers; resource nationalism and inadequate legal protection of investors and investments rights; to mention but a few have exacerbated legal risks relating to solid minerals investment in Africa.[142] Furthermore, lack of accountability, arbitrary confiscation of traditional and indigenous lands without compensation, siting of resource processing projects in poor and vulnerable communities, criminalization and persecution of rights advocates, and the absence of judicial and quasi-judicial remedies for victims of the above-mentioned problems.[143] As far back as 1980, the Organization of African Unity, the precursor to the African Union (AU), adopted the Lagos Plan of Action for the economic development of Africa summarized the major problems confronting Africa in the field of natural resource development as follows:

> Lack of information on natural resource endowment of large and unexplored areas…; lack of adequate capacity (capital, skills and technology) for the development of these resources; a considerable dependence on foreign transnational corporations for the development of a narrow range of African natural resources selected by these corporations to supply raw material needs of the developed countries; the inadequate share in the value added generated by the exploitation of natural resources of member States …; non-integration of the raw materials exporting industries into the national economies of the member States thus impeding backward and forward linkages; the extremely low level of development and utilization of those natural resources of no interest to foreign transnational corporations; and the disappointingly low general contribution of natural resources endowment to socio-economic development.[144]

The above analyses unfortunately remain true in contemporary Africa several years after. Apart from environmental, human rights and regulatory challenges, the geometric rise of artisanal and small-scale mining (ASM), and their impacts on commercial earnings that could have been utilized by government, have remained economic concerns in many parts of Africa. ASM, as defined in the Nigerian Minerals and Mining Act, means artisanal, alluvial and other forms of mining

[141] See Elbra (2017), pp. 18–25.

[142] Ibid.

[143] See generally Olawuyi (2015a, b); Oluduro (2014), p. 165 (providing an account of the nature and scope of the problems discussed); see also Human Rights Watch (1999), p. 47 (discussing the impact of governmental corruption in the forms of systematic kickbacks for the award of contracts, special bank accounts in the control of the presidency, and fraudulent allocation of oil or refined products to political favorites, on sustainable resource governance in Nigeria).

[144] See United Nations Economic Commission for Africa (2011), p. 5.

operations involving the use of low-level technology or application of methods not requiring substantial expenditure for the conduct of mining operations within a small scale.[145] African Mining Vision (AMV) reports that between 15 and 20% of the world's non-fuel minerals, 18% of Africa's gold, and almost all African gemstones except diamonds are produced by ASM. In Africa, an estimated eight million people are employed by the ASM sector; against 3 and 3.7 million in 1999.[146] Increased participation in ASM across the continent has been linked to factors such as poverty, conflict, natural disasters, political crises, decline in agriculture and climate change. Countries such as South Africa have sought to regularize and recognize the ASM sector as a vehicle for social and economic development, particularly for disadvantaged communities that are often excluded from participating in the national mining economy.[147] While South Africa and Nigeria have recognised ASM in mining laws, a number of other African countries still grapple with how best to regulate ASM.

As noted in Chap. 1, another key concern facing the overall outlook of the extractive sector in Africa is the near neglect of mineral development in several African countries, especially oil producing countries such as Nigeria. Despite the vast amount of mineral resources in countries such as Nigeria, DRC, Botswana, Kenya, and Tanzania, mineral exploration has taken the back seat while government expend several million on oil field exploration. The volumes of domestic trade deficit and foreign exchange losses resulting from the deficiency in exploiting the solid minerals sector are colossal. The availability of these minerals opens up myriad of opportunities, which include exports and use in domestic industries for generation of foreign exchange and internal revenue; economic diversification; and increased employment, particularly in the rural areas where the minerals are found. The development of the solid minerals sector could also contribute to technology transfer and development; and the development of infrastructure, especially in rural areas. The new and renewed emphasis on the mining sector in Africa is both timely and important.

The extractives industry must be viewed and developed holistically as one industrial sector that can provide complementary economic and social benefits. To achieve a sustainable and holistic development of the African extractives industry, African countries must address law and governance gaps that reduce sustainable resource management. There is an urgent need to put in place proper legislation and legal and regulatory reform to address some of the obstacles and bottlenecks, which prospective investors in Africa's mining sector often encounter. A starting point is for national authorities to ensure the issuance and administration of mining titles proceed in an objective, efficient and transparent manner.

[145] Nigerian Minerals and Mining Act No. 20 of 2007.

[146] International Labour Organisation (1999), Benkenstein (2012) and Buxton (2013).

[147] Minerals and Mining Policy for South Africa 1998.

2.6 Conclusion

The economic potential of Africa's solid minerals sector is enormous and can completely transform the economic fortunes of several African countries if sustainably and accountably harnessed. However, if lessons from the environmental, human rights and accountability challenges associated with petroleum activities in the continent are anything to go by, there is a need for African countries to revitalize and reassess extant extractive sector laws and institutions to make them more effective, and to avoid some of the problems associated with oil and gas investments in the continent. The holistic development and diversification of the African extractives industry to focus less on oil and gas, and promote solid mineral development, will depend on how well African countries can develop adequate law and governance structures that can effectively address recurrent problems of corruption, resource theft, environmental pollution and human rights abuses amongst others.

There is also a need for African governments to provide the right investment climate and environment that could attract foreign investment in the African extractive market. For example, in an increasingly globalized and competitive world, it will remain difficult and complex for African countries to leverage their geological affluence to attract foreign participation in the solid minerals and petroleum sectors, without adequate extractive infrastructure. This will include robust network of pipelines, safe road networks functioning rail systems, stable electricity and telecommunication facilities, all of which are currently unreliable across the continent.

Even though Africa's extractive potentials and opportunities have increasingly been rehashed in various international fora, legal and institutional frameworks for addressing current infrastructure gaps in the continent have not been exhaustively explored and communicated. To design effective legal and institutional response frameworks, an understanding of the intricacies of the various physical and associated infrastructure required to expand and stimulate extractive investments in Africa must, as a starting point, be clarified and examined. The next chapter therefore provides detailed examination of the current state of extractive sector infrastructure in Africa.

References

Alberta Energy. What is Oil Sands? http://www.energy.alberta.ca. Accessed 12 Apr 2018

American Association of Petroleum Geologists. What is Petroleum. http://www.aapg.org. Accessed 10 June 2018

American Gem Society. 4 Cs of Diamonds. http://www.americangemsociety.org/page/4cs. Accessed 2 May 2018

Ayoade M (2002) Disused offshore installations and pipelines: towards sustainable decommissioning. Kluwer, Alphen aan den Rijn

Bazhenova K (2009) Evolution of oil and gas generation in the Earth's history and petroleum prediction in sedimentary basins. Russ Geol Geophys 50(4):308–319

BBC News (2015) World's most expensive coloured gem sells for $30m. http://www.bbc.com/news/world-europe-32717029. Accessed 2 May 2018
Benkenstein A (2012) Artisanal and small-scale mining in Africa: opportunities and challenges. SAIIA, Johannesburg. http://www.saiia.org.za/opinion-analysis/artisanal-and-small-scale-mining-in-africa-opportunities-and-challenges. Accessed 10 Oct 2015
Bhardwaj A (2013) Challenges and solutions in an upstream and downstream oil and gas operation. http://globalenergy.pr.co/65678-challenges-and-solutions-in-an-upstream-and-downstream-oil-and-gas-operation. Accessed 5 May 2018
British Petroleum (2017) BP statistical review of World Energy. https://www.bp.com/content/dam/bp/en/corporate/pdf/energy-economics/statistical-review-2017/bp-statistical-review-of-world-energy-2017-full-report.pdf. Accessed 5 May 2018
Buxton A (2013) Responding to the challenge of artisanal and small-scale mining. How can knowledge networks help? IIED, London http://pubs.iied.org/pdfs/16532IIED.pdf. Accessed 10 Oct 2017
Cairn Energy (2010) Oil and gas project life cycle. http://www.cairnenergy.com/files/reports/responsibility/cr2010/overview/oil-and-gas-project-life-cycle.html. Accessed 10 June 2018
CM Enterprises. Minerals. CM Enterprises. http://www.cm-enterprises.org. Accessed 3 May 2018
Commission of the European Communities (2008) Is there a speculative bubble in commodity markets? Report of the Task force on the role of speculation in agricultural commodities price movement. SEC (2008)2971
Conoway C (1999) The petroleum industry: a non-technical guide. PennWell, Tulsa
Control Risks (2012) A new frontier: oil and gas in East Africa. http://www.cailaw.org/media/files/IEL/ConferenceMaterial/2013/OilfieldServices/ASekandi-paper.pdf. Accessed 21 Apr 2018
Daily Nation (2015) Titanium earns Kenya Sh9bn, surpasses soda ash. http://www.nation.co.ke/business/Titanium-earns-Kenya-Sh9bn-surpasses-soda-ash/996-2703628-j3cp47z/index.html. Accessed 11 June 2018
Diamond Facts. An Estimated 65% of the world's diamonds come from African countries
Diamond Facts. Conflict Diamonds. http://www.diamondfacts.org/index.php?option=com_content&view=article&id=118&Itemid=156&lang=en. Accessed 10 June 2018
Eckert A (1997) The World of Opals. Wiley, Hoboken
Economides M, Oligney R (2002) The color of oil: the history, the money, and the politics of the World's biggest business. Round Oak, Katy
Egypt Oil and Gas Web Portal. More than A 100-year journey. http://www.egyptoil-gas.com/publications/more-than-a-100-year-journey
Elbra PA (2017) Governing African gold mining: private governance and the resource curse. Springer, Berlin
Electric Power Research Institute (1999) Enhanced oil recovery scoping study. Final Report, No. TR-113836. Palo Alto, CA
Environment Canada. Petroleum Refining. http://www.ec.gc.ca. Accessed 5 May 2018
Extractive Sector Transparency Measures Act of Canada, S.C. 2014, c. 39, s. 376
Fatoye FB, Gideon YB (2013) Geology and occurrences of limestone and marble in Nigeria. J Nat Sci Res 3(11):60–65
Foraminifera Market Research (2013) Barite ore mining and milling in Nigeria: how viable? http://www.foramfera.com. Accessed 12 Apr 2018
Foraminifera Market Research. Iron ore mining in Nigeria: the opportunities. http://www.formafera.com. Accessed 12 Apr 2018
Foster NH, Beaumont EA, Bacheller WD, Peterson RM (1991) Hassi Messaoud Field—Algeria Trias Basin, Eastern Sahara Desert. In: Foster NH, Beaumont EA (eds) Treatise of petroleum geology Atlas of oil and gas fields, structural traps. American Association of Petroleum Geologists, Tulsa
Gao Z (1998) Environmental regulation of oil and gas. Kluwer Law International, Alphen aan den Rijn

Government of Canada, Stages of Mineral Exploration & Development in Northwest Territories. http://publications.gc.ca/collections/collection_2008/inac-ainc/R2-466-2007E.pdf. Accessed 12 Apr 2018
Grayson LE (1981) National oil companies. John Wiley & Sons, Chichester
Hsu T, Lucas A, Pardieu V (2016) Gem fossicking: recreational mining in Australia. https://www.gia.edu/gia-news-research/gem-fossicking-recreational-mining-australia. Accessed 12 Apr 2018
Human Rights Watch (1999) The price of oil: corporate responsibility and human rights violations in Nigeria's oil producing communities. Human Rights Watch, Washington, DC. hrw.org/reports/1999/nigeria/nigeria0199.pdf. Accessed 12 May 2017
International Council on Mining and Metals (2014) The role of mining in national economies. ICMM, London http://www.icmm.com/romine/index. Accessed 10 June 2018
International Energy Agency (2015) Africa energy outlook special report. IEA, Paris
International Energy Agency (IEA) (2016) Medium-term coal market report. IEA, Paris
International Labour Organisation (ILO) (1999) Social and labour issues in small-scale mines. Geneva, Switzerland
International Peatland Society (2015) What is a Peat? http://www.peatsociety.org/peatlands-and-peat/what-peat. Accessed 10 June 2018
Kazeem Y (2017) Two investigations into corruption in resource-rich DR Congo show why it stays so poor. Quartz Africa. https://qz.com/1038143/two-investigations-into-corruption-in-resource-rich-dr-congo-show-why-it-stays-so-poor/. Accessed 12 Apr 2018
KPMG (2014) Nigerian mining sector. KPMG, Lagos
Llewellyn J, Hansen B, Llewellyn P (2013) The changing face of the oil industry: White Paper Llewellyn consulting
Magloire PR (1970) Geology of giant petroleum fields, Triassic Gas Field of Hassi R'Mel.Algeria, American Association of Petroleum Geologists Memoir 14
Martin P, Kramer B (2009) Williams & Meyers manual of oil and gas terms. Lexis Nexis, New York
Mineral and Petroleum Resources Development Act No. 28 of 2002 of South Africa
Mineral Councils South Africa. http://www.chamberofmines.org.za/sa-mining/coal. Accessed 11 June 2018
Minerals and Mining Act No. 20 of 2007 of Nigeria
Minerals and Mining Policy for South Africa 1998
Mining Act of Kenya, No 12 of 2016
Mining Facts (2012) What happens to mine sites after a mine is closed? http://www.miningfacts.org/environment/what-happens-to-mine-sites-after-a-mine-is-closed/. Accessed 12 May 2017
Mining Greece. Ten most expensive minerals in the World. http://www.miningreece.com/miningreece-the-10-most-expensive-minerals-in-the-world/. Accessed 2 May 2018
Nassau K (1978) The origins of color in minerals. Am Mineral 63:219–229
Natural Resource Governance Institute (2015a) The mining industry. https://resourcegovernance.org/sites/default/files/documents/nrgi_primer_mineral-industry.pdf. Accessed 10 June 2018
Natural Resource Governance Institute (2015b) The oil and gas industry: overview and trends. https://resourcegovernance.org/sites/default/files/documents/nrgi_primer_oil-and-gas-industry.pdf. Accessed 10 June 2018
Natural Resources Canada (2013) Exploration and mining guide for aboriginal communities. https://www.nrcan.gc.ca/sites/www.nrcan.gc.ca/files/mineralsmetals/files/pdf/abor-auto/mining-guide-eng.pdf. Accessed 12 May 2018
Nigerian Investment Promotion Council. Solid minerals. https://www.nipc.gov.ng/?wpfb_dl=11. Accessed 04 June 2018
O'Brien R (2009) Australia's iron ore product quality. Geoscience Australia, Canberra
Olawuyi D (2015a) Increasing relevance of right-based approaches to resource Governance in Africa: shifting from regional aspiration to local realization. McGill Int J Sustain Dev Law Policy 11(2):113–158

Olawuyi D (2015b) Principles of Nigerian environmental law. Afe Babalola University Press, Ado-Ekiti
Olawuyi D (2018) Regulation of upstream oil & gas. Gulf legal advisor. Lexis Nexis, New York
Oluduro O (2014) Oil exploitation and human rights violations in Nigeria's oil producing communities. Afrika Focus 25:2
OPEC (2013) I need to know: an introduction to the oil industry & OPEC. http://www.opec.org/opec_web/static_files_project/media/downloads/publications/ChildrenBook2013.pdf. Accessed 21 Apr 2018
OPEC. Profile of Member Countries. http://www.opec.org/opec_web/en/about_us/25.htm. Accessed 21 Apr 2018
Petroleum (Exploration and Production) Act of Kenya, Cap 308
Petroleum (Production) Act 1934 of the United Kingdom
Petroleum Act of Nigeria, No. 1 of 1969, Cap P10, Laws of the Federation of Nigeria (2004)
Petroleum. In: Hugh Chisholm (ed) Encyclopædia Britannica, 11th edn. Cambridge University Press
Rossman G (2005) New discoveries of Painite in Myanmar (Burma). Gems Gemol 41(4):356
Shor R, Weldon R (2009) Ruby and Sapphire production and distribution: a quarter century of change. Gems Gemol 45:236–257
Sigam C, Garcia L (2012) United Nations Conference on Trade and development. In: Extractive industries: optimizing value retention in host Countries. UNCTAD, Geneva
Social and Economic Development Group, Middle East and North Africa Region (2006) Socialist people's Libyan Arab Jamahiriya Country Economic Report. http://siteresources.worldbank.org/INTLIBYA/Resources/libyacountryeconomicreport.pdf. Accessed 10 May 2018
South Africa Government. Mineral resources. http://www.gov.za/about-sa/minerals. Accessed 12 Apr 2018
The Citizen (2013) Nigeria's wasting bitumen wealth
Traut MW, Boote DRD, Clark-Lowes DD (1998) Exploration history of the Palaeozoic petroleum systems of North Africa. Geol Soc 132:69–78
Tullow Oil (2014) Tullow in Ghana. https://www.tullowoil.com/operations/west-africa/ghana. Accessed 19 Dec 2017
United Nations Economic Commission for Africa (UNECA) (2011) Minerals and Africa's development: the international study group report on Africa's mineral regimes. UNECA, Addis Ababa
United States Department of the Interior (2015) The mineral industry of Africa: 2012 Minerals Yearbook Africa. https://minerals.usgs.gov/minerals/pubs/country/2012/myb3-sum-2012-africa.pdf. Accessed 10 June 2018
US Department of Energy (2002) Energy and the environmental profile of the US mining industry, lead and zinc (Chapter 6). http://www.energy.gov
US Department of Energy- Office of Fossil Energy. Petroleum reserves. https://www.energy.gov/fe/services/petroleum-reserves. Accessed 5 May 2018
US Energy Information Administration (2015) Country analysis brief: Nigeria. http://www.eia.gov/beta/international/analysis_includes/countries_long/Nigeria/nigeria.pdf. Accessed 10 June 2018
US Energy Information Administration (2016a) Transportation sector energy consumption. In: International energy outlook. US EIA, Washington, DC
US Energy Information Administration (2016b) Industrial sector energy consumption. In: International energy outlook. US EIA, Washington, DC
US Energy Information Administration (2018) What are petroleum products, and what is petroleum used for? https://www.eia.gov/tools/faqs/faq.php?id=41&t=6. Accessed 10 June 2018
US Energy Information Administration. FAQs. https://www.eia.gov/tools/faqs/faq.php?id=41&t=6. Accessed 10 June 2018
US Geological Survey (2017) Mineral commodity summaries. https://minerals.usgs.gov/minerals/pubs/mcs/. Accessed 10 June 2018

US. Energy Information Administration. *International Energy Statistics*. Accessed 3 May 2018
Visual MD. Minerals. The Visual MD. http://www.thevisualmd.com. Accessed 10 June 2018
Wolf COH (2009) Petroleum sector value chain. World Bank Group, Washington, DC http://siteresources.worldbank.org/INTOGMC/Resources/noc_chapter_1.pdf. Accessed 10 June 2018
World Bank (2018a) Libya's economic report. https://data.worldbank.org/indicator/NY.GNP.PCAP.CD?locations=LY. Accessed 10 June 2018
World Bank (2018b) Nigeria's economic report. https://data.worldbank.org/indicator/NY.GNP.PCAP.CD?locations=NG. Accessed 10 June 2018
World Diamond Council. The Diamond Industry Fact Sheet. Diamond Facts. http://www.diamondfacts.org/pdfs/media/media_resources/fact_sheets/Diamond_Industry_Fact_Sheet.pdf. Accessed 12 Apr 2018
World Gold Council. Gold mining. https://www.gold.org/about-gold/gold-supply/gold-mining. Accessed 12 Apr 2018
Yager TR, Bermúdez-Lugo O, Mobbs PM, Newman HR, Taib M, Wallace GJ, Wilburn DR (2012) The mineral industries of Africa. In: Minerals yearbook. U.S. Geological Survey, Reston
Yglesias M (2014) Why speculators are stashing vast quantities of crude oil on tanker ships. Vox. https://www.vox.com/2014/12/25/7443569/contango-oil-storage. Accessed 5 May 2018
Zimbabwe Herald (2016) Fresh lithium deposit to transform Zimbabwe's Economy. http://www.herald.co.zw/fresh-lithium-deposits-to-transform-zim-economy/. Accessed 2 May 2018

Chapter 3
Extractive Sector Infrastructure

Abstract This chapter discusses the importance of a comprehensive and functional infrastructure network as a fundamental requirement for optimizing and maximizing the benefits of Africa's vast extractive resources. It unpacks the priority infrastructure required for extractive operations; the nature and extent of infrastructure gaps and the opportunities for infrastructure investments in Africa; and the legal framework for constructing, maintaining and operating extractive infrastructure in African countries. It discusses the need for appropriate legal frameworks that support the development and financing of shared use infrastructure projects across Africa.

3.1 Introduction

The surge in exploration and production of petroleum and solid mineral resources across Africa has created large scale demand for infrastructure to connect extracted resources with refineries, processing plants, and ultimately to end users.[1] Consequently, the African midstream sector, hitherto projected to enter into full maturity stage by 2050, has entered a high growth phase requiring significant amounts of capital investment to meet growing infrastructure needs, and providing investment opportunities for international extractive companies (ECs).

Extractive infrastructure refers to physical infrastructure (rail, roads, ports, bridges, pipelines, refineries, gathering systems) and associated infrastructure (telecommunications, electricity, logistics, and the institutions) required to extract, produce, refine, transport, store and distribute extractive resources across the extractives industry value chain.[2] As noted in Chap. 1, many resource rich African countries have significant infrastructure gaps that can make it more challenging to bring

[1] Olawuyi (2018), pp. 1–2.

[2] Ramdoo (2015). See also the Nigerian case of *Shell Petroleum Development Company of Nigeria Ltd. vs. Abel Isaiah and Ors.* (2001) 1 N.W.L.R. Part 723 page 169, where the court held that installation of pipelines, producing, treating and transmitting of crude oil to the storage tanks is a fundamental part of petroleum mining operations.

D. S. Olawuyi, *Extractives Industry Law in Africa*,
https://doi.org/10.1007/978-3-319-97664-8_3

extractive resources to market and transform their geological affluence into long-term development.[3] According to the World Bank's Africa Infrastructure Country Diagnostic, the cost of redressing Africa's infrastructure deficit is estimated at around US$75 billion, with the current funding gap being around US$35 billion annually.[4] Consequently, poor infrastructure remains as a major constraint to doing business and is found to depress firm productivity in Africa by around 40%.[5] Due to many years of corruption, sabotage, neglect and lack of proper maintenance, some of the essential infrastructure -such as pipelines, barge, cargo ship, refineries and road networks—that are essential for transporting, refining, storing and processing extracted resources are either non-existent, most especially across the West African region, or are preforming below optimal levels.[6] Similarly, pipeline networks have over the last decade faced heightened and persistent disruptions and sabotage.[7] In Nigeria for example, about over 4000 incidences of pipelines vandalization was reported since 2015 by NNPC as against 1121 cases in 2000.[8] The perennial vandalization of extant energy infrastructure has have further stifled the security and supply of natural gas required to power electricity plants in the country.[9]

Africa's largest infrastructure deficit is in the power sector. According to the World Bank, the 48 countries of Sub-Saharan Africa (with a combined population of 800 million) generate roughly the same amount of power as Spain (with a population of 45 million), barely enough to power one 100-watt light bulb per person for 3 h a day.[10] In Nigeria for example, an investor in the extractive sector faces logistical

[3] Ramdoo (2015).

[4] Foster (2008).

[5] Ibid, also Escribano et al. (2008).

[6] A Escribano, Ibid. Collier (2011), p. 48.

[7] Sabotage is the wilful act of destroying, damaging, or obstructing petroleum installations and transportation infrastructure thereby hindering petroleum production and distribution processes. Section 1 of the *Nigerian Petroleum Products and Distribution (Anti-Sabotage) Act of 1975* LFN 2004 c P12, outlaws the willful obstruction or prevention of petroleum production and distribution as an offence in Nigeria. It criminalizes all acts that disrupt petroleum production and distribution such as pipe vandalization, blow ups, obstruction of petroleum production or distribution; obstruction of the procurement of petroleum products; or the obstruction of vehicles distributing petroleum products such as motor spirits, gas oil, diesel oil, automotive gas oil, fuel oil, aviation fuel, kerosene, liquefied petroleum gases and any lubricating oil or greases or other lubricant. Under this Act, the offence of sabotage is punishable by a death penalty or a term of imprisonment not exceeding 21 years. In addition to the principal offender, the Act also outlines punishments for anyone who aids; incites, counsels or procures any other person, to commit an act of sabotage, whether or not that other person actually does the act in question. See also s. 7 of *The Miscellaneous Offences Act of 1983* Cap. M17 Laws of the Federation of Nigeria, 2004, which criminalizes the willful or malicious breaking, damage, obstruction, disconnection or tampering with the free flow of any crude oil or refined petroleum product through any oil pipeline. Any one found guilty of pipeline tampering and destruction shall be punishable by imprisonment for life.

[8] Punch Newspaper (2016).

[9] Punch Newspaper (2018).

[10] The report also notes that Africa's firms report losing 5% of their sales as a result of frequent power outages; this rises to 20% for informal sector firms unable to afford backup generation facilities. Foster (2008), pp. 2–3.

challenges relating to inadequate electricity supply; ageing refineries; timeworn pipeline infrastructure; and lack of motorable access roads to sites of mineral deposits.[11] With the exception of South Africa,[12] these problems are also prevalent in Kenya, Ghana, Tanzania, Mozambique, Zimbabwe and across the entire extractive industry in the continent.[13] As capacity increases with new investments in the generation, transmission and distribution sectors, these challenges can themselves provide opportunities for investors to spearhead infrastructural rehabilitation and expansion projects that are required to connect extracted resources with refineries, processing plants, and ultimately to end users.[14] The expansion potential of the African extractive industry, especially the midstream sector, can be enhanced if supported by robust and dynamic legal and governance frameworks that simplify and incentivize investment opportunities for potential investors. The International Energy Agency, in its latest report, states that $44tn is needed for investment in energy infrastructure assets worldwide with 60% of this relating to electricity-generating assets (including both fossil fuel and low-carbon energy infrastructure, and for mining the energy sources for use).[15] Bloomberg's annual report states that $10.2tn is needed for new electricity-generating infrastructure.[16] For Africa's extractive sector to be competitive, a significant of these investments must occur in Africa.

This chapter discusses the importance of a comprehensive and functional infrastructure network as a fundamental requirement for optimizing and maximizing the benefits of Africa's vast extractive resources. It also discusses the need for legal frameworks that support the development and financing of shared use infrastructure projects across Africa. Rather than the prevailing approaches in which investors have had to build small power plants, generators, single use pipeline networks, modular refineries and other build-as-you require forms of infrastructure investments to suit current project needs, African governments can stimulate and facilitate new infrastructure projects that meet long term, shared-use and more than one objective.[17]

After studying this chapter, you should be able to understand the priority infrastructure required for the extractive operations in Africa; appreciate the nature and

---

[11] See KPMG (2017), pp. 6–10.

[12] Southern Africa is one of the most developed regions in Africa, from an infrastructure and intra-regional connectivity perspective. Perhaps most notable is the widespread use of rail as a key mode of transport for freight transportation in Southern Africa—an option which is currently heavily under-utilized in the rest of Sub-Saharan Africa. According to the International Energy Agency (IEA), South Africa has the most advanced energy infrastructure in Sub Saharan Africa. See IEA (2014), pp. 30–34.

[13] See Ramdoo (2015) and Foster and Briceno-Garmendia (2010). See also Natural Resource Governance Institute (2015) pp. 2–3, discussing how the logistical challenges of getting coal to market in Mozambique from this remote location have impacted the profitability of the mine and left large stores of coal waiting outside the mine.

[14] OECD (2012).

[15] IEA (2016a, b, c).

[16] Bloomberg (2017).

[17] See Natural Resource Governance Institute (2015).

extent of infrastructure gaps in your country (if any) and the opportunities for infrastructure investments; understand and appraise the legal framework for constructing, maintaining and operating extractive infrastructure; and the key institutions tasked with approving or reviewing the construction of extractives infrastructure in an African context.

Forms of Extractive Infrastructure As earlier noted, extractive infrastructure can be categorized into two: physical infrastructure (rail, roads, ports, bridges, pipelines, refineries, gathering systems); and associated infrastructure (electricity, water, telecommunications, information technology, logistics, and the institutions). Each of these are discussed below.

3.2 Physical Infrastructure

Physical infrastructure refers to the basic structures required to move petroleum and solid minerals from one point to the other. Physical extractive infrastructure, such as rail, roads, ports, bridges, pipelines, refineries, gathering systems, serve important purposes in the transportation, storage and distribution of petroleum and mineral products.[18] A robust and functional physical infrastructure allows the seamless flow of produced resources between the various fixed points, from the points of production to the points of consumption that are distinctly spatially separated. An inexpensive, high quality transportation also encourages an indirect form of competition by making goods available to a market that normally could not withstand the high cost of transportation. This is an important issue, especially in determining the prices of extractive products.[19]

Given their different characteristics, petroleum and solid minerals are transported almost differently. Solid minerals are often transported through railways, marine ports, airstrips, concrete pipeline, roads and conveyor belts.[20] Each mode has its relative advantages and weaknesses with reference to carrying capacity and distance. Besides distance, transportation mode is also dependent on the mineral characteristics. In the case of low value bulk minerals, transportation can represent a high percentage of the consumer final cost. Minerals are shipped in dry bulk vessels for long distances, and in barges in rivers and channels. Some mineral ores such as coal may also be transported by pipelines. Crushed coal can be mixed with oil or

[18] See Lucas and Thompson (2016).

[19] Abduljabbar and Tahar (2012); also Ehimomen and Adeleke (2012).

[20] Some differences in transportation needs occur with mines that produce large volumes of commodities such as base metals.

water (the mixture is called a slurry) and sent by pipeline to an industrial user.[21] As vessels and barges can ship different bulk commodities such as coal, iron ore and bauxite, the rates and availability can vary a lot and are heavily dependent on the conditions of the different mining markets.

Truck transportation is used to move ore to a transloader, for placement onto a water or rail carrier, or for direct shipment to the customer in the case of refineries. Trucks have the advantage of routing flexibility and modest capital requirements. They are used mainly for shorter distances due to the high unit cost of moving generally low-value products in relatively small quantities (compared to rail or barges). Storage is necessary at several points in the mining supply chain. Because minerals are transported mainly in batches (e.g., a unit train or a vessel), rather than moved continuously through a network like natural gas, the supply chain must accommodate surges and lulls in demand at the mine; at the origin and receipt dock or port for water shipment; and at the end user, such as a power plant in the case of coal. Some minerals must be carefully handled, and storage must consider their specific characteristics to avoid problem to the environment or the hazards of damage to the quality of the mineral. For example, improperly stored coal can oxidize (weather), causing a loss of heat content, and if heat is allowed to build up in a stagnant coal pile (or in a vessel, barge, or railcar) the coal can self-ignite.[22]

Road transportation of solid minerals, *i.e.* movement of solid minerals via trucks and vehicles, is flexible, easy to access and readily available in many countries, although the quality and safety of roads can vary from one African country to another. Rail transport on the other hand, is comparatively cheap, more expensive than roads to construct but cheaper to maintain, relatively safer in terms of accident occurrence, and has a greater carrying capacity when compared with road transport. Ports are also highly useful in transporting large volumes across longer distances and routes. They are however more expensive to construct and maintain. Generally, the various modes of transporting solid minerals complement each other on numerous occasions, although in some cases they stand in competition against each other. Each of these transportation activities is a source of air pollutants. The construction and maintenance of access roads and of loadouts and transfer points disturbs the ground and discharges particulate matter into the air. Utilization of the roads and loadouts for transport releases particulates because of disturbance of the ground. The engines of the trucks, trains or conveyors used to transport the ores also produce pollutants.[23]

Oil on other hand is normally transported by one of four options: pipelines, rail, trucks and ships. The two main modes that are used for the long-distance transportation of petroleum are pipeline and ports, while rail and road may become highly advantageous when relatively short distances are involved. Road transport forms the major linkage between the depots and the bulk consumers and retails outlets (petrol/gas stations and warehouse) to the petty consumers. The cost per unit of transporting

[21] US Department of Energy.

[22] See Ehimomen and Adeleke (2012).

[23] See Olawuyi (2015a, b), pp. 28–30.

the products by road depends both on the road tanker vehicles in use and the marketers.[24] Since trucks can go virtually anywhere on land, trucks are often used for the first leg of crude deliveries. Oil pumped from a well might go straight into a truck, which then delivers it to a central facility for loading onto a pipeline, boat or railcar. Trucks are also faster to arrange when other forms of transportations are not available.[25] However, trucks tend to be the least preferred method in terms of safety. More than any other mode of oil transport, they operate in proximity to the public and can therefore be catastrophic when never there is an accident.[26] Oil spills from tankers is also be a major source of environmental pollution in many oil producing countries in Africa.[27] Similarly, road transportation can be at great risk of sabotage and attacks from hoodlums, in which case oil theft, vandalization and economic loss can almost be unavertable.

Ships, barges and boats can be very advantageous to transfer large volumes of oil. A barge has a cargo capacity of around 1.3 million gallons—and there can be several barges per tow—while the largest transoceanic tankers can carry around 84 million gallons.[28] A truck can move only about 9000 gallons, while a train of 100 cars can transport about 3 million gallons. Consequently, one barge can carry the load of hundreds of rails and trucks. Additionally, even with their great capacity, barges do not face the same kind of traffic and other logistical issues as ground transport.[29] However, if boats have spills, the environmental damage, especially to water bodies, can be much greater than that from other modes of transport. And while most spills dump very little oil, spills from ships and boats can be very significant and are a principal source of marine pollution.

Trains tend to spill a smaller amount of oil than other forms of transport. An International Energy Agency study noted that from 2004 to 2012 there were six times as many rail spills as pipeline spills, but the average pipeline spill was far graver.[30] Rail transport can be speedier, too.[31] However, oil transportation by rail can be very risky and catastrophic especially for local communities in cases of accidents.[32] Train accidents potentially threaten lives and can cause widespread property damage. Trains often travel straight through many cities at street level—as opposed to pipelines, which tend to be located underground and often far from populated

[24] Ehimomen and Adeleke (2012).

[25] Ibid.

[26] Olawuyi (2015a, b).

[27] Ibid. There have been several fatal incidents of oil tanker spillage in Africa. A recent example is the 2018 oil tanker disaster in Lagos, Nigeria. See Usman (2018).

[28] Molinski (2015).

[29] Molinski (2015); also TTI (2007), pp. 1–10.

[30] Tencer (2013).

[31] Molinski (2015).

[32] There is a long history of fatal accidents involving oil trains. Over 31 train accidents including oil trains were recorded in the US and Canada between 2013 and 2015. For example on July 5, 2013, a train that had been left unattended derailed, spilling oil and catching fire inside the town of Lac-Megantic in Quebec. About 1.6 million gallons of oil was spilled, with 47 people killed and 30 buildings burned down. See Associated Press (2016).

areas. A rail accident can therefore result in explosions and oil spills that could result in loss of lives and properties in monumental proportions.[33]

The most commonly used form of oil transportation is oil pipelines.[34] Given the strategic and unique importance of pipelines in the petroleum sector value chain, it is extensively discussed below.

3.2.1 Pipelines

A pipeline has been aptly defined as "a tube for the transportation of crude oil or natural gas between two points, either offshore or onshore."[35] Oil and natural gas fields are often located in long distances that are far from the market and could range from a few 100 km to more than 1000 km. Consequently, the transportation system for oil and natural gas consists of a complex network of pipelines, designed to safely, quickly and efficiently transport petroleum products from points of origin, to areas of demand. Pipelines are the essential link points for moving crude oil from the point of production (wellhead) to gathering and processing facilities and from there to refineries and tanker loading facilities.[36] Pipelines are also used to transport solids suspended in liquids, such as coal slurry, which consists of powdered coal suspended in water.

Pipelines can be characterized as interstate, intrastate or cross border. Interstate pipelines carry oil and natural gas from one state or province, within a country, to another. A good example is the Esravos-Lagos pipeline system, built in 1989, that transports natural gas from Escravos region in Niger Delta area to Egbin power station near Lagos in Nigeria.[37] Intrastate pipelines, on the other hand, transport natural gas internally within a particular state or province. Cross border pipelines are pipelines that transport processed oil and gas from a terminal situated in a producing country to a terminal situated in a consuming country. In essence, they transport oil and gas from one country to another.[38] A good example is the Pande-Temane Gas Pipeline, a $721 million, 865 km pipeline designed to transport natural gas from Mozambique to South Africa.[39] There is also the East African Crude Oil Pipeline, a $3.55 billion, 897-mile pipeline that will transport crude oil between Uganda and Tanzania. Once completed, the East African Crude Oil Pipeline is expected to transport 200,000 barrels of oil per day.[40] Similarly, the Chad-Cameroon Oil Pipeline (CCO Pipeline) is another cross-border pipeline project in Africa that has since

[33] Ibid, also Vertabedian (2015).

[34] Kilbourn (1970), p. vii.

[35] PricewaterhouseCoopers (2006), p. 60.

[36] See Lucas and Thompson (2016).

[37] Onyegbule (2012).

[38] Vinogradov (1999), pp. 75–80.

[39] World Bank (2015).

[40] Pipeline and Gas Journal (2017).

2003 transported oil from three oil fields in South-Western Chad to the Cameroon port of Kribi on to a floating storage and offloading vessel on the coast of Cameroon.[41]

Another key example is the $1billion West African Gas Pipeline (WAGP) a 421-mile long regional high-pressure gas transmission system, built to transport gas from the Niger Delta area in Nigeria to Ghana, Benin and Togo.[42] The WAGP is the first cross border pipeline to be installed in sub-Saharan Africa and is owned and operated by the West African Gas Pipeline Company. Cross border pipeline projects are more complex to construct and regulate given the myriads of international law questions that must be addressed in each of the countries that the pipeline will pass through.[43] For example, the WAGP was stalled for many years, after local communities in Nigeria complained it would damage land, destroy livelihoods and pollute fishing areas.[44] The social, economic and environmental issues that arise in the construction and financing of pipelines vary and differ depending on the resource to be transported in the pipeline. Various issues such as the flow characteristic of the resource, environmental hazards and costs implications and security risks must be carefully considered and mitigated when planning pipeline projects.

3.2.1.1 Types of Pipelines

Oil Pipelines The Nigerian *Oil Pipeline Act of 1969* defines an oil pipeline as 'a pipeline for the conveyance of mineral oils, natural gas and any of their derivatives or components, and also any substance (including steam and water) used or intended to be used in the production or refining or conveying of mineral oils, natural gas, and any of their derivatives or components.'[45] Even though this definition does not distinguish between pipelines used to transport oil, and those utilized to transport natural gas, condensates and other oil and gas derivates, the practical reality is that different configurations of pipelines are required to transport each of these products

[41] Energy Charter Secretariat 2015.

[42] World Bank (2014), pp. 1–10.

[43] See, for example, significant work by the United Nations Economic Commission for Europe (UNECE); World Bank (2003), "Cross Border Oil and Gas Pipelines: Problems and Prospects" noting that the reasons for the complexities include: different parties, each with different interests, are involved; there is no overarching legal regime that can be used to police and regulate activities and contracts; and profit and rent are to be shared between the various parties and mechanisms exist to encourage one or other party to seek a greater share of that profit and rent. See also Dulaney and Merrick (2005) discussing the various legal issues that are encountered in cross-border oil and gas pipeline projects and how treaties, intergovernmental agreements and host government agreements can effectively manage some of those legal issues.

[44] See Olawuyi (2015a, b).

[45] S. 11 (2) of the *Oil Pipelines Act*, Chapter 338 Laws of the Federation of Nigeria 1990. Similarly, in *Nigerian AGIP Oil Company vs. Kemmer* (2001) 8 N.W.L.R. Part 716 page 506, where the meaning of oil pipeline was considered. The court held that: "An oil pipeline means a pipeline for the conveyance of oil minerals, natural gas and any of their derivatives or components, and also any substance (including steam and water) used or intended to be used in the production or refining or conveying of mineral oils, natural gas, and any of their derivatives or components."

given their distinctive characteristics, such as heaviness, viscosity and flow. These differences in physical properties between oil and natural gas dictate that they move on different pipelines. Given that oil is less bulky, oil pipelines can transport about five times the volume of what a gas pipeline can transport. Oil pipelines are therefore usually cheaper to construct and operate, and have less risks than gas pipelines. For oil pipelines, the main risk is the risk of oil spill during transportation. The process of cleaning up oil spills can be more straightforward and cost efficient.[46]

The pipeline construction industry has grown from a support industry to a key industry in the midstream sector of the oil and gas value chain. Early pipelines in the United States were installed in 1834 in Millville, New Jersey, to carry manufactured gas for gas lighting purposes in major cities.[47] After oil was discovered in Pennsylvania in 1858, a new use for pipe and pipelines soon emerged. The first successful oil pipeline, a 2-1/2-mile-long 2-inch-diameter pipeline was laid in 1863 and it moved 800 barrels (33,600 gallons) of oil per day.[48] Since that time, there have been significant advances in oil pipeline manufacturing, construction, corrosion control, and testing and inspection aimed at reducing the possibilities of leakage and spills. Generally, once crude oil is refined into petroleum products, the products may be transported to the final users using other less expensive forms of transportation such as trailers and trucks and rail. However, oil pipelines are still very important and often utilized to transport refined oil to tank farms and to different retail and distribution depots. Oil pipelines run through specific route created for it by the national oil company or supervising Ministry for petroleum. In Nigeria, most of the current oil pipelines are owned and operated by the Federal Government through the Petroleum Pipelines and Marketing Company (PPMC), a subsidiary of the national oil company.[49] A number of international oil companies (IOCs) however own and operate their own oil pipelines following approval from NNPC and the Ministry for Petroleum Resources.

Gas Pipelines Like oil, the most commonly used form of transportation for gas is the gas pipeline. Transporting gas is far more expensive than transporting oil. The gaseous and bulky nature of natural gas means that it is difficult to transport gas by road, rail or through other forms unless if the gas has been liquefied. Similarly, unlike oil pipelines, gas pipelines need to be operated at high pressure. Consequently, high pressure gas pipelines are the main options for transporting natural gas in its gaseous form. High pressure gas pipelines have a higher construction cost and maintain, including the high cost of compressing the gas.

---

[46] Olawuyi (2015a, b).

[47] Kiefner and Trench (2001).

[48] Ibid.

[49] See the case of *Shell Petroleum Development Company vs. Burutu Local Government Council* (1998) 9 N.W.L.R. Part 565 page 318, on the issue of ownership of installations for petroleum operations by the NNPC.

The composition of the natural gas defines how it will be processed for transport. When in the reservoir, natural gas is found in three states: *non-associated*, where natural gas is not in contact with, or dissolved in, crude oil in the reservoir; *gas cap*, where it is overlying an oil reserve; and *associated gas*, where natural gas is mixed with oil in underground reservoirs.[50]

For gas pipelines, the main risk is the risk of explosion which can be catastrophic and difficult to deal with or clean up.[51] In addition to the environmental and health implications of gas pipeline explosion, explosions are difficult to contain and could result in monumental loss to the gas company. Similarly, the consequences of disruption in the operation of a gas pipeline may be more serious and complex than an oil pipeline. Methane is highly flammable; undetected leaks could lead to accidents and explosions. Additionally, natural gas is predominantly methane, a potent greenhouse gas. Undetected natural gas pipeline leaks could be a contributor to climate change.[52] There are also environmental concerns on the impacts of gas pipeline leakage for gas pipelines crossing environmentally sensitive areas, for example delicate natural habitats, near communities, and close to water aquifers.[53]

Furthermore, unlike oil pipelines that can be used to transport different blends and grades of oil, gas pipelines are tailored to pressure profile, surges and cycles and can therefore not be easily converted or used to transport other gas products. Appurtenances such as leak detection and monitoring systems, pump stations, compressor stations, tank farms and In-Line Inspection (ILI) launching/receiving facilities, flow meters, strainers, liquid separators, corrosion control devices, leak detection devices, control valves and sectionalizing valves may need to be altered.[54]

Like oil pipelines, gas pipelines run through specific route created for it by the national oil company or supervising ministry for petroleum resources. In Nigeria, most of the current gas pipelines are owned and operated by the Federal Government through the Nigeria Liquefied Natural Gas Company (NLNG), a limited liability company mainly owned by NNPC and established pursuant to the *NLNG Act,* to harness Nigeria's vast natural gas resources and produce Liquefied Natural Gas (LNG) and Natural Gas Liquids (NGLs) for export.[55] A number of international oil companies however own and operate their own gas pipelines following approval from NNPC and the Ministry for Petroleum Resources.

LNG Transportation Systems LNG refers to gas that has been converted to liquid form for ease and safety of storage or transport. It is 'an odourless, colourless, non-corrosive and non-toxic product of natural gas consisting primarily of methane (CH_4) that is in liquid form at near atmospheric pressure.'[56] Converting and shipping

[50] PwC (2006).

[51] Reuters (2018).

[52] Olawuyi (2015a, b), pp. 108–110.

[53] Ibid.

[54] Olawuyi (2015a, b), pp. 108–110.

[55] The Nigeria LNG (Fiscal Incentives Guarantees and Assurances) Act Cap N87 2004.

[56] PwC (2006), p. 47.

gas in liquid form is often preferred for international transportation because, in a liquid form, natural gas takes up less volume and does not ignite, making it easier and safer for shipment and storage. Where natural gas cannot be delivered by pipeline, it can be liquefied and delivered by ship. However, transporting LNG requires a complex and expensive infrastructure system that includes use of pipelines, liquefaction plants, tankers and ships and regasification plants.[57] LNG infrastructure includes a gas pipeline leading to the seaside, gas liquefaction plant, storage facilities and an LNG terminal for shipment.[58]

Natural gas is liquefied by lowering or cooling the temperature of natural gas to approximately −162 °C (−260 °F).[59] This temperature drop liquefies the methane present in the natural gas therefore shrinking the volume of the gas 600 times and making transportation at atmospheric pressure in the form of LNG possible.[60] LNG is then transported internationally via tankers and ships in its liquid state. In its liquid state, LNG is safer to transport as it will not ignite. Additionally, LNG takes up significantly less space as a liquid than a gas.[61] LNG is kept in its liquid form through auto refrigeration and then introduced into specially insulated tankers and transported around the world. When LNG reaches its destination, it is turned back into a gas at regasification plants, for distribution and sale. The gas is then transported to end users via gas pipelines.

Constructing and maintaining the LNG liquefaction system is by itself specifically complex, extremely expensive, and could take several years to complete. The cost of a LNG pipeline project, including gas field development, liquefaction plant, special LNG tankers, and the regasification plant, in the past typically would be quoted at US$9–12 billion, therefore reducing the revenue benefit to the governments concerned.[62] The process of liquefaction is also highly energy intensive, with around 15–18% of the gas effectively wasted in producing the liquefaction process.[63]

3.2.1.2 Flow Process and Systems

Generally, in order to travel from point of production to refinery and then to the end users, oil and gas must pass through three major categories of pipeline systems: the gathering system, the interstate pipeline system, and the distribution system.

---

[57] According to the EIA, Qatar is currently the largest exporter of LNG in the world. Qatar started exporting LNG in 1997, and currently exports over 4.3tn cubic feet (tcf) with 31% of market share in 2014. Qatar is also at the forefront of gas-to-liquids (GTL) production, and the country is home to the world's largest GTL facility.

[58] See Shell, *Liquefied Natural Gas (LNG)*.

[59] Ibid.

[60] Ibid.

[61] LNG is approximately 1/600th the volume of the same amount of natural gas. Ibid.

[62] See World Bank (2003).

[63] Ibid.

The primary function of the gathering system (also known as the collecting system or gathering facility) is to gather crude oil or natural gas from different offshore fields and sources of production at the wellsite, and bring it to a main storage facility, a processing plant, or a shipping point for treatment and/or further transportation.[64] The gathering system consists of low pressure, small diameter pipelines that transport raw natural gas from the wellhead to the processing plant. Most gathering systems are owned by ECs or the national oil company and are designed, constructed, and installed over many years, to interconnect the wellhead to the mainline transmission company. After the gathering system delivers the oil or natural gas to the common collection point, it may then be delivered to a gas transmission system (for natural gas) and the oil trunk lines (for oil) for onward movement to another processing or storage terminal.

Once gas is processed, natural gas pipeline operators ship the natural gas through large gas transmission systems to different delivery points. Natural gas must be highly pressurized to move it along the pipeline. To ensure that the natural gas remains pressurized, compressor stations are placed in intervals along the pipeline. The natural gas enters the compressor station, where it is compressed by either a turbine, motor, or engine. Metering stations are also installed throughout the pipeline network to monitor for pressure, flow and leaks.[65] For oil, large trunk lines consisting of pipelines are used to transfer crude or refined oil across countries or within states to the delivery point. The delivery point may be an entity or utility that used high volumes of gas, such as a power station, or it may be a distribution system that is located in a state or province that will then split and transmit the gas in smaller quantities to the end users.

The distribution system is therefore a network of low pressure pipelines, storage tanks, and fuel lines, interconnected to facilitate and monitor the flow of oil and gas from the transmission system to the individual end user. Each of these systems have important roles to play in the petroleum sector value chain and must be closely and properly maintained to avoid sudden catastrophic leaks.

Rupture of or leakage from fuel distribution systems can threaten environmental quality and can also result in disruption of power, water and all other associated infrastructure in a country. A number of laws have therefore been put in place in many resource producing countries to ensure the safety and integrity of pipeline networks and systems.

3.2.1.3 Regulation of Pipelines

Pipeline design, construction, approval, maintenance, safety, pricing, environmental issues, and other aspects of pipeline operation are all highly regulated. To ensure the security and safety of oil and natural gas supply, oil pipelines are often regulated by a legislation by the resource producing country. In Nigeria for example, oil pipelines

---

[64] See de Silva and Bremen (2008), p. 149.

[65] See de Silva and Bremen (2008), p. 149.

are regulated by the *Oil Pipelines Act*.[66] Under the Act, the permission of the Minister for petroleum resources is expressly required to build and operate a pipeline together with any ancillary installations necessary for the operation of oil pipelines, including storage tanks, pumping stations and loading terminals in Nigeria. *The Oil and Gas Pipelines Regulations and the Department of Petroleum Resources' Guidelines and Procedures for Construction Operation and Maintenance of Oil and Gas Pipelines* prescribe, at length, standards to be adhered to in the design, construction, operation, maintenance and upgrade of petroleum pipelines in Nigeria.

There are two types of permits that can be obtained under the Act: a permit to survey routes for oil pipelines and license to construct, maintain and operate oil pipelines.[67]

Permit to Survey Routes for Petroleum Pipelines A permit to survey entitles the holder to enter upon a defined land area to access and survey routes for the construction of oil pipelines. This includes the right to enter, together with officers, agents, workmen or other servants and with any necessary equipment or vehicles, to survey and take levels of the land; dig and bore into the soil and subsoil; cut and remove such trees and other vegetation as may impede the purposes specified in this subsection and; pass over land adjacent to such route to the extent that such may be necessary or convenient for the purpose of obtaining access to land upon the route specified; and do all other acts necessary, to ascertain the suitability of the land for the establishment of an oil pipeline or ancillary installations.[68]

The power to grant permit to survey is vested in the Minister of Petroleum, pursuant to Section 3 of the Oil Pipelines Act which states that the Minister may, grant permits to survey routes for oil pipelines and licenses to construct, maintain and operate oil pipelines. Consequently, prior to the installation of any pipeline, an under the ground permit to survey the land must be obtained. The survey permit precedes the oil pipeline licence. Each license shall be issued in respect of and authorise the construction, maintenance and operation of one pipeline only. A survey would have to be carried out on the land on which the oil pipeline will be laid in order to ascertain its fitness for the oil pipeline.

The Oil Pipelines Regulations provides that no oil pipeline license shall be granted or renewed unless the route of the pipeline has been surveyed; or in the case of a renewal, the pipeline has been re-surveyed. Furthermore, no survey or re-survey shall be carried out under Regulation 1 of the regulations unless a permit to survey has been obtained in accordance with the provisions of sections 4 and 5 of the Act. An application for a permit to survey shall be accompanied with ten copies of the topographical map of the proposed route of the pipeline or alternative routes proposed. The Minister may grant the permit to survey on payment of the fees required by section 31 of this Act; or may refuse to grant the permit to survey. If the Minister

[66] *Oil Pipelines Act of 1956*, Laws of the Federation of Nigeria 2004, c O7.

[67] Ibid, s. 4.

[68] Ibid, s 5.

refuses to grant the permit to survey the applicant will be notified in writing with the reasons for the refusal.

The Act contains notice provisions which mandate the holder to give not less than 14 days' notice to the owner or the lawful occupiers of the land through which a proposed pipeline would run. This provision protects the property and land rights of existing land owners who must be duly consulted prior to the construction of pipeline projects. Furthermore, upon obtaining the prior consent of such owners or lawful occupiers and the payment of adequate compensation to them, the holder of a permit to survey must also take all reasonable steps to avoid necessary damage to any land entered upon and any buildings, crops, or profitable trees thereon, and shall make compensation to the owners or occupiers for any damage done under such authority and not made good.[69] Upon the successful completion of a survey, the holder may apply to the minister for an oil pipeline license.

License to Construct, Maintain and Operate Oil Pipelines (Petroleum Pipeline License) A Petroleum Pipeline License (PPL) typically gives the holder the right to construct and operate a pipeline on a designated pipeline land. In Nigeria for example, pursuant to Section 7 of the Oil Pipelines Act, the holder of a permit to survey may make an application to the Minister for the grant of an oil pipeline license in respect of any oil pipeline. The oil pipeline must be the one for which the applicant has completed the survey of the route. An applicant for a license shall deliver to the Minister an application stating the terminal points and giving a description of the pipeline. The application shall also be accompanied by a plan of the proposed route of the pipeline sufficient to identify the land affected thereby and the proposal of any pumping stations, tanks or other ancillary installations.[70] The application must be accompanied by a statement indicating the services to be rendered by the pipeline; the specification of the pipeline; the characteristics of the fluids to be conveyed through the pipeline; and the total estimated cost of construction of the pipeline.

The Minister shall upon receipt of the application appoint a date not less than 6 weeks ahead for the hearing of any objections, The Minister will also nominate the person or persons by whom and the place(s) at which any such objections will be heard. The applicant must notify the general public, in a Federal Gazette, in the Gazette of each state concerned; and in newspapers circulating in the areas through which the route of the projected pipeline passes about the project. Also, holders of exclusive prospecting licenses, mining rights, oil exploration licenses and oil prospecting licenses, lessees of mining leases, temporary mining leases or oil mining leases must be notified.

The Minister may approve the application or refuse to grant the license for reasons which he/she considers sufficient. If the Minister refuses to grant the license, the applicant shall be notified in writing of such refusal and the reasons for such refusal. Where satisfied that the pipeline can be operated efficiently, the minister will approve the application and the parties will agree conditions for use of the

---

[69] Ibid, s 6(1) (2) (3) (4).

[70] Ibid, s 8.

pipeline, failing which, the minister may determine such conditions.[71] The terms and conditions of access are to be determined by agreement between the parties and failing, such agreement shall be determined by the Minister. The Minister may impose such requirements that are considered necessary to secure the access right of the applicant and regulating the access charge.

No person other than the holder of a license shall construct, maintain or operate an oil pipeline. Every person who acts in contravention of the of the Act provisions with respect to obtaining the licence shall be guilty of an offence and shall be liable on conviction to a term of imprisonment or to a fine or to both imprisonment and fine. The Minister may require any person who is convicted of an offence under that section to have the pipeline and any ancillary installation removed. An offender who is required by the Minister to have a pipeline or any ancillary installation removed shall be liable for making good any damage done to any land by such removal.

Section 17 (1) of the Oil Pipelines Act stipulates that a licence may be granted for such period not exceeding 20 years as the Minister may direct. Where any person other than the owner of the pipeline seeks to have a right to convey any petroleum product in an oil pipeline constructed, maintained and operated in pursuance of a licence granted under the Act he must make an application to the Minister.[72] Every such application shall be made in the prescribed manner and form containing the prescribed particulars. To renew an existing license, there is a mandatory requirement to resurvey an existing pipeline and obtain a permit to the survey as the preconditions for the renewals.[73] The duration of such pipeline renewal should not be less than the residue of the oil mining lease.

The holder of an oil pipeline licence may terminate the licence, in respect of the whole route or part thereof at any time during the currency of the term granted, by the giving of not less than 3 months prior notice of his decision (to so terminate) to the Minister (in accordance with the provisions contained in section 17 (3) of the Pipelines Act.

Similarly, the Minister may terminate or revoke a pipeline license on three grounds. First is for the effluxion of time, which occurs at the end of the term granted and of any renewals where such renewals had been granted. Second, is for cessation of oil and natural gas operations from which petroleum products is transported through the pipeline. Given that the cessation of petroleum operations would make the continued retention of the license and the maintenance of the pipeline by the holder meaningless, the Minister may revoke the pipeline license. Practically, the license holder may apply to covert the pipeline for other usage. For example, an oil pipeline may be converted to an natural gas pipeline and vice versa. In such a case, the holder shall apply to the DPR as provided in Regulation 22 of the 1995 Oil and Gas Pipelines Regulations. The third ground is revocation for cause, which occurs if the holder of an oil pipeline licence is in default of any of the terms or conditions upon which the licence was granted, and fails to cure such default after

[71] Ibid.

[72] Ibid, s 18 (1).

[73] See Regulation 1 of the Oil and Gas Pipelines Licence Regulations 1995.

due written notice thereof has been given.[74] The Minister may by notice to such holder, proceed to revoke the licence. Revocation may also be effected by the Minister if the holder fails to comply with certain requirements imposed by the Minister relating to the use of the pipeline by persons other than the licence.[75] Upon the termination or expiration of an oil pipeline licence, the holder shall within 3 months after such termination be free (upon giving notice of 3 weeks to the Minister) to remove such pipeline and the ancillary installations to the extent that the Minister does not elect to purchase such pipeline and ancillary installations or any part thereof. In the event of a disagreement as to the purchase price, the price payable therefore would be determined by arbitration.[76]

Except for cases of sabotage, the licence holder is generally under obligation to compensate any person who suffers damage as a consequence of any breakage of or leakage from the oil pipeline or any ancillary installations. However, the licensee does not have any obligation for any losses and or damage that are as a result of sabotage or by a person not responsible to the licence holder or whom the licensee is not expected to control.[77]

A number of other secondary legislation and regulations apply to pipeline installation and operation. There are regulations that require an environmental impact assessment (EIA) be carried out prior to approval being obtained for the project and subsequent execution. For example, under the Nigerian EIA Act, pipelines more than 50 km require an EIA approved by the Ministry of Environment.[78] Also, pursuant to the Nigerian Land Use Act, the use of land for the construction of gas pipelines may be regarded as use for an overriding public interest for which the Government may compulsorily acquire land.[79] Furthermore, the *Oil Terminal Dues Act* contains relevant provisions on the export of crude oil that must be complied with. It provides for the levying and payment of terminal dues by vessels evacuating crude oil at terminals in Nigeria.[80] The Act also prohibits the installation and operation of oil terminals, except with the written approval of the minister of petroleum, among other requirements.[81] Crude oil is evacuated at oil terminals by ocean tankers that must comply with the Crude (Transportation and Shipment) Regulations which,

---

[74] Ss 27 (1) and 27 (2) of the Oil Pipelines Act.

[75] Ibid, section 18 (9).

[76] Ibid, s. 28 (1).

[77] Ibid, s 11 (5) c.

[78] *Environmental Impact Assessment Act of 1992*, Laws of the Federation of Nigeria 2004, c E12.

[79] S. 28 of the *Land Use Act of 1979*, Laws of the Federation of Nigeria 2004, c L5 which provides that that land may be appropriated for "overriding public interests." This includes "the requirement of the land for mining purposes or oil pipelines or for any purpose connected therewith." For an excellent and detailed critique of this section and the Land Use Act in general, see Ako (2009).

[80] *Oil Terminal Dues Act* 1969 No. 9 (Laws of the Federation of Nigeria 1990, Cap. 339).

[81] Ibid, in *Seleba vs. Mobil Producing (Nigeria) Unlimited* (2002) 12 N.W.L.R. Part 995 page 643 the Nigerian court defined "Oil Terminal" to mean an oil loading terminal, pumping or booster station, or other installation (or structure associated with a terminal, including its storage facilities) other than a terminal situated within a port or any approaches within the meaning of this Port Act.

among other things, require all declarations regarding the capacity of a vessel in which crude oil is carried to be verified by the appropriate government authority.[82]

A petroleum exporter will also require export permits issued by the Department of Petroleum Resources and the Federal Ministry of Industry, Trade and Investment to export oil or gas. *The Pre-shipment Inspection of Exports Act* subjects exports from Nigeria (including crude oil exports) to inspection by pre-shipment inspection agents.[83] The Act also requires exporters of goods, including petroleum products, to open, maintain and operate a foreign currency domiciliary account in Nigeria, into which all export proceeds must be paid. *The Coastal and Inland Shipping (Cabotage) Act* restricts the use of foreign vessels for the carriage of goods within coastal territorial inland waters, or any point within the waters of Nigeria's exclusive economic zone.[84] Furthermore, the Act restricts vessels, tugs or barges that are not beneficially owned by a Nigerian citizen from the carriage of materials or supply services to and from oil rigs, platforms and installations, whether offshore or onshore or within any ports in Nigerian waters.[85] It is a criminal offence in Nigeria to export, import or distribute petroleum or petroleum products without a valid permit or license from the petroleum ministry.[86]

---

[82] *Crude Oil (Transportation and Shipment) Regulations (S.I. 44 of 1984).* These Regulations prescribe rules for the transportation of crude oil by ships. It prohibits, among other things: (a) the loading or storage in tanks other than those designated for the purpose; (b) loading of crude oil in an unauthorized location; (c) transportation of crude oil without the required documentation; and (d) the making of false declarations regarding the capacity of the ship used for the transportation of crude oil.

[83] *The Pre- shipment Inspection of Export. Act,* no 10 of 1996, Cap P26 Laws of Federation of Nigeria, 2004. The aim of the Act is primarily to address issues relating to under valuation of oil exports, delays in issuance of invoices and payments for goods, manipulation of crude oil prices, illegal swaps, deliberate illegal over loadings, topping-up of tanker loads by as much as 1% (against International Standard of 0.1(10a) and other disturbing leakages and lapses that were prevalent in the Nigerian oil sector. The aim of the Act is to instil probity and integrity in oil shipment to reduce losses due to sharp practices of employees at the oil terminals, monitor the export activities of the Joint Venture Oil Companies and their agents, to enforce effective supervision of loading and lifting of the Nigerian Crude Oil and Non-oil export based on quantity, quality and value of the product.

[84] *Coastal and Inland Shipping (Cabotage) Act* 2003 (No. 5 of 2003). The objective of the Act is primarily to reserve the commercial transportation of goods and services within Nigerian coastal and inland waters to vessels flying the Nigerian flag and owned by persons of Nigerian citizenship. It encourages indigenous shipping lines owned to participate in domestic shipping and cabotage trade.

[85] See ss 3–5, ibid.

[86] *The Miscellaneous Offences Act of 1983* M17 Laws of the Federation of Nigeria, 2004, criminalizes any illegal imporation, export or purchase of minerals without the permission of the appropriate national authority. Section 17 prohibits the illegal importation, export, sale or distribution of crude oil, petroleum or petroleum product in Nigeria, while s.18 criminalizes petroleum adulteration.

3.2.1.4 Pipeline Operating and Transportation Agreements

In the past, the roles of pipeline companies were to facilitate the transfer of oil and gas from a producer to the purchaser. However, increasingly over the last decade, this role has metamorphosed with oil and gas producers selling directly to end users. Many oil and gas producing countries and IOCs have increasingly put in place their own pipeline infrastructure and pipeline companies that enable them to supply gas directly to local distribution companies (LDC), industrial facilities, and to other end users. As a result, pipeline companies increasingly serve as both the producer and the transporter of extractive products. Pipelines companies provide transportation services to their customers under contracts that are in accordance with established pipeline tariffs. These contracts have historically been long-term in nature, typically more than 20 years, but can also range anywhere from 30 days to 10 years depending on the commercial intention of the pipeline owner and the customer. As these long-term contracts expire, replacement contracts are negotiated for shorter average periods. The transportation contract specifies the type of service to be provided, the points of receipt and delivery, the transportation fees, and other contractual terms such as maximum daily quantities, quality of gas, amongst others.[87]

There are four key types of agreements that have been adopted for pipeline construction, operation and investment: the pipeline construction and operating agreement (PCOA); transportation agreements; operational balancing agreements (OBAs); and the ship or pay contracts.

Pipeline Construction and Operating Agreement (PCOA) A PCOA sets out the terms and conditions under which one or more parties agree to jointly develop, finance, operate and maintain a pipeline. A PCOA is a form of joint venture project often put in place by one or more producers, or by a producer and a long term consumer, to enable the joint financing and operation of a pipeline, as well as to ensure the availability of a reliable infrastructure that will guarantee stability of energy supply.[88] A good example is the Dolphin Gas Pipeline project, a 364 kilometer undersea pipeline and the longest gas pipeline in the Middle East, which was constructed by a consortium of public and private sector entities, to transport natural gas from Qatar's North Field to regional markets in Oman and the United Arab Emirates.[89] The 48-in. (1200 mm) pipeline has capacity of 33 billion cubic metres (1.2 trillion cubic feet) of natural gas per year.[90] This project demonstrates how two or more entities can come together to finance and operate a cost intensive pipeline project for long term mutual benefits.

A PCOA sets put the technical specifications and design of the pipeline, in addition to the roles of each of the joint venturers. It outlines detailed rules on which the

[87] Wright and Gallun (2008).

[88] See de Silva and Bremen (2008), p. 149.

[89] The pipeline was designed by UAE Offsets Group (UOG) and is operated by Dolphin Energy Limited. See Dargin (2011), pp. 317–320.

[90] Ibid.

joint venture will operate throughout this joint relationship. A PCOA is designed to last for the life of the project, which may be as much as 30 years, and the parties will remain together unless termination, withdrawal, assignment or default occur. Much like a JOA used in oil and gas exploration, the PCOA sets out the percentage interest held by each party to the pipeline project and the roles allocated to each party. One or more of the parties may be designated as the operator. The operator is responsible for the day-to-day management and operation of the pipeline. A PCOA will also establish a management committee to oversee the construction and operation of the pipeline. Each of the partners will normally be represented in the management committee and decisions are made by means of 'pass mark' voting (the aggregate percentage interest required to agree a proposal).[91] These pass marks will differ according to the phase of operations and voting practices are likely to vary from one PCOA to another. Fundamental changes or decisions such as the dismissal of the Operator would normally require a higher pass mark than many other decisions.

Other key provisions included in a PCOA, to protect the parties' interests, include assignment clauses (to specify if and when one of the parties may transfer its interest to a new participant); use of the pipeline by third parties, governing law; dispute resolution clause; and sole risk provisions which allows some parties to proceed with a proposal which does not meet with majority approval. A termination clause is also important to outline how the project may be discontinued. The PCOA parties may separate in one of three ways: (1) due to fault or default; (2) they may agree to separate by mutual consent; or (3) a party may choose to transfer its interest. Default may arise as a result of a failure by a party to pay its percentage interest share of cash calls within the time set for payment. Parties in default will normally lose their rights under the PCOA and ultimately may lose their interest; non-defaulting parties will have to pay pro rata the amounts that are outstanding and may ultimately acquire the defaulting party's interest. Defaults could arise in connection with decommissioning obligations if a part sought to escape this liability. It may also arise in connection with liability for significant environmental damage from spills or explosion.

Transportation Agreements A transportation agreement (TA) is the primary contract for oil and gas transportation services. It defines the terms and conditions under which a transportation services provider (mostly pipeline owners, the "Transporter") agrees to transport oil and gas on behalf of a private or commercial client that owns the product to be transported (the "Shipper").[92] In this case, the shipper owns the petroleum product and can be a producer, marketer, broker, or end user. The pipeline company agrees to receive the shipper's product and deliver it at designated points along the pipeline system. Delivery points are locations on oil and gas pipelines routes where gathering systems connect (*receipt*) to the pipeline and where local

[91] Typically, a management committee resolution is adopted and binding upon the joint venture when one or more members representing more than 50% of the participating interests have voted in favour of the proposal.

[92] See de Silva and Bremen (2008), p. 149.

distribution systems and other end-users connect to the pipeline (*delivery*). A TA must clearly specify the exact delivery point for the purpose of passing risks from the transporter to the shipper.

Given the different characteristics of oil and gas, TAs can be tailored for gas—called the gas transportation agreement, or the crude oil transportation agreement (COTA) for oil. For example, an IOC may enter into a COTA for a year with a pipeline owner under which the IOC secures the right to pump its oil through the pipeline for the duration of the contract. The pipeline company under a COTA guarantees the oil company a secured form of transportation for the oil via the pipeline for a specified fee. Similarly, under a GTA a gas pipeline operating company typically agrees to provide natural gas transportation services to an IOC.

The GTA sets out the rates and terms under which a transporter will transport gas on behalf of a shipper. Generally, TAs include provisions that specify conditions surrounding the nature of service to be provided; volume or quantity of the throughput (defined as the volume of oil or gas flowing or transported through a pipeline); quantity of service; delivery point, scheduled payments and term or duration of the agreement.[93] It includes general provisions dealing with notices, assignments, contract termination, governing law and dispute resolution. Generally, most pipeline owners and operators have standard terms and conditions that they offer to shippers. Once both parties agree on these conditions, they do not change unless renegotiated in accordance with the TA. For example, if the oil company contracted to use the pipeline to transport its oil for 1 year but only transported it for 9 months, it must still pay the full amount of the contract, as the pipeline has been reserved specifically for that oil company to use for 1 year.

The pipeline owner generally has the obligation to maintain, repair, renew and operate the pipeline system. Having some or all of this obligation on the shipper must therefore be carefully negotiated. In such cases, it is important to specify which facilities are to be maintained by the customer, standard of maintenance, whether this will result in a reduction in the charges payable for transportations services, the requirement for the transporter to minimize disruption of services due to maintenance or repair works as quickly and as it reasonably can, and the routine for carrying out such maintenance (scheduled or unscheduled).

Operational Balancing Agreements (OBA) An OBA is mostly required for the transportation of gas. It is an agreement between pipelines and parties at delivery or receipt points, whereby the parties agree to specified procedures for balancing discrepancies between the nominated levels of service and the actual quantities.[94] Practically, the amount of gas received and delivered cannot be precisely controlled as physical gas activity measured through a gas meter does not always match the expected gas flow. This mismatch creates imbalances between the pipeline and the customers. OBAs are used to balance the difference between measured and expected gas flow at a particular meter. The natural gas delivered to a meter allocated among

93 See de Silva and Bremen (2008), p. 149.

94 PwC (2006), p. 58.

the various shippers that had designated deliveries at that point based on the amount of gas that they requested, or nominated at the meter. The remaining difference of actual measured gas flow at the meter is then allocated to the OBA. OBAs can help facilitate more efficient operations, accounting, and systems management at delivery points. They can also reduce litigations and disputes relating to discrepancies in gas measurement and delivery.

Ship or Pay Contracts A ship or pay contract or clause is an agreement between a transporter and a natural gas producer or shipper, that requires the shipper to pay for the transportation of petroleum products even in case such product is not transported.[95] They are stand-alone contracts or at times clauses in transportation contracts that provide an option for the shipper to either use the transportation service to which the contract relates, or pay for it anyway.[96] It places an obligation on the shipper to pay for the service (either as well as, or instead of, taking up the service by injecting gas).[97] As such, the shipper's obligation is in the alternative—it can be satisfied in either of the two ways (1) by shipping, and paying the contract price for, a minimum contract quantity of commodity that must be shipped each year (the "SOP Quantity"); or (2) pay the applicable contract price for such SOP Quantity if it is not shipped during the applicable year. Such clauses are designed to guarantee a transporter with an assured revenue stream that ensures an adequate return on the usually significant project capital investment and risks invested in developing the transpiration infrastructure.[98]

While some have argued that a SOP contract operates as a penalty which will be contrary to general common law rules that prevent the enforcement of clauses that operate as a penalty against the party in default.[99] The better view arguably is that courts should and will generally enforce such clauses to give effect to commercial contracts freely entered into between sophisticated commercial parties of comparable bargaining power, and will and only rarely finds ship or pay clauses to be unenforceable penalties.[100] The UK supreme court in *Cavendish Square Holdings BV v Talal El Makdessi* explained the basis and enforceability of such contracts as follows:

> where a contract contains an obligation on one party to perform an act, and also provides that, if he does not perform it, he will pay the other party a specified sum of money, the obligation to pay the specified sum is a secondary obligation which is capable of being a

---

[95] PwC (2006), p. 68.

[96] See Holland (2016), pp. 443–453, also *Total Gas Marketing Ltd v Arco British Ltd and Others* [1998] 2 Lloyd's Rep 209 HL; [1998] UKHL 22; and *Amoco (UK) Exploration Company v Teesside Gas Transportation Limited and Another* [2001] 1 All ER (Comm) 865; [2001] UKHL 18.

[97] See Holland (2016), Ibid.

[98] Ibid.

[99] See *Alfred McAlpine Capital Projects Ltd v Tilebox Ltd* [2005] EWHC 281 (TCC), [2005] BLR 271 (QBD).

[100] See Holland (2016).

> penalty; but if the contract does not impose (expressly or impliedly) an obligation to perform the act, but simply provides that, if one party does not perform, he will pay the other party a specified sum, the obligation to pay the specified sum is a conditional primary obligation and cannot be a penalty.[101]

The foregoing confirms that a SOP contract must be carefully constructed to operate as primary obligation and not as a penalty which might make it unenforceable in court. Also, implications of *force majeure* events on the SOP contract must be carefully structured to avoid the possibility that a customer may be required to pay for transportation that it failed to make due to a *force majeure* event that prevented performance. Since the shipper's obligations under a SOP clause are couched in the alternative, the occurrence of force majeure may excuse the customer's failure to transport the SOP Quantity, but does it excuse the customer from paying the pipeline operator for such transportation not made? As long as transportation payment can be made, the pipeline operator can strongly argue that the shipper can fully perform its contract obligations by making payment of the applicable SOP deficiency amount. Absent clear and specific elaboration of how force majeure affects the customer's SOP obligation, both the pipeline operator and customer may find themselves testing the effectiveness of the contractual dispute resolution clause when a force majeure event occurs and the parties have different views as to whether payment remains due.

3.3 Associated Infrastructure

Associated infrastructure refers to systems and structures that are required for, or support, the effective functioning of physical infrastructures. These include telecommunications systems, information and communication technology systems, electricity, water systems, logistics, and institutions such as ports agencies. In the extractive context, associated infrastructure refers to systems and processes that provide secondary support for the extraction, processing, storage and transportation of extractive resources from point of extraction to the end user. For example, without reliable electricity supply, functional water systems, and dependable telecommunication systems, use of pipeline and port infrastructures might be difficult or even impossible.

As earlier noted, Africa lags significantly when it comes to important associated infrastructure especially electricity. According to the World Bank, more than 1.4 billion people worldwide, mostly in Africa do not have access to electricity, thereby living in energy poverty.[102] In Nigeria for example, many homes and businesses go several weeks without electricity. The overall power generation in the country in

[101] [2015] UKSC 67 [14] (Lord Neuberger and Lord Sumption).

[102] According to the World Energy Outlook 2016, more than 95% of those living without electricity and modern cooking fuels are in countries in sub-Saharan Africa and Asia. See International Energy Agency (2016a, b, c). See also Escribano et al. (2010).

2016 was 3000–4000 MW, an abysmally low figure for a country of over 150 million people. Estimates show that based on current demands in Nigeria, the country must generate at least 40,000 MW to meet power demands and to address power deficiency.[103] Power generation in Nigeria dates back to 1896 when the first power generation plant was constructed in Ijora, Lagos. Subsequently, many other power generation plants and grids were constructed, for example in Kainji and Shiroro to name but a few. However, many years of oppressive military rule, corruption and neglect have resulted in inadequate investments in infrastructural facilities in the power sector and a dilapidation of many of the existing power plants. Currently, only about 40% of the country's population has access to electricity that is erratic and unsteady. Many rural communities in Nigeria are not even connected to national electricity supply grids. The problem of inadequate power supply has generally stifled the profitability and ease of conducting extractive operations in many sub-Saharan countries.[104] According to the World Bank's Africa Infrastructure Country Diagnostic published in 2008, the cost of redressing Africa's infrastructure deficit is estimated at around US$75 billion, with the current funding gap being around US$35 billion annually.[105] Many African governments are currently unable to afford the cost of addressing these associated infrastructure gaps. For example, The Nigerian government identified this in the National Energy Policy which notes that "The power sector is very capital intensive. Government, with its many responsibilities in other sectors of the economy, cannot fund its development. Therefore, there is a need to reform the sector to attract and encourage private sector participation."[106]

The lack of associated infrastructure, such as electricity, roads, water systems mean that ECs have to allocate extra budgets for generating their own water and electricity which often mean that the private infrastructure expenditure often dwarfs the direct cost of constructing mines or conducting petroleum operations themselves. Inadequate access to energy in Africa is also a critical contributor to social challenges, such as poverty, food scarcity, inadequate access to clean water, poor health, stunted economic growth, and climate change.[107] The United Nations' Sustainable Energy for All and the associated Sustainable Development Goal 7 therefore encourage countries to develop projects and initiatives aimed at providing access to energy to the over one billion people who do not have access to reliable energy.[108] To achieve these targets, African countries must scale up investments in infrastructure development projects aimed at expanding energy access ("energy access infrastructure").

---

[103] See Igikiowubo (2008) and Ibitoye and Adenikinju (2007).

[104] See Adenikinju (2005).

[105] Foster (2008), pp. 1–5.

[106] Report of the Presidential Task Force on Power (2011).

[107] See International Energy Agency (2016a, b, c).

[108] SDG 7 aims to achieve three goals by 2030: Ensuring universal access to modern energy services; doubling the global rate of improvement in energy efficiency; and doubling the share of renewable energy in the global mix. *See* Sustainable Energy for All.

There is need for dynamic governance innovations and approaches to attract the private sector participation and involvement to address these associated infrastructure gaps. One key public private sector participation approach is for governments and extractive companies to jointly develop and finance shared infrastructure systems to jointly address multifaceted infrastructure problems across Africa.[109] There is an opportunity for governments to leverage extractive-related infrastructure to address associated infrastructure gaps across all other sectors. For example, an independent power project designed to power a mine can indeed provide electricity to community members and to small scale industries located in proximity to the mine. Governments have prominent roles to play in coordinating the delivery of shared extractive infrastructure as a strategic tool for achieving national infrastructure expansion and renewal plans, as well as for promoting energy security and eliminating poverty. Facilitating shared use of extractive infrastructure between the extractive company and third parties is the key to addressing energy poverty in local communities, and a platform for distributing the benefits of extractive-related investments more widely.

To leverage the development of extractive associated infrastructure projects as energy access projects in accordance with a country's long-term infrastructure strategy, national governments must put in place the right mix of policies and fiscal incentives that will accentuate the development of shared and integrated infrastructure by ECs. One key strategy will be to incorporate development of the energy access projects as a critical indicator of local content in African countries. Currently, local content policies in African countries focus extensively on economic benefits such as employment and opportunities for local industries. In a continent in dire need of associated infrastructure, local content policies can accentuate the importance of social objectives such as delivery of shared energy access infrastructure. For example, ECs can be required to develop and share their associated infrastructure assets with local residents and communities, especially local miners or oil companies in the area at a reasonable tariff rate.

To achieve this, the energy access objective must be incorporated into the negotiation and award of extractive licenses. Issue such as geographic location of the extractive project and the local communities that can benefit from energy access infrastructure projects must be identified and cooperatively integrated into the licensing process. A good example of this approach was the granting of a mining license in 2013 by the Nigerian government to a Chinese-Nigerian consortium, HTG-Pacific Energy, for the exploitation of the Ezinmo coal block in Enugu State.[110] The license included a memorandum of understanding between the Nigerian government and the licensee, to the effect that the licensee was to develop a 1000 MW coal-fired power plant near the mine site. Power generated from the power plant is expected to supply the national grid and contribute to power access in the region and country. This project is still at the initial stages, but it clearly demonstrates how extractive development activities can be leveraged as basis for addressing energy

[109] See Norton Rose Fulbright (2015).

[110] Ibid.

poverty and associated infrastructure gaps in the continent. This shared infrastructure approach can be successful if carefully negotiated and supported by the national government through fiscal incentives, such as lower taxes, that can make such energy access infrastructure beneficial to the EC both in terms of cost recovery and the commercial profitability of the extractive operations.

Implementing a shared infrastructure approach also requires greater coordination and alignment of policy objectives between various agencies and ministries. Different government ministries, organizations and state parastatals play prominent roles in approving extractive and power projects in Nigeria. For example, the Ministry of Finance is responsible for approving currency importation and approval of payment instruments, while the Ministry of Internal Affairs is needed for granting approvals for project execution. The Corporate Affairs Commission will be needed to grant approvals for foreign agencies to carry out investment activities. Other ministries such as power, energy and transportation will have to be involved in approving all projects to be executed under them. It could be herculean task for investors to get approval from each of these ministries owing to weak information link and coordination between these agencies and ministries. This weak link between government ministries and agencies often makes them work at cross-purposes and request their separate approval processes from investors. For example, while the ministry of finance may be pushing for greater revenue through increased tax, the ministry of power is seeking to increase power generation in the national grid, while the ministry of labor may be reviewing a project for social welfare and employment perspectives. Without strategic coherence and harmonization between the diverse ministries the EC may not achieve a commercially viable or profitable outcome. Failure of governments to foster policy coordination between relevant ministries and agencies could make shared infrastructure development and implementation projects in Africa less attractive to investors, thereby posing an institutional barrier infrastructure development. There is a need for African countries to put in place adequate regulatory frameworks and institutions to coordinate the design, approval, financing and implementation of shared infrastructure projects in the extractive industries. The legal and regulatory framework is a major determinant of a country's ability to attract and benefit from foreign investment. Before committing funds to any project, investors consider whether laws and contracts are likely to be properly enforced and whether their rights and responsibilities are well defined and likely to be respected. The legal and regulatory framework for issuing licenses or concessions should define the rights and obligations of utilities, clarify pricing mechanisms and establish procedures for dispute resolution. The establishment of a strong and autonomous regulatory agency is essential to attractive private sector investments in shared energy infrastructure projects.[111] With appropriate intergovernmental coordination, granting and getting approvals for projects will be less cumbersome and straightforward. Such coordination will also simplify the information flow between government ministries, thereby eliminating regulatory overlaps.

[111] See Olawuyi (2013), pp. 110–112.

3.4 Conclusion

Extractives industries across Africa currently face significant shortage of physical and associated infrastructure necessary to link and commercialize extractive resources from the point of extraction to international and regional markets. This infrastructure shortage is a hindrance to sustainable resource utilization, economic and social development, as well as addressing energy poverty in the continent. To address infrastructure deficits in the continent, there is need for dynamic legal solutions and innovations that leverage the expansion of extractive projects as an opportunity to promote energy access infrastructure to local communities, local industries and operators close to locations of extraction activities. Governments must as a starting point put in place adequate policies and incentives that can make the development of shared associated infrastructure projects attractive to ECs both in terms of cost recovery and ease of executing such projects. Governments must also ensure collaboration and coordination between the multiple stakeholders involved in such projects to reduce risk exposure from such projects.

From an EC standpoint, the key objective is to increase probability of an extractive venture by eliminating direct and indirect legal risks. The next section of the book therefore delves into the legal and regulatory framework applicable to the extractives sector in Africa with the aim of highlighting potential legal risks points and the legal tools for managing such risks. The next chapter kicks of this inquiry with an assessment of the legal and jurisprudential viewpoints on resource control and ownership in the continent.

References

Abduljabbar WK, Tahar R (2012) A case study of petroleum transportation logistics: a decision support system based on simulation and stochastic optimal control. Afr J Bus Manag 6(12):4350

Adenikinju A (2005) Analysis of the cost of infrastructure failures in a developing economy: the case of electricity sector in Nigeria. African Economic Research Consortium AERC Research Paper, Nairobi

Ako R (2009) Nigeria's land use act: an anti-thesis to environmental justice. J Afr Law 53(2):289–304

Alfred McAlpine Capital Projects Ltd v Tilebox Ltd [2005] EWHC 281 (TCC), [2005] BLR 271 (QBD)

Amoco (UK) Exploration Company v Teesside Gas Transportation Limited and Another [2001] 1 All ER (Comm) 865; [2001] UKHL 18

Associated Press (2016) A timeline of recent oil trains crashes in the US and Canada. https://www.apnews.com/84b1e8273d854697b34af57bc60badc2. Accessed 11 June 2018

Bloomberg New Energy Finance (2017) New energy outlook 2017: long-term projections of the global energy sector. Executive Summary. www.bloomberg.com/company/new-energy-outlook. Accessed 23 Jan 2018

Coastal and Inland Shipping (Cabotage) Act *2003* (No. 5 of 2003)

Collier P (2011) Building an African infrastructure. Finance Dev 48(4):18–21

Crude Oil (Transportation and Shipment) Regulations (S.I. 44 of 1984)
Dargin J (2011) Qatar's gas revolution. In: Fattouh B, Stern J (eds) Natural gas markets in the Middle East and North Africa. Oxford University Press, Oxford
De Silva L, Bremen J (2008) Oil and gas transportation agreements. In: Jennings A (ed) Oil and gas production contracts, 1st edn. Sweet & Maxwell, p 149
Dulaney M, Merrick R (2005) Legal issues in cross-border oil and gas pipelines. J Energy Nat Resour Law 23(3):247–265
Ehimomen C, Adeleke A (2012) An assessment of the distribution of petroleum products in Nigeria. J Bus Manag Econ 3(6):232–241
Environmental Impact Assessment Act of 1992, Laws of the Federation of Nigeria 2004, c E12
Escribano A, Guasch JL, Pena J (2008) Impact of infrastructure constraints on firm productivity in Africa. Working Paper, World Bank, Washington, DC
Escribano A, Guasch JL, Pena J (2010) Assessing the impact of infrastructure quality on firm productivity in Africa: cross country comparisons based on investment climate surveys from 1999 to 2005. Policy Research Working Paper 5191. World Bank, Washington, DC
Foster V (2008) Africa infrastructure country diagnostic: overhauling the engine of growth: infrastructure in Africa. World Bank, Washington, DC
Foster V, Briceno-Garmendia CM (2010) Africa's infrastructure: a time for transformation. World Bank, Washington, DC
Holland B (2016) Enforceability of take-or-pay provisions in English law contracts – resolved. J Energy Nat Resour Law 34(4):443–453
Ibitoye F, Adenikinju A (2007) Future demand for electricity in Nigeria. Appl Energy 84:492–504
Igikiowubo H (2008) Nigeria – A Nation in Darkness. Vanguard Newspaper
International Energy Agency (2014) World Energy Outlook 2014 special report: African Enehrgy Outlook. IEA, Paris
International Energy Agency (IEA) (2016a) Energy access projections. World Energy Outlook 2016. http://www.worldenergyoutlook.org/resources/energydevelopment/energyaccessprojections/. Accessed 15 Aug 2017
International Energy Agency (IEA) (2016b) Modern energy for all. World Energy Outlook 2016. http://www.worldenergyoutlook.org/resources/energydevelopment/. Accessed 15 Aug 2017
International Energy Agency (IEA) (2016c) World Energy Outlook 2016. OECD/IEA
Kiefner JF, Trench CJ (2001) Oil pipeline characteristics and risk factors: illustrations from the decade of construction. American Petroleum Institute's Pipeline Committee, Washington, DC
Kilbourn W (1970) Pipeline. Clarke Irwin, Toronto
KPMG (2017) Nigerian mining sector brief. https://assets.kpmg.com/content/dam/kpmg/ng/pdf/advisory/ng-Nigerian-Mining-Sector.pdf. Accessed 9 June 2018
Land Use Act of 1979, Laws of the Federation of Nigeria 2004
Lucas AR, Thompson CB (2016) Infrastructure, governance and global energy futures: regulating the oil sands pipelines. J Environ Law Pract 28(3):355–394
Miscellaneous Offences Act of 1983 M17 Laws of the Federation of Nigeria, 2004
Molinski D (2015) How to transport oil more safely. Wall Street J
Natural Resource Governance Institute (2015) Extractives-linked infrastructure exploring options for shared use of infrastructure projects. https://resourcegovernance.org/sites/default/files/nrgi_Extractives-Linked-Infrastructure.pdf. Accessed 10 June 2018
Nigerian AGIP Oil Company vs. Kemmer (2001) 8 N.W.L.R. Part 716, p 506
Norton Rose Fulbright (2015) Sharing mining infrastructure. http://www.nortonrosefulbright.com/knowledge/publications/127581/sharing-mining-infrastructure. Accessed 11 June 2018
Oil and Gas Pipelines Licence Regulations 1995
Oil Pipelines Act of 1956, Laws of the Federation of Nigeria 2004
Oil Terminal Dues Act 1969, Laws of the Federation of Nigeria 1990
Olawuyi DS (2013) Power generation through renewable energy sources: an analysis of the legal barriers and potentials in Nigeria. J Resour Energy Dev 10(2):105–112

Olawuyi D (2015a) Fostering accountability in large scale environmental projects: lessons from CDM and REDD+ projects. In: Wouters J, Ninio A, Doherty T, Cisse H (eds) Improving delivery in development: the role of voice, social contract, and accountability. The World Bank Legal Review, Washington, pp 127–147
Olawuyi D (2015b) Principles of Nigerian environmental law. Afe Babalola University Press, Ado-Ekiti
Olawuyi D (2018) Midstream oil and gas operations. Gulf legal advisor. Lexis Nexis, New York
Onyegbule E (2012) An overview of the Nigerian downstream gas market. National Electricity Regulatory Commission. https://pubs.naruc.org/pub.cfm?id=5385E110-2354-D714-512F-1FED305A7575. Accessed 11 June 2018
Organisation for Economic Co-operation and Development (OECD) (2012) Mapping support for Africa's infrastructure investment. OECD, Paris
Petroleum Products and Distribution (Anti-Sabotage) Act of 1975 of Nigeria
Pipeline and Gas Journal (2017) Uganda and Tanzania reach pipeline construction agreement. https://pgjonline.com/2017/05/22/uganda-tanzania-reach-pipeline-construction-agreement/. Accessed 11 June 2018
PricewaterhouseCoopers LLP (PWC) (2006) LNG: a glossary of terms, 3rd edn. Petroleum Economist Ltd, London
Punch Newspaper (2016) Nigeria records 1,600 pipeline vandalism cases. Punch Newspapers
Punch Newspapers (2018) Blackout looms as explosion rocks Escravos gas pipeline again. Punch Newspapers
Ramdoo I (2015) Synergising and optimising mineral infrastructure in regional development strategies. E15 Expert Group on trade and investment in extractive industries
Report of the Presidential Task Force on Power (2011) https://www.esi-africa.com/wp-content/uploads/Beks_Dagogo.pdf. Accessed 12 Dec 2017
Reuters (2018) Explosion hits major gas pipeline in Nigeria. Reuters, London. https://www.reuters.com/article/us-nigeria-gas-blast/explosion-hits-major-gas-pipeline-in-nigeria-idUSKBN1F01PG. Accessed 23 Apr 2018
Seleba vs. Mobil Producing (Nigeria) Unlimited (2002) 12 N.W.L.R. Part 995, p 643
Shell Petroleum Development Company of Nigeria Ltd. vs. Abel Isaiah and Ors (2001) 1 N.W.L.R. Part 723, p 169
Shell Petroleum Development Company vs. Burutu Local Government Council (1998) 9 N.W.L.R. Part 565, p 318
Shell. Liquefied Natural Gas (LNG). Shell. https://www.shell.com/energy-and-innovation/natural-gas/liquefied-natural-gas-lng.html. Accessed 12 May 2017
Sustainable Energy for All. http://www.se4all.org/our-mission. Accessed 15 Aug 2017
Tencer D (2013) Pipeline spills release three times as much oil as rail spills: IEA. HuffPost. https://www.huffingtonpost.ca/2013/05/14/pipeline-oil-spills-rail-spills_n_3273725.html. 11 June 2018
Texas Transport Institute (TTI) (2007) A modal comparison of domestic freight transportation effects on the general public. TTI, Houston
The Nigeria LNG (Fiscal Incentives Guarantees and Assurances) Act, Cap N87, Laws of Federation of Nigeria, 2004
The Pre-shipment Inspection of Export. Act, no 10 of 1996, Cap P26 Laws of Federation of Nigeria, 2004
Total Gas Marketing Ltd v Arco British Ltd and Others [1998] 2 Lloyd's Rep 209 HL; [1998] UKHL 22
U.S. Department of Energy. https://fossil.energy.gov/education/energylessons/coal/coal_mining.html. Accessed 21 Apr 2018
Usman E (2018) Many feared dead, 67 cars burnt as tanker explodes on Lagos Otedola bridge. https://www.vanguardngr.com/2018/06/many-feared-dead-67-cars-burnt-tanker-explodes-lagos/. Accessed 12 Aug 2018

Vertabedian R (2015) Why are so many oil trains crashing? Track problems may be to blame. Los Angeles Times. http://www.latimes.com/nation/la-na-crude-train-safety-20151007-story.html. Accessed 21 Apr 2018

Vinogradov S (1999) Cross-border pipelines in international law. Nat Resour Environ 14(2):74

World Bank (2003) Cross-border oil and gas pipelines: problems and prospects. UNDP/World Bank Energy Sector Management Assistance Programme. http://siteresources.worldbank.org/INTOGMC/Resources/crossborderoilandgaspipelines.pdf. Accessed 11 June 2018

World Bank (2014) Ghana - the West African gas pipeline project (English). World Bank Group, Washington, DC

World Bank (2015) Pande-Temane gas fields: SASOL – CMH – IFC PPA UJV. https://www.igu.org/sites/default/files/3-3%20IFC%20-%20Katia%20Daude%20Goncalves%20-%20Gas%20Competence%20Seminar%20-%20September%2022%202015.pdf. Accessed 11 June 2018

Wright and Gallun (2008) Fundamentals of oil and gas accounting. PennWell Books, Houston

Part II
Legal Theories, Contracts and Fiscal Regimes on Extractives Investments in Africa

Chapter 4
Ownership and Control of Extractive Resources

Abstract This chapter surveys the applicable theories and regimes underpinning the ownership and control of extractive resources in Africa. In doing so, it deals with several complicated matters relating to the extent of interest and rights that any person, individual or body has in petroleum and mineral resources. First, it contextualizes ownership debates and contentions from the perspectives of dominial and non-dominial jurisdictions. It then reviews the diverse theories that have been put forward to clarify whether petroleum and minerals are capable of being owned: namely the absolute ownership theory, qualified interest theory, non-ownership theory, the rule of capture and ownership puzzles relating to the tragedy of the commons.

4.1 Introduction

Ownership is generally the absolute right to possess, enjoy or freely dispose of a resource or commodity.[1] To own a thing is to have full and complete dominion over that thing.[2] In practice however, this absolute right is limited. The issue of who owns and controls petroleum or minerals has been a contentious issue, and a subject of extensive litigation, in many parts of Africa. Ownership rights over extractive resources vary from one African country to the other, depending largely on the country's socio-political culture, historical and colonial background, the enduring legal regime and development framework. Consequently, when investing in Africa, it is important to understand the specific resource ownership structure in the target country: who owns and controls oil and gas resources? Is it the central or federal government? Is it provincial or state government? Or, is there private ownership of oil and gas resources? Answers to these questions are highly varied from one African

[1] Graulau (2011), pp. 14–15.

[2] Njotini (2017), pp. 14–19.

D. S. Olawuyi, *Extractives Industry Law in Africa*,
https://doi.org/10.1007/978-3-319-97664-8_4

country to the other and may at times fall outside the traditional ownership norms in other foreign jurisdictions.

Jurisprudential interpretations and theories have evolved in the process of trying to unpack who owns or controls these important economic resources in Africa.[3] In order to clarify the extent of interest and rights that any person, individual or body has in petroleum and mineral resources, this chapter surveys the applicable ownership theories and regimes underpinning the African extractive industry. In doing so, it deals with several complicated matters. It contextualizes the issue of ownership of extractive resources from the perspective of dominial and non-dominial jurisdictions. It reviews the diverse theories that have been put forward as to whether petroleum and minerals are capable of being owned: namely the absolute ownership theory, qualified interest theory, non-ownership theory, the rule of capture and the theory of ownership relating to the tragedy of the commons.

After studying this chapter, you should be able to understand the concept of ownership of oil and gas and mineral resources, especially in Africa; appreciate the underpinning theories of ownership of petroleum and mineral resources and how they have evolved; and finally develop a robust understanding of how ownership of natural resources is conceptualized in your local jurisdiction, especially from the dominial and non-dominial theoretical viewpoints.

4.2 The Elusive Ontology of Resource Ownership

In the literature on extractive resource management, defining the idea of resource ownership is itself a contested subject.[4] It has therefore been said that defining ownership is either so simple as to need no explanation or so elusive as to defy definition.[5] In many countries of the world, companies and individuals are only granted certain rights over extractive resources, making it difficult to lay claim to absolute ownership of those resources. Practically therefore, what ownership translates to is the ultimate residual bundle of rights that remains in law over a thing. It is the totality of, or the bundle of rights that a person has, over and above every other person in relation to a thing.[6] An owner is a person who has sole dominion over a property, including the right of possession and control of that property.[7]

[3] Olawuyi (2015), pp. 247–265.

[4] Njotini (2017), see also Bouckaert (1990).

[5] Bouckaert, ibid.

[6] Njotini (2017).

[7] See Clark (2002), stating that ownership [here implies] "the legal right that a legal system grants to an individual in order to allow him or her to exercise the maximum degree of formalized control over a scarce resource. This idea can be derived from the civil law concept of dominium, the greatest right in property to use and dispose of a thing in the most absolute way…"

This bundle of rights enjoyed by a legal owner includes: the power to use and enjoy a property; the right to deal with, produce or destroy as the owner pleases; the power of possession, which also includes the right to exclude others; the power to sell, rent or alienate; the power to bequeath by will; as well as the power to charge the property as security.[8] The other incidentals of this right of ownership includes the right to protect and defend the property against intrusion or trespass, and the right to dispose of the property, provided that it is within the confines of the law and does not infringe on the rights of others. Summarily, ownership entails all of the above incidental and bundle of rights.[9] The idea of ownership is therefore simply nature, scope and extent of legal rights, interest and control that a legal system grants to a person over a property.

Ownership rights over extractive resources takes various forms, and varies from country to country, or jurisdiction to jurisdiction. These changes are often dependent on a country's socio-political, historical background, colonial history if any, legal system, the laws, which directly impact natural resources, as well as contractual arrangements that are in existence between governments and third parties. The legal and regulatory regime in place in a given country, will determine the rules and procedure, which govern the allocation, maintenance, transfer and cancellation of oil and gas rights, as well as the rights and obligations of contractual titleholders.[10]

In dominial regimes, ownership of natural resources is vested in the government, whether federal, state or province where the natural resource is located. While in non dominial regimes, private ownership of natural resources by the surface landowner is recognized and allowed. This section examines the ownership theories or modes of ownership of natural resource in dominial and non-dominial jurisdictions. It also offers examples of jurisdictions, which fall into dominial or non-dominial categories.

4.2.1 Ownership in Dominial Jurisdictions

Under the dominial system of ownership, the entire ownership and control of all petroleum and mineral resources in, under or upon any lands is generally vested in the State. Dominial control of resources is the more typical system of ownership of natural resources, as all countries of the world, with the exceptions of the United States and Canada, retain absolute control and sovereign rights over natural resources found in their territories. All African countries exercise dominial control an ownership over natural resources.[11] This right is often enshrined in constitutions

[8] See Omorogbe (2001), pp. 30–31. Also, Badenhorst et al. (2006).

[9] See generally, Omorogbe and Oniemola (2010), pp. 116–117. Also Badenhorst et al. (2006).

[10] Omorogbe and Oniemola (2010), p. 115.

[11] Ibid, also Momodu (1988), p. 268.

or principal natural resource laws of these countries.[12] In African jurisdictions, ownership rights over natural resources are vested absolute in the sovereign—that is the state or governing authority. This gives the state the power and prerogative to explore and exploit its natural resources either by itself, or by granting exploration and development rights to other third parties, based on stipulated conditions.[13] Such dominial control typically extend to all land: including land covered by water which is in the State; the territorial waters of the State; part of the continental shelfs; and part of the Exclusive Economic Zone.[14]

This system of ownership in which natural resources belong to, or are controlled by the state has its origin in the system known as the 'regalian system', which originated under Roman law.[15] The regalian doctrine is anchored on the principle that the

---

[12] For example, the ownership and control of all minerals, mineral oils and natural gas in Nigeria is vested in the Federal Government of Nigeria by virtue of the Constitution of the Federal Republic of Nigeria, Laws of the Federation of Nigeria 2004, c 23 [*CFRN 1999*]:

> also Section 1 of the Petroleum Act of Nigeria and S. 1 of the *Nigerian Mines and Mineral Act, 2007*, Cap. F 34 LFN 2004.
>
> See s 44 (3) of the *CFRN 1999,* stating that:
>
> ... entire property in and control of all minerals, mineral oils and natural gas in under or upon any land in Nigeria or in, under or upon the territorial waters and the Exclusive Economic zone of Nigeria shall rest in the Government of the Federation and shall be managed in such manner as may be prescribed by the National Assembly.
>
> See also S. 1 of the *Nigerian Mines and Mineral Act, 2007*, Cap. F 34 LFN 2004 stating that:
>
> The entire property in and control of all mineral resources in, under or upon any land in Nigeria, its contiguous continental shelf and all rivers, streams and watercourses throughout Nigeria, any area covered by its territorial waters or constituency and the Exclusive Economic Zone is and shall be vested in the Government of the Federation for and on behalf of the people of Nigeria.

[13] See Omorogbe (2001), pp. 34–35. See also Law No. 10/04 of 12 November 2004 of ANGOLA which vests absolute ownership of resources in Angola in the State. Also Article 3 of the Constitution of Angola, 2010 vests in the State, the exclusive jurisdiction and rights of sovereignty over the conservation, development and use of natural, biological and non-biological resources in the contiguous zone, the exclusive economic area and on the continental shelf, under the terms of the law and international law. SOUTH AFRICA's Mineral and Petroleum Resources Development Act (MPRDA) Act No 28 of 2002 provides that "South Africa's mineral and petroleum resources belong to the nation and that the State is the custodian thereof" (see preamble and ss 2, 3). The notion of custodianship under South African law was extensively discussed in *Agri SA v Minister for Minerals and Energy* 2013 (4) SA 1 (CC), stating that the MPRDA vests the absolute custodianship of coal and other mineral and petroleum resources in the state on behalf of the people of South Africa. However, as custodian, the state does not acquire ownership to these mineral and petroleum resources, but has become "a facilitator or a conduit through which broader and equitable access to mineral and petroleum resources can be realised." Para 68–69. See also *Minister of Mineral Resources and Others v Sishen Iron ore Company (Pty) Ltd and Another* 2014 (2) SA 603 (CC) para 63.

[14] See for example Section 1 of the Petroleum Act of Nigeria.

[15] Jolowicz and Barry (1972), p. 140; also Davies and Fouracre (1995), pp. 4–15.

King or State is the owner of all conquered lands, waters and mineral resources.[16] Under the regalian system, the King owned all lands and natural resources found in such lands and can therefore determine who shall be granted rights and privileges over public lands and natural resources, as well as under what terms they may be granted such privilege.[17] The King, represented by a national authority, is therefore responsible for granting permits, licenses and leases for the exploration, and exploitation of mineral resources, in exchange for royalties, taxes or fees by the grantee.

Natural resources laws in dominial systems continue this tradition of the regalian doctrine. The dominial law system vests ownership of all lands, territories and natural resources in the sovereign, meaning that there is no private ownership of subsurface ownership of natural resources. The state vests control over natural resources in itself, while landowners only have a right of compensation for the loss of surface rights. The rights of other third parties are usually limited to contractual rights of use, management, control, or enjoyment of benefits derived from the resource. This means that these third parties do not have the legal right to extract or sell any natural resources without first obtaining prior authorization from the government.[18] This authorization is usually in the form of permits, leases or concessions, as we will see in subsequent chapters.

4.2.2 Ownership in Non-Dominial Jurisdictions

In non-dominial jurisdictions, private ownership of natural resources by the surface landowner is possible and recognized. In such jurisdictions, ownership of mineral resources could be both freehold and leasehold (i.e. a majority owned by the government, while some are privately owned). A good example is the Province of Alberta in Canada which recognizes some form of private ownership of petroleum and natural gas. In Alberta, approximately 81% of all mineral rights are owned by the Crown, which is administered by the Alberta Department of Energy. The remaining 19% are freehold mineral rights, which encompass rights owned privately by individuals or companies.[19] Individuals who are the assigns or descendants of Alberta's original homesteaders are the registered owners of the petroleum and/or natural gas beneath approximately 6 million acres (4% of the Province's surface area). Freehold leases confer ownership over production reduced to possession by the lessee, while the freehold estate owner typically retains a royalty interest. The terms of these leases vary and are independently negotiated with the freehold estate owners, although standard form agreements drafted by industry organisations, most the Canadian Association of Petroleum Landmen, are typically used. In such cases,

[16] Davies & Fouracre, ibid.

[17] See Omorogbe and Oniemola (2010), p. 120.

[18] Ibid, 115–116, 120. See also South Africa's MPRDA, section (s) 70, 74,76,79 and 83.

[19] Plourde (2010), pp. 10–13.

the role of public authorities is to provide the legal framework that guides negotiations and settlements between the right holders.

Similarly, in many parts of the United States, ownership of mineral resources is vested in private owner of the land surface, or the subsoil where the petroleum is located. For example in 1915, the Texas court in the case of *Texas Company v. W. H. Daugherty et al*,[20] held that oil and gas beneath the surface of the earth, belonged to the person who owned the land, in keeping with the Latin maxim, "*cuius est solum eius usque ad coelum et ad inferos*", which means "whomsoever owns the soil, owns up to the sky and down to the depths."[21] This theory of land ownership, also referred to as the *Ad Coelum* theory of ownership, has evolved over time due to more knowledge and information about the nature of oil and gas reserves and their formation. We will examine in the next part the prevailing theories of natural resource ownership.

In Africa and the Middle East, private ownership of land and natural resources is generally not recognized.[22] Government ownership and control over all lands, territories and natural resources is the prevailing system across the African continent. Furthermore, in many parts of Africa, ownership and control of natural resources is often vested in the central government at the national level, as opposed to the system in Canada and the United States in which States or provinces own and control natural resources. In Canada for example, oil and gas resources are owned and controlled by the provinces, as constitutionally recognized in section 109 of the Constitution Act, 1867 which explicitly vests ownership rights to extractive resources with the provinces in which they are located.[23] This distinction is unique and important because in many parts of Africa, the central government maintains radical title and ownership to all petroleum and natural gas within the country. As such, ownership of oil and gas by provincial or state government is not widespread. Secondly, freehold ownership is not widespread in African countries as all oil and gas resources are typically owned by the Federal Government.

In Africa therefore, investors have to deal directly with the government or national oil companies. Knowledge of this seemingly negligble fact may save the cost and time of dealing with wrong agencies and unscrupulous individuals who may lay claim to freehold ownership. Furthermore, in some African countries, even

[20] See Texas Company v. W. H. Daugherty et al., 107 Tex. 226 (Tex. 1915).

[21] See Omorogbe and Oniemola (2010), p. 118.

[22] For example, Kuwait, Qatar, Iran, United Arab Emirates and Saudi Arabia are all dominial jurisdictions. Under Qatar Law No. 3/2007 on the Exploitation of Natural Resources, which governs petroleum operations in the State of Qatar, the State owns all oil and gas deposits. Similarly, under Saudi Arabia's Basic Law of Governance 1992, all the states natural resources are vested in the state. It provides that All God's bestowed wealth, be it underground, on the surface, or in national territorial waters, on the land or maritime domains under the State's control, all such resources shall be the property of the State as defined by the Law. The Law shall set forth the means for exploiting, protecting and developing such resources or the benefit, security and economy of the State. Saudi Arabia explores and produces oil through Saudi Aramco, which is fully owned by the Saudi government. See Olawuyi (2018), pp. 1–2.

[23] See Plourde (2010).

after an oil exploration license is granted by the national oil company, the grant must be formally ratified by parliament before it becomes effective. What this means is that oil exploration work cannot not start until an oil license is granted parliamentary approval. This requirement is not typical in many foreign jurisdictions. In Canada for instance, there is no legal requirement for the federal or provincial parliament to pass a vote on a crown lease granted by the responsible energy regulator. The required parliamentary approval in many African countries may take several months or in extreme cases years to come through. Knowledge of such seemingly minute requirements may save an investor from the embarrassment of making investment commitments or plans too early, before actual parliamentary approval is received.

Many African regimes also confer absolute and often excessive powers on local energy regulators and national oil companies, which can be a source of legal risk to unaware investors. For example, local enactments and constitutions in a number of African countries often confer excessive powers on national oil companies to unilaterally cancel or amend petroleum contracts and licences. For example, in countries such as Guinea, Tanzania, Democratic Republic of Congo and Sierra Leone, there have been cases of: increased government participation; arbitrary confiscation and nationalization of Oil Mining Licences; unilateral modification of the oil license or contract; forced contract renegotiation, retroactive change in legislation and licensing regime after exploration rights have been granted; and breach of contract by national oil company.[24] These resource nationalism trend in many parts of Africa often poses financial risks to investors that have committed resources to establishing oil operations. It is therefore important to understand the ownership regime in a country, the long-term validity of an exploration license or permit, and the rights that come with such permits. It should also be an essential component of due diligence search to investigate and determine the ownership structure in specific countries; the nature of legislative power conferred on the national oil company; and the nature of investment protection and tenure security guaranteed to potential investors. During contract negotiations, it is also pertinent to ensure that stabilization, adaptation and foreign arbitration clauses are established as part of any agreements with the host state.[25]

---

[24] For example, in September 2011, Guinea adopted a new mining code which vests the state a 15% stake in all new mining projects with the right to purchase a further 20%. This led to the exit of investors such as BHP Billiton from Guinea and a negative rating for the country as an investment destination. In the 2013 Resource Nationalism Index, Guinea was classified as being an 'extreme risk' country. Guinea was also ranked last in competitiveness and investor protection in the World Economic Forum's 2014–2015 Annual Competitiveness Index (WEF 2014).

[25] Olawuyi (2015).

4.3 Theories of Resource Ownership and Control

When the potential commercial value of petroleum and mineral resources first became apparent during the 1870s, ownership disputes increasingly arose and progressively became prevalent. Furthermore, the advance of modern technology led to the discovery of resources below the surface that were unimaginable when the depths to which people could go were limited by what manual labour could achieve. The nature of subsurface natural resources and the extent of ownership claims that could be made by a landowner presented the courts of the late 1800s with great difficulties. Therefore, different theories and principles evolved in efforts to explain who controls subsurface natural resources and the extent of such control. In the oil and gas context, there are several theories that mainly evolved from the fact that oil and natural gas are not solid and do not remain in one place. Thus, theories of oil and gas law ownership are slightly different from the ownership theories that apply to the ownership, leasing, and mining of solid minerals. This section elaborates the various theories of ownership and control of extractive resources, especially the *ad ceolum* doctrine and the rule of capture amongst others. Although these theories have been essentially developed and applied to explain ownership of oil and gas resources, they can be equally useful and applicable in analysing ownership and control of solid mineral resources.

4.3.1 The Ad Ceolum Doctrine

The *ad coelum* doctrine is the earliest legal theory developed to explain the nature of subsurface natural resources. For centuries, the well-known Latin brocard—*cujus est solum, ejus est usque ad coelum et ad inferos*—the owner of the surface owns everything from the skies to the centre of the earth—applied to ownership disputes. This theory postulates that, *prima facie*, the owner of the surface is entitled to the surface itself and everything below it down to the centre of the earth (the ground below), including any minerals or natural resources found in between. Under this theory, one was totally free to exclude any trespassers. The principle was recognized and firmly established in common law by Edward Coke in *Bury* v. *Pope* (1587) where the court noted that "and lastly, the earth hath in law a great extent upwards, not only of water as hath been said, but of air, and all other things even up to heaven, for *cujus est solum ejus est usque ad coelum,* as it is holden."[26] This principle has been applied in a number of US and UK cases.

[26]The phrase also appeared in *Blackstone's Commentaries*, Book 2, Chapter 2, p. 18: "Land hath also, in its legal signification, an indefinite extent, upwards as well as downwards. *Cujus est solum, ejus est usque ad coelum,* is the maxim of the law, upwards; therefore no man may erect any building, or the like, to overhang another's land: and, downwards, whatever is in a direct line between the surface of any land, and the center of the earth, belongs to the owner of the surface; as is every day's experience in the mining countries. So that the word "land" includes not only the face of the

In *Mitchell v Mosley,* Cozens Hardy MR stated that the grant of the land includes the surface and all that is *supra*—houses, trees and the like—and everything that is *infra*—mines, earth and clay, etc. Agreeing with him, Swinfen Eady and Phillimore LJJ noted that this was a recognised rule of law.[27] Furthermore in *Kelsen v Imperial Tobacco Co (of Great Britain and Northern Ireland) Ltd*, a mandatory injunction was granted ordering the defendants to remove a sign which projected 8 inches over the plaintiff's property on the ground that, applying the brocard, this was a trespass.[28]

However, with increased technological advancements, the relevance and utility of the *ad coelum* doctrine of property began to face greater scrutiny, most especially with the development of airplanes that could go very high beyond the surface of the land, and with oil and gas drilling facilities and technology that could penetrate deeper into the earth's subsurface.[29] For example, in the US case of *United States v Causby* the respondent had argued that there was a taking of their property, within the meaning of the Fifth Amendment, by frequent and regular flights of army and navy aircraft over their land at low altitudes.[30] The US Supreme Court held that the *ad ceoleum doctrine* has no place in the modern world.[31] Similarly, in *Willoughby Hills v Corrigan* the court noted that the *ad coelum doctrine*, to the effect that the ownership of land extends to the periphery of the universe, has no place in the modern world.[32]

Furthermore, when commenting on the question whether an owner of land may prevent airlines from flying over his or her land, Griffiths J in *Bernstein of Leigh v Skyviews & General Ltd* held that "I can find no support in authority for the view that a landowner's rights in the air space above his property extend to an unlimited

---

earth, but everything under it, or over it. And therefore if a man grants all his lands, he grants thereby all his mines of metal and other fossils, his woods, his waters, and his houses, as well as his fields and meadows."

[27] [1914] 1 Ch 438, 450; see also *Rowbotham v Wilson* (1860) 8 HL Cas 348, 360, per Lord Wensleydale; *Bowser v Maclean* (1860) 2 De G F & J 415, 419, per Lord Campbell LC; *Pountney v Clayton* (1883) 11 QBD 820, 838, per Bowen LJ; *Elwes v Brigg Gas Co* (1886) 33 Ch D 562, 568, per Chitty J; and *National Carriers Ltd v Panalpina (Northern) Ltd* [1981] AC 675, 708, per Lord Russell of Killowen. The proposition that prima facie everything below the surface belongs to the surface owner is often linked to the proposition that everything above it belongs to him too: "everything up to the sky", as Sir William James VC put it in *Corbett v Hill* (1870) LR 9 R 671, 673, or "everything under the sky" in the words of Bowen LJ in *Pountney v Clayton.*

[28] [1957] 2 QB 334.

[29] The soundness of this brocard as a proposition of law was questioned in *Commissioner for Railways v Valuer-General* [1974] AC 325. Describing the doctrine that 'land' means the whole of the space from the centre of the earth to the heavens, as "so sweeping, unscientific and unpractical a doctrine is unlikely to appeal to the common law mind."

[30] 328 US 256 (1946).

[31] Per Douglas J: The air is a public highway, as Congress has declared. Were that not true, every transcontinental flight would subject the operator to countless trespass suits. Common sense revolts at the idea. To recognise such private claims to the airspace would clog these highways, seriously interfere with their control and development in the public interest, and transfer into private ownership that to which only the public has a just claim."

[32] 278 NE2d 658, 664 (Ohio 1972).

height." [33] He noted that although the *ad coelum* phrase was 'colourful', the rights of a land owner over his or her land extend only to a height 'necessary for the ordinary use and enjoyment of the land and structures upon it'. These cases confirm the view that the *ad ceolum* doctrine has ceased to apply to the use of airspace above a height, which may interfere with the ordinary user of land.

In the extractive context however, the *ad ceolum* doctrine is still relied upon as good law, and as a valuable tool for analyzing the extent of ownership enjoyed by the land owner, with respect to the strata beneath it, including the minerals found therein. For example, a neighbor could drill a well on his side of the fence and drain away a landowner's oil or gas. A common question then is how could someone be said to own the oil or gas found on his or her land if they cannot prevent a neighbor from draining or interfering with what he/she owned. Furthermore, it increasingly became difficult to determine to what extent the owner's claim over resources extends. Does it extend down indefinitely to the strata below the surface or is there a sopping point or limit to the depth in which a subsurface owner may exercise ownership rights? In *Chance v BP Chemicals Inc,*[34] the Supreme Court of Ohio, following *Willoughby Hills*, took the airspace approach to subsurface ownership rights when it noted that just as a property owner must accept some limitations on the ownership rights extending above the surface of the property, there are also limitations on property owners' subsurface rights, and as such absolute ownership of subsurface rights, just like absolute ownership of air rights, is a doctrine which has no place in the modern world.

The issue of how much further into the earth's crust can ownership go was a key question before the Supreme Court of the United Kingdom in *Star Energy Weald Basin Limited and Anor. v Bocardo SA*.[35] In this case, Star held a licence to search, bore for and extract petroleum from the Palmers Wood oil field in Surrey, a naturally occurring reservoir of petroleum and petroleum gas, which extends beneath the Oxted Estate of which Bocardo is the owner. To reach the oil, Star's predecessors, Conoco (UK) Ltd, drilled three diagonal wells from neighbouring land owned by them and only entered the sub-strata of Bocardo's land at a depth of 800 ft. The drilling and existence of the wells did not affect Bocardo's use or enjoyment of their land, however Bocardo argued that the drilling of the three wells under its land, without obtaining the landholder's agreement to do so was an actionable trespass by Star.

Bocardo's case was that it is trite law that a conveyance of land includes the surface and everything below it. Star does not dispute this proposition as a general rule, but maintains that Bocardo's ownership right does not extend indefinitely, neither does it extend to the depth at which the operations were and are being carried out in this case. According to Star, the wells and their tubes and casing were too deep for the landowner's interest in his land to be affected in any meaningful way at all. Star also argued that the *ad coelum* doctrine was no longer relevant to land ownership.

---

[33] [1978] QB 479. See also the Alberta Court of Appeal decision in *Didow v Alberta Power Ltd* [1988] 5 WWR 606, 613 which quoted the decision in *Bernstein of Leigh (Baron).*

[34] 670 NE2d 985 (Ohio 1996).

[35] *Bocardo SA v Star Enegy UK Onshore Ltd and another* [2010] UKSC 35.

The Supreme Court unanimously affirmed the *ad coelum* doctrine and upheld both the High Court and Court of Appeal's decisions that deviational drilling constituted trespass. Whilst acknowledging that subterranean ownership could not extend indefinitely, the Supreme Court nevertheless upheld the claimant's right to claim for trespass at depths of 250–400 m below the surface.[36] This decision confirms that although property ownership does not extend downward 'indefinitely', the owner of the land owns everything below it including the minerals found therein.

The decision however continued to raise questions on the indeterminacy and lack of a logical approach for determining how far that ownership could stretch. The UK parliament responded by enacting the *Infrastructure Act 2015* which, in section 43, permits the use of 'deep level land' for the purposes of exploiting petroleum or deep geothermal energy, without liability for trespass. Deep-level land is defined as any land at a depth of at least 300 m below surface level. Pursuant to this Act, only intrusion beyond deep level would constitute actionable trespass.

The *ad ceolum* doctrine has remained a useful theoretical support for subsurface ownership of all natural resources found within a portion of land.[37] In dominial regimes of Africa, this doctrine will reinforce the right of a license holder to conduct exploration activities within the portion of land covered by lease and to search form, win, obtain and transport extractive resources by drilling, mining, etc, and to conduct all activities incidental to such operations in accordance with the terms and conditions of the license. The application of this theory may however face uncertainties when two license holders have rights over adjoining lands from which one right holder may utilize advanced drilling technologies to drain gaseous or liquid natural resources from the other right holder. The uncertainties surrounding the doctrine has led to the evolution of what is now known as the 'non-ownership' or 'qualified ownership' theory.

---

[36] Para 27–28:

> There must obviously be some stopping point, as one reaches the point at which physical features such as pressure and temperature render the concept of the strata belonging to anybody so absurd as to be not worth arguing about. But the wells that are at issue in this case, extending from about 800 feet to 2,800 feet below the surface, are far from being so deep as to reach the point of absurdity. Indeed the fact that the strata can be worked upon at those depths points to the opposite conclusion. I would hold therefore that the appellant's title extends down to the strata through which the three wells and their casing and tubing pass.

[37] The ownership in place theory enjoys wide recognition and application in Arkansas, Colorado, Kansas, Maryland, Michigan, Mississippi, Montana, New Mexico, North Dakota, Pennsylvania, Tennessee, Texas, Washington, and West Virginia have all adopted the ownership in place theory. Martin and Kramer (2012), p. 201.

4.3.2 Non-Ownership Theory

Under the non-ownership theory, ownership of land does not immediately translate to ownership of oil and gas underneath the land. Given that oil and gas in the ground is migratory or "fugacious" in nature, it is incapable of being owned until it is produced and reduced to possession. It follows therefore that surface owner's ownership of petroleum underneath his land is qualified, that is they belong to the land owner and are part of it, so long as they are on or in it, and are subject to his control; but when they escape, and go into other land, or come under another's control, the title of the landowner is gone. This much was stated by the US court, as far back as in 1889, in the case of *Westmoreland & Cambria Natural Gas Co. v. Dewitt*:

> ...as minerals *ferae naturae*. [Like] animals, they have the power and the tendency to escape without the volition of the owner. Their 'fugitive and wandering existence within the limits of a particular tract was uncertain,' . . . They belong to the owner of the land, and are part of it, so long as they are on or in it, and are subject to his control; but when they escape, and go into other land, or come under another's control, the title of the former owner is gone. Possession of the land, therefore, is not necessarily the possession of the [oil and] gas. If an adjoining, or even a distant, owner, drills his own land, and [produces gas from another's land], so that it comes into his well and under his control, it is no longer [owned by the owner of the other land], but [becomes] his control, it is no longer [owned by the owner of the other land], but [becomes] his[38]

The court, in its above reasoning, likened oil to other fugacious or migratory things like air, birds and sunshine, which even though may be present on a landowner's property, and can be used by the landowner while across or over the land, are not necessarily the property of the land owner. This line of reasoning was followed in *State v. Ohio Oil Co* in which the court held that:

> To say that the title to natural gas vests in the owner of the land in or under which it exists today, and that tomorrow, having passed into or under the land of an adjoining owner, it thereby becomes [the property of that adjoining owner], is no less absurd, and contrary to all the analogies of the law, than to say that wild animals or fowls, in 'their fugitive and wandering existence,' in passing over the land, become the property of the owner of such land, or that fish, in their passage up or down a stream of water, become the property of each successive owner over whose land the stream passes. [Hence, the court reasoned that to hold otherwise will be] as unreasonable and untenable as to say that the air and the sunshine which float over the owner's land is a part of the land, and is the property of the owner of the land. [The court] therefore [held] that the title to natural gas does not vest in any private owner until it is reduced to actual possession.[39]

The court rightly noted that ownership of land does not automatically vest ownership of oil and gas found in the land, unless the landowner extracts or reduces the resources to actual possession. Similarly, in *Kelly v. Ohio Oil Co*, the court endorsed the non-ownership theory when it noted that:

---

[38] 130 Penn. 235, 18 A. 724 (1889).

[39] *State v. Ohio Oil Co.*, 49 N.E. 809, 812 (1898). See also Martin and Kramer (2012), p. 203 (quoting *State v. Ohio Oil Co.*, 49 N.E at 812).

> Petroleum oil is a mineral, and while in the earth it is part of the realty, and, should it move from place to place by percolation or otherwise, it forms part of that tract of land in which it tarries for the time being, and, if it moves to the next adjoining tract, it becomes part and parcel of that tract; and it forms part of some tract until it reaches a well, and is raised to the surface, and then for the first time it becomes the subject of distinct ownership, separate from realty, and becomes personal property, the property of the person into whose well it came.[40]

Under this approach, whoever removes the extractive resource owns it. The landowner has no absolute property in oil, gas, and water, found underneath the land, but by virtue of his or her proprietorship, may bore wells for the purpose of extracting these substances. Until these substances are actually reduced by him to possession, he has no title whatever to them as owner.[41] The principle that oil and gas cannot be owned absolutely until found and reduced to possession is recognized in all oil and gas producing jurisdictions of the United States[42] and in Canada.[43]

The ownership in place theory is based on the principle of 'fee simple determinable'. Under fee simple determinable, a person has an estate in property, *i.e.* an ownership interest, unless a stated condition occurs. If and when that condition occurs, they lose their interest in the property. In its application to oil and gas ownership, the owner of the oil and gas is the person on whose land the resource is located. However, if the oil or gas migrates to another person's property, then the original owner no longer owns the resource. The primary condition for ownership is that the resource must be on the owners' property. As such, the only limitation to this ownership theory is where the mineral estate has either been severed from the surface

---

[40] Martin and Kramer (2009).

[41] Ohio Oil Company v. Indiana U.S.S.C. [1900] 44 L. Ed. 729, par. 64, stating that "Although in virtue of his proprietorship the owner of the surface may bore wells for the purpose of extracting natural gas and oil, until these substances are actually reduced by him to possession, he has no title whatever to them as owner." That is, he has the exclusive right on his own land to seek to acquire them, but they do not become his property until the effort has resulted in dominion and control by actual possession.

[42] See *Strother v. Mangham* (1915) 138 La. 437, also *Rich v. Donaghey* (1918) Okla. S.C., 3 A.L.R. 352, p. 355.

[43] This principle was incorporated into Canadian oil and gas law by the 1953 decision of the Judicial Committee of the Privy Council in *Borys v. CPR and Imperial Oil Limited* J.C.P.C. [1953] 7 W.W.R. (NS) pp. 550–551 where Lord Porter, for the Privy Council, stated:

> The substances are fugacious and are not stable within the container although they cannot escape from it. If any of the three substances (petroleum, gas or water) is withdrawn from a portion of the property which does not belong to the appellant but lies within the same container and any oil or gas situated in his property thereby filters from it to the surrounding lands, admittedly he has no remedy. So, also, if any substance is withdrawn from his property, thereby causing any fugacious matter to enter his land, the surrounding owners have no remedy against him. The only safeguard is to be the first to get to work, in which case, <u>those who make the recovery become owners of the material which they withdraw from any well which is situated on their property or from which they have authority to draw.</u> [emphasis added]

right,[44] or where ownership of the resource is lost by virtue of the operation of the rule of capture if neighbors begin to drain the resource.

4.3.3 The Rule of Capture

A different ownership theory, known as the rule of capture, developed in the State of Texas, USA, and has been adopted in a number of oil producing jurisdictions in and outside of the US. Under this theory, a landowner may drill as many wells on his or her land and may drain as much oil as possible, without being liable to the adjacent landowners whose lands are drained as a result of such operations, so long as the producing well does not trespass. The rule of capture gives the landowner a right to produce all of the oil and gas that flow out of the well on his or her land, even though it is proved that part of the oil migrated from adjoining lands.[45]

This rule was applied in the US, in the 1886 case of *Wood County Petroleum Co v. West Virginia Transportation Co*,[46] and also in the 1889 case of *Westmoreland & Cambria Natural Gas Co v. De Witt*.[47] Furthermore in *Jones v. Forest Oil Co.*,[48] a Pennsylvania court, citing English common law, held that the rule of capture accords each owner an unlimited right of withdrawal. Also in *Barnard v. Monongahela Gas Company*, the court explained that

> An oil and gas well may draw its product from an indefinite distance and in time exhibit a large space. The vagrant character of the mineral and the porous rock in which it is found and though which it moves fully justify... the right of every land owner to drill a well on his own land at whatever spot he may see fit. What, then, can the neighbor do? Nothing, only go and do likewise.[49]

The rule of capture has also become a well-established doctrine in jurisdictions that recognize a landowner's rights to produce the oil and gas in place beneath their

---

[44] Chapter 7 of this book examines the different types of interests in oil and gas, including a mineral interest.

[45] See Hardwicke (1935). See also *Brown v. Humble Oil & Refining Co.* (1935) Tex. S.C., 83 S.W. (2d) 935, where the Court at p. 940, stated as follows:

> Owing to the peculiar characteristics of oil and gas, the foregoing rule of ownership of oil and gas in place should be considered in connection with the law of capture. <u>This rule gives the right to produce all of the oil and gas that will flow out of the well on one's land; and this is a property right.</u> And it is limited only by the physical possibility of the adjoining landowner's diminishing the oil and gas under one's land by the exercise of the same right of capture. [emphasis added]

[46] *Wood County Petroleum Co v. West Virginia Transportation Co* [(1886) 28 W Va 10 (W Va).

[47] *Westmoreland & Cambria Natural Gas Co v. De Witt* 130 Pa 235 (1889); 18 Atl 724 (pa).

[48] *Jones v. Forest Oil Co* 194 Pa. 379 (1900).

[49] *Barnard v. Monongahela Gas Company* 216 Pa., 362; 65 Atl. 801 (1907).

land, as well as the oil and gas which flows to the land as the result of physical conditions and natural laws relating to the migratory nature of oil and gas.[50]

The policy aim of this theory is to provide an economic imperative for landowners to drill their land and produce oil as rapidly as possible, otherwise risk the oil and gas resources in the land being drained by the owners of adjoining properties. In reality however, the application of this rule has been plagued by widespread waste of oil and gas resources in large part due to the drilling of unnecessary wells to prevent adjoining owners from capturing the resource.[51] This often leads to a race to the finish, where operators raced to be the first to drill, and exploit the high natural reservoir pressures that are responsible for driving the oil to the surface. Increasingly across the US, large number of wells were drilled almost simultaneously, resulting in large amounts of oil being left in the ground as a result of loss of well reservoir pressure occasioned by indiscriminate drilling. The drilling also resulted in large scale flaring of gas. Consequently, the unchecked production resulted in environmental concerns, colossal fall in oil prices, storage issues, as well as the premature shutting in of productive wells that cost more to exploit.[52]

Following heightened risk of damage to the environment resulting from the application of this rule, efforts were made to eliminate the unnecessary competitive and wasteful aspects of the rule, by requiring that oil reservoirs must be developed as a unit, regardless of the surface land divisions. This eliminated the need for competitive drilling, which occasioned some of the problems highlighted above. Despite these efforts, the rule of capture remains in full application in its original form in many parts of the U. It is only replaced where there are conservation rules or regulation that promote unitization or unit operations in place. Conservation statutes exist in many states to place limits to the rule of capture, by introducing conditions such as where drilling can take place, production methods and the amount of production.

The rule of capture has also found application in other parts of the world, especially in Romania, Poland, Ukraine and Russia. Its influences are also found in the French Civil Code of 1804. In common law England, the rule of capture was the default property law principle to be applied by the courts when ruling on oil and gas disputes. It is important to note that even in its current application, the rule of capture is not an absolute rule but is subject to exceptions. One exception relates to natural gas that has already been produced. Natural gas maybe stored by injecting it into underground storage areas.[53] This natural gas which has already been 'cap-

---

[50] *SWEPI, L.P. v. Camden Res., Inc.*, 139 S.W.3d 332, 341 (Tex. App. – San Antonio 2004, rev. denied).

[51] See *State of Alabama v U.S. Department of Interior*, 84 F. 3d. 410, 413, 133 O&G R. 437 (11th Cir. 1996) stating that the application of the rule is causing "(1) economic waste in drilling unnecessary wells; (2) a corresponding heightened risk of damage to the environment; and (3) physical waste of the oil and gas itself because the faster production occurs, the lower the long-term recovery will be from the reservoir."

[52] See Daintith (2010), p. 141.

[53] This is very similar to the technique that is utilized with carbon sequestration.

tured' remains the property of the party who initially captured it even though it has been injected underground for storage, except in situations where it is abandoned by the owner. Another exception is the doctrine of correlative rights.

4.3.4 Correlative Rights or Qualified Ownership Theory

The doctrine of correlative rights states that the right to take oil and gas from another party's land under the rule of capture does not apply where the removal of the oil and gas is done negligently or causes waste. A correlative right is the right the landowner has to produce oil and gas from their own property provided they are not negligent or wasteful in the production. Each owner of land lying over the common reservoir has certain correlative rights (and duties) in the common pool, including the duty not to waste the oil and gas, and not produce it in such manner as to damage the reservoir formation and reduce the ultimate recovery.

This theory, developed as one of the exceptions to the rule of capture, was first introduced in the case of *Eliff v. Texon Drilling Company*.[54] Similarly, in Eliff's case, Texon, in draining oil wells from its property also drained the oil from Eliff's property. However, Texon was negligent in the operation of one of its well's causing the well to blow out and a large amount of gas, distillate, and oil from the shared reservoir blew into the air. The blow out caused large quantities of oil and gas under Eliff's land to be drained away and destroyed. Texon was held liable because its negligence caused Eliff to suffer a loss.[55] Generally, landowners whose tracts overlay a producing formation have correlative rights and duties not to damage or waste the oil and gas in the formation.[56]

Furthermore, the rule of capture does not provide a total shield from liability for removing oil and gas from another person's property. It does not exempt a trespasser, who trespasses unto another's land and illegally removes oil and gas. It also does not exempt persons who utilize directional drilling techniques, such as 'slant drilling' to remove oil and gas from the property of another. Slant drilling, which is done by setting up a drilling rig close to the border of two pieces of land, but instead of drilling straight down onto their own property, they drill at a slant, angling the operation to drill into a reserve that is under a neighbouring property, has been described as a form of trespass.[57]

---

[54] See *Eliff v. Texon Drilling Company* 146 Tex. 575, 210 S.W. 2d 558, 1948 Tex. 4 A.L.R. 2d 191. The correlative rights theory is followed in US states of California, Indiana, Louisiana and Oklahoma.

[55] See generally, *Eliff v. Texon* 146 Tex. 575, 210 S.W. 2d 558, 1948 Tex. 4 A.L.R. 2d 19.

[56] Martin and Kramer (2012), 203.2.

[57] See generally, Oil and Gas Law, Delmar Learning.

4.3.5 *Ownership of Strata Theory*

This theory was developed as another exception to the rule that the landowner does not own oil and gas in place in the land. This theory posits that although that the landowner does not own oil and gas in place in the land, the landowner owns the strata or formations containing the oil and gas. The landowner owns the sedimentary layer containing the oil and gas within the limits of the vertical planes representing the boundaries of his track.

With increased exploitation of coal bed methane and coalbed gas, this theory has played a significant role in clarifying the extent of rights that a landowner has over the strata or seams found in the land. Under this theory, gas present in coal seams, as long as it remains within the coal found on the land, belongs to the owner of the coal.[58] As noted in the leading Pennsylvania case of *United States Steel Corp. v Hoges*, coalbed gas contained in coal is, *ab initio*, property of the coal owner so long as it remains within his property and subject to his exclusive dominion and control.[59]

4.3.6 *Servitude Theory (Profit à Prendre)*

The servitude theory of ownership is mostly applied to solid minerals. A mining license typically grants the holder the right to enter the land, to which such right relates, to prospect, mine, explore or produce minerals for which such right has been granted and to remove and dispose of any such mineral found during the course of prospecting, mining, exploration or production, as the case may be.[60] In mining law however there has always been conflict between the holder of a title deed who, in most cases, is the surface landowner and the mineral right holder who has the rights to access and sever the minerals beneath the surface by virtue of the licenses granted.[61] Given that solid minerals are non-migratory, many of the foregoing rules of ownership for oil and gas are unhelpful in clarifying this and other conflicts relating to ownership of solid minerals.

The servitude theory posits that the holder of a mineral license, applicable to another's land, enjoys a mineral servitude and has no ownership right over the land. A mineral servitude is defined in *Frost-Johnson Lumber Co. v. Salling's Heirs* as a right to go upon land and reduce its minerals to possession.[62] The mineral servitude

---

[58] Martin and Kramer (2009), p. 683.

[59] 85 W. Va. L. Rev. 803 (1982–1983).

[60] See for example Section 5 of the South African Mineral and Petroleum Resources Development Act 28 of 2002.

[61] See for example the South African case of *Minister of Minerals and Energy v Agri South Africa* (Centre for Applied Legal Studies as amicus curiae) [2012] 3 All SA 266 (SCA).

[62] 150 La. 756, 91 So. 207 (1922).

grants a mineral right holder the right to enter private property to search for minerals on the basis that the nature of mineral rights is separated from the ownership of the land.[63] The servitude allows the person to unto the land and win mineral resources therein.

A mineral servitude is therefore the right of enjoyment of land belonging to another for the purpose of exploring for and producing minerals and reducing them to possession and ownership.[64] A mineral servitude may be created by a landowner or anyone who owns the right to explore for and produce minerals when the servitude is created. It entitles the holder thereof to go upon the property to which they relate to search for minerals, and, if the holder finds any, to sever them and carry them away.

Pursuant to the servitude theory, when there is difference between the landowner and the owner of the minerals extracted from land, the owner of minerals has a non-possessory (incorporeal) right, something akin to an easement or a *profit à prendre*.[65] Profit à prendre is a right to take from the land owned by another person, part of the natural produce grown on that land or part of the soil, mineral, earth or rock comprising the land. The doctrine was described in the Canadian case of *National Trust Co. v. Bouckhuyt*:

A *profit à prendre* is a right to take something off the land of another person. A *profit* includes the power and privilege to acquire, through severance, ownership of some part of the physical substance included in the possession of the land that is subject to the *profit*, or by reduction to possession, ownership of some substance which, were it not for the existence of the *profit*, could be appropriated only by the possessor of the land that is subject to the *profit*.[66]

Under the servitude theory, a mineral right holder may enter into land to reduce mineral resources to possession without necessary acquiring any right or interest over the land. In such cases, such mineral right holder has a *profit à prendre*, which enables him/her to take the mineral resources off the land of another person.[67] The owner of a mineral servitude is under no obligation to exercise it. If he does, he is entitled to use only so much of the land as is reasonably necessary to conduct his operations. He is obligated, insofar as practicable, to restore the surface to its original condition at the earliest reasonable time. The owner of a mineral servitude must conduct his operations subject to the restrictions that apply to a landowner. He must

[63] Per Innes CJ in the South African case of *Van Vuren and Others v Registrar of Deeds* 1907 TS 289 at 294;

[64] See Article 21 of the Louisiana Mineral Code [R. S. 31. 21].

[65] Van Vuren and Others v Registrar of Deeds, *n. 65.*

[66] 1987 39 DLR 4th 60 (1987, Ontario). See also *Nywening v. Melton Holdings Ltd.* (1998, ABQB) 58 Alta LR (3d) 333: "A *profit à prendre* is a right to take something off the land of another person. It may be more fully defined as a right to enter the land of another person and to take some profit of the soil, or a portion of the soil itself, for the use of the owner of the right".

[67] See Article 21 of the Louisiana Mineral Code.

also protect his right against interference or damage by all of the means available to a landowner.[68]

While the foregoing ownership theories have been helpful in explaining ownership and control in less dominial jurisdictions such as Canada and the United States, ownership theories have found little application in the African context. As noted earlier, African countries exercise dominial control over oil and gas resources. The ownership and control of all minerals, mineral oils and natural gas in is vested in the State by virtue of constitutional or statutory provisions.[69] Given that there is no private ownership of oil and gas *in situ*, private individuals or companies can only obtain rights to prospect, explore and mine for oil and gas under the express terms and conditions stipulated in a lease, license or permit.

4.4 Conclusion

Given that African countries maintain dominial control over all natural resources, the extent to which resources can be exploited and captured are often expressly outlined in the applicable petroleum or mineral development law, and in the specific licenses and leases granted. In most cases, concerns or disputes relating to the extent and limits of exploration are within the purview of the state, or the responsible national authority which oversees, manages, and authorizes extractive operations.

When developing investment plans for African extractive markets, it therefore makes business sense for an investor to carefully examine the nature, extent and scope of rights and interests that are conveyed to a right holder under the various licenses and permits granted by African States to avoid legal risks. Extensive due diligence, which includes a careful combination of surveying and understanding applicable legal regimes, negotiating protective contractual provisions, informed deal structuring and active management of political risk, and adequate insurance protection for political risks are highly important.[70]

The next chapter explores the network of laws, regulations and guidelines that have been developed to regulate the exploration and development of petroleum and mineral resources in frontier African jurisdictions.

---

[68] ibid. Articles 21-79.

[69] See *Constitution of the Federal Republic of Nigeria*, Laws of the Federation of Nigeria 2004, c 23. See also the *Petroleum Act 1969* which vests entire ownership and control of all petroleum in Nigerian lands, territorial waters or the continental shelf in the Nigerian government.

[70] See Olawuyi (2015).

References

Badenhorst PJ, Pienaar JM, Mostert H (2006) Silberberg & Schoeman's the law of property, 5th edn. Lexis Nexis, South Africa

Barnard v. Monongahela Gas Company 216 Pa., 362; 65 Atl. 801 (1907)

Blackstone's Commentaries, Book 2, Chapter 2

Bocardo SA v Star Enegy UK Onshore Ltd and another [2010] UKSC 35

Borys v. CPR and Imperial Oil Limited J.C.P.C. [1953] 7 W.W.R. (NS)

Bouckaert B (1990) What is property? Harv J Law Public Policy 13:781

Bowser v Maclean (1860) 2 De G F & J 415, 419, per Lord Campbell LC

Brown v. Humble Oil & Refining Co. (1935)

Clark B (2002) Migratory things on land: property rights and a law of capture. Electron J Comp Law 6(3):1–29

Commissioner for Railways v Valuer-General [1974] AC 325

Constitution of the Federal Republic of Nigeria, Laws of the Federation of Nigeria 2004, c 23

Corbett v Hill (1870) LR 9 R 671, 673

Daintith T (2010) The *Rule of Capture*: the least worst property rule for oil and gas. In: McHarg A, Barton B, Bradbrook A, Godden L (eds) Property and the law in energy and natural resources. Oxford University Press, Oxford

Davies W, Fouracre P (eds) (1995) Property and power in the early middle ages. Cambridge University Press, Cambridge

Didow v Alberta Power Ltd (Alberta Court of Appeal)

Eliff v. Texon Drilling Company 146 Tex. 575, 210 S.W. 2d 558, 1948 Tex. 4 A.L.R. 2d 191

Elwes v Brigg Gas Co (1886) 33 Ch D 562, 568, per Chitty J

Graulau J (2011) Ownership of mines and taxation in castilian laws, from the middle ages to the early modern period: the decisive influence of the sovereign in the history of mining. Continuity Change 26:13–44

Hardwicke RE (1935) The rule of capture and its implications as applied to oil and gas. Tex Law Rev 13:391

Jolowicz H, Barry N (1972) Historical introduction to the study of roman law, 3rd edn. Cambridge University Press

Jones v. Forest Oil Co 194 Pa. 379 (1900)

Louisiana Mineral Code [R. S. 31. 21]

Martin P, Kramer B (2009) Williams & Meyers manual of oil and gas terms. Lexis Nexis

Martin P, Kramer B (2012) Williams & Myers oil and gas law. Lexis Nexis

Mineral and Petroleum Resources Development Act 28 of 2002 of South Africa

Minister of Minerals and Energy v Agri South Africa (Centre for Applied Legal Studies as amicus curiae) [2012] 3 All SA 266 (SCA) (South Africa)

Momodu MK (1988) Legal aspects of ownership of natural gas in Nigeria. J Energy Nat Resour Law 6:268–270

National Carriers Ltd v Panalpina (Northern) Ltd [1981] AC 675, 708, per Lord Russell of Killowen

Nigerian Mines and Mineral Act, 2007, Cap. F 34 LFN 2004

Njotini M (2017) Examining the 'objects of property rights' - lessons from the Roman, Germanic and Dutch legal history. De Jure 50(1):136–155

Nywening v. Melton Holdings Ltd. (1998, ABQB) 58 Alta LR (3d)

Ohio Oil Company v. Indiana U.S.S.C. [1900] 44 L. Ed. 729

Olawuyi D (2015) Legal strategies and tools for mitigating legal risks associated with oil and gas investments in Africa. OPEC Energy Rev 39(3):247–265

Olawuyi D (2018) Regulation of upstream oil & gas. Gulf Legal Advisor. Lexis Nexis

Omorogbe Y (2001) Oil and gas law in Nigeria. Malthouse Press Limited, Lagos, Nigeria

Omorogbe Y, Oniemola P (2010) Property rights in oil and gas under dominial regimes. In: McHarg A, Barton B, Bradbrook A, Godden L (eds) Property and the law in energy and natural resources. Oxford University Press, Oxford

Petroleum Act 1969 of Nigeria

Plourde A (2010) Oil and Gas in the Canadian Federation. University of Alberta, Working Paper No. 2010-01

Pountney v Clayton (1883) 11 QBD 820, 838, per Bowen LJ

Rich v. Donaghey (1918) Okla. S.C., 3 A.L.R. 352

Rowbotham v Wilson (1860) 8 HL Cas 348, 360, per Lord Wensleydale

State of Alabama v U.S. Department of Interior, 84 F. 3d. 410, 413, 133 O&G R. 437 (11th Cir. 1996)

State v. Ohio Oil Co., 49 N.E. 809, 812 (1898)

Strother v. Mangham (1915) 138 La. 437

SWEPI, L.P. v. Camden Res., Inc., 139 S.W.3d 332, 341 (Tex. App. – San Antonio 2004, rev. denied)

Tex. S.C., 83 S.W. (2d) 935

Texas Company v. W. H. Daugherty et al, 107 Tex. 226 (Tex. 1915)

Van Vuren and Others v Registrar of Deeds 1907 TS 289 (South Africa)

Westmoreland & Cambria Natural Gas Co v. De Witt 130 Pa 235 (1889); 18 Atl 724 (pa)

Wood County Petroleum Co v. West Virginia Transportation Co [(1886) 28 W Va 10 (W Va)]

World Economic Forum (2014) Annual competitiveness. WEF

Chapter 5
Sources of Extractive Industry Law

Abstract This chapter provides an overview of the different sources of law that govern the extractives industry. It discusses the roles of international, regional and national laws in shaping extractive sector activities and investments across the African continent. First, it identifies some of the basic characteristics of the different sources of law that are applicable to extractive operations. It then analyses the character, status and force of the different sources including the interrelationships between them.

5.1 Introduction

To participate in extractive operations in Africa, foreign companies must, in most cases, satisfy incorporation, registration and procurement formalities which are found in several locally applicable corporate, business organizations, foreign investment, energy, environmental, trade, currency importation, competition, tax and finance laws. To avoid legal risks relating to noncompliance, it is very pertinent for prospective participants in Africa's extractive sector to understand the diverse applicable legislation, institutions and frameworks that are applicable to extractive operations.

This chapter provides an overview of the different sources of law that govern the extractive industry. By sources, it is meant the genres of law that underpin and govern extractive sector activities and investments. Generally, every stratum of the overall framework of laws have direct application in the extractive sector, be it international law, environmental law, criminal law, administrative law, tort law, or constitutional law. This chapter explores how these bodies of law contribute to extractive sector activities and investments across the African continent. This chapter will lay the foundation for subsequent discussions on the licensing and permit regime for extractive investments, as well as the conduct of petroleum and solid mineral operations.

At the end of this chapter, you should be able to: understand where extractive sector law comes from, at international, regional and national levels; identify some of the basic characteristics of the different sources of extractive law in your country;

D. S. Olawuyi, *Extractives Industry Law in Africa*,
https://doi.org/10.1007/978-3-319-97664-8_5

and understand the character, status and force of the different sources including the interrelationships between them.

5.2 Sources of Extractive Industry Law

There are five key sources of extractive industry law: international law; domestic legislation; common law; judicial decisions; and scholarly publications.

5.2.1 International Law on Resource Extraction

International law governs relations between states. One of the key aims of international law is to provide an underlying legal framework that governs the interactions between states with respect to the exploration, exploitation, management, exchange, sale and use of natural resources found in the respective territories of countries. International energy and natural resources law is therefore a field of international law that regulates the behaviour of states and international organizations with respect to the management and use of natural resources, including petroleum and solid mineral resources. As the preamble to the United Nations Charter notes "the peoples of the United Nations aim to… establish conditions under which justice and respect for the obligations arising from treaties and other sources of international law can be maintained".[1] International law on natural resources aims to reinforce and support the sustainable development, management and utilization of natural resources found in a country's territory.

Since the early 1970s, widespread concern about the need for sustainable use and management of natural resources, especially in developing countries, has gained international attention. Prior to this awareness, international law traditionally regulated the conduct of states in regard to war and peace. For example, at the time when United Nations Charter was drawn up in 1945, natural resource management were not at the forefront of global attention and discussions, the main focus was world peace. Today there are nearly 3000 international legal instruments that are either primarily directed to sustainable resource management or have implications for the extractive industry.

Pursuant to Article 38 of the Statute of the International Court of Justice the sources of international law, and by extension international natural resources law, are: international conventions and treaties; international custom; general principles of law; and judicial decisions and teachings of highly qualified publicists.[2]

[1] Charter of the United Nations (Adopted June 26 1945, entered into force on 24 October 1945).

[2] Statute of the International Court of Justice 3 Bevans 1179; 59 Stat. 1031; T.S. 993; 39 AJIL Supp. 215 (1945).

5.2.1.1 International Conventions/Treaties

One of the main sources of international law are "international conventions, whether general or particular establishing rules expressly recognized by the contesting states".[3] Treaties are viewed more or less the same way as contracts in domestic law, *i.e* having effect and binding only as between the parties that signed them.

According to the 1969 Vienna Convention of the Law of Treaties ("VCLT"), which sets out the laws that govern treaties in international law, a treaty is 'an international agreement concluded between states in written form and governed by international law, whether embodied in a single instrument or in two or more related instruments and whatever its particular designation.'[4] Treaties are the major mechanism employed by states in the conduct of their relations with one another, and can also be known as conventions, protocols, covenants, charters, pacts, engagement, accord, exchange of notes, Memorandum of Understanding etc., and regardless of their characterization, once they fit the above particulars, they are recognized as treaties. They are essentially written or oral agreement between states or between states and international organizations, which are governed by international law. The aim of treaties generally is the resolution of matters of international importance. They are the major mechanism employed by states to regulate conduct between states. They are also the most frequently used instruments for creating multilateral rules that apply to the environment. Every responsible state is expected to carry out the terms of its obligations under an international agreement. This is based on the legal principle of *pacta sunt servanda* (let the agreement be kept). According to Article 18 of the VCLT:

> A State is obliged to refrain from acts which would defeat the object and purpose of a treaty when: (a) it has signed the treaty or has exchanged instruments constituting the treaty subject to ratification, acceptance or approval, until it shall have made its intention clear not to become a party to the treaty; or (b) it has expressed its consent to be bound by the treaty, pending the entry into force of the treaty and provided that such entry into force is not unduly delayed.[5]

Treaties are not on the face binding on parties unless: (a) the intention be binding is clearly expressed and (b) the concerned states clearly accept the benefits and obligations imposed by the treaty, mostly by the process of signature and ratification. However, as we will see below, when treaty provisions have evolved into customary international law, they can be binding on non-parties. While there are no fixed rules prescribing the form of treaties, the VCLT provides rules that apply to treaties concluded after its entry into force in 1980.[6] Depending on their composition, and the number of parties to the treaties, they are sometimes described as framework or bilateral or multilateral among others.

[3] Art. 38, ibid.

[4] Vienna Convention on the Law of Treaties, 23 May 1969, 1155 UNTS, 331.

[5] Ibid.

[6] Ibid.

The general classifications of treaties are briefly highlighted below.

Bilateral/Multilateral Treaties Bilateral treaties are between two countries, while multilateral treaties are agreements between more than two countries. Bilateral investment treaties (BITs) are increasingly adopted in the extractive sector to create obligations between states with respect to the treatment of investors from the other treaty party in the cultivation, extraction or exploitation of natural resources. The aim of BITs is to encourage and increase investments between investors of one State Party into the territory of the other State Party, in a manner that supports the sustainable development of each Party, and in particular the Host State where an investment is to be located.[7] They may be standalone agreements between two states, they may also be regional agreements, such as the Southern African Development Community (SADC) region and the pending Common Market for Eastern and Southern Africa (COMESA) agreement. Investment treaties may also be found as parts of bilateral or regional free trade treaties. To date, there are some 3000 investment treaties in force globally.

BITs typically contain substantive provisions on investment protection and dispute settlement mechanisms. The substantive provisions found in most investment treaties include: the *national treatment clause,* which contains the basic requirement that investors from the other country will be assured treatment "no less favourable" than investors of the host state.[8] The *most favoured nation (MFN) treatment clause* is also utilized in BITs to guarantee that a government cannot treat an investor from the state of the treaty partner any less favourably than it treats any investor from any other third country.[9] The International Law Commission Study Group has identified four core elements of MFN clauses.[10] First, under such a provision each State agrees to grant a particular level of treatment to the other State or States, and to persons and entities in a defined relationship with that State or those States.[11] Second, the level of treatment provided by an MFN provision is determined by the treatment given by the State granting MFN to third States ("no less favourable").[12] Third, an MFN commitment applies only to treatment that is in the same category as the treatment granted to the third State ("*ejusdem generis*").[13] Fourth, the persons or entities enti-

---

[7] See Article 1 of the SADC Model Bilateral Investment Treaty Template of 2012.

[8] See OECD (2004).

[9] MFN clause has been defined as: "treatment accorded by the granting State to the beneficiary State, or to persons or things in a determined relationship with that State, not less favourable than treatment extended by the granting State to a third State or to persons or things in the same relationship with that third State." Art. 5 of the 1978 Draft Articles of the MFN Clause, *Yearbook of the International Law Commission 1978*, vol. II, Part Two, p. 21.

[10] International Law Commission, Most-Favoured-Nation Clause Final Report of the Study Group on the Most-Favoured-Nation Clause (2015) Yearbook of the International Law Commission para 35.

[11] Ibid.

[12] Ibid.

[13] Ibid.

tled to the benefit of MFN treatment are limited to those in the same category as the persons or entities of the third State that are entitled to the treatment being claimed.[14] By requiring national treatment and most favoured nation treatment in extractives investments, these provisions can help provide less stringent requirements and favourable terms, with respect to market entry, fund repatriation, free carry shares and performance requirements, for investors from the treaty states. A state party is required to accord investments of investors of the other State Party protection and security no less favourable than that which it accords to investments of its own investors or to investments of investors of any third State.

BITs can also include *expropriation clauses* that protect foreign investors from direct, indirect and regulatory expropriations of an extractive investment or operation by the host state without compensation to the investor.[15] BITs often recognize exceptions, especially when national expropriations are (a) in the public interest; (b) in accordance with due process of law; and (c) on payment of fair and adequate compensation within a reasonable period of time.[16] Fair and adequate compensation will normally be assessed in relation to the fair market value of the expropriated investment immediately before the expropriation took place.[17] The fair and equitable treatment clause places an obligation on the host state to provide accountability mechanisms by improving the transparency, efficiency, independence and accountability of their legislative, regulatory, administrative and judicial processes in accordance with their respective domestic laws and regulations.

Another substantive clause is the *umbrella clause* which often stipulates that a government must adhere to any other commitments it has made to an investor and

[14] International Law Commission, Most-Favoured-Nation Clause Final Report of the Study Group on the Most-Favoured-Nation clause (2015) Yearbook of the International Law Commission para 35.

[15] Direct expropriation, where an investment is nationalized or otherwise directly expropriated through formal transfer of title or outright seizure, while indirect expropriation, is where an action or series of actions by a Party has an effect equivalent to direct expropriation without formal transfer of title or outright seizure. See Dominican Republic-Central America Free Trade Agreement (CAFTA-DR) (2006) in annex 10-C. See also Harvard Draft Convention on the International Responsibility of States for Injury to Aliens (1961) and the Third Restatement of Foreign Relations Law of the United States (1987). In *Starrett Housing v. Iran*, Interlocutory Award No. ITL 32-24-1, 19 December 1983, 4 Iran-United States Claims Tribunal Reports 122, p. 154, the tribunal noted that it is recognized under international law that measures taken by a State can interfere with property rights to such an extent that these rights are rendered so useless will constitute expropriation, even though the State did not purport to expropriate them and the legal title to the property formally remains with the original owner.

[16] This exception is found in the Dominican Republic-Central America Free Trade Agreement (CAFTA-DR) (2006), Egypt-Germany BIT (2005), Mexico-United Kingdom BIT (2006), Japan-Lao People's Democratic Republic BIT (2008), Investment Agreement for the Common Market for Eastern and Southern Africa (COMESA) Common Investment Area (2007) and in the Turkey model BIT (2009), all stating that investments of investors of the contracting parties "shall not be nationalized or expropriated, either directly or indirectly through measures having effect equivalent to nationalization or expropriation ("expropriation") except for a public purpose, on a non-discriminatory basis, in accordance with due process of law and against compensation…."

[17] Nikièma (2013), pp. 4–10.

that such obligation cannot be avoided by the change of laws or the introduction of legal changes that were not in place as of the time of investment.[18] Most BITs also include *personnel and employment clauses* that allow the foreign investor maximum flexibility to engage staff from its country of origin or other countries. Under this clause, a State Party shall not require an Investor to appoint, to senior management positions for its Investment, individuals of any particular nationality. This provision is particularly important in order to waive the application of domestic local content laws that may stipulate preferential employment for nationals or work force at the host state. A well-crafted personnel clause provides an investor the maximum flexibility to engage highly-skilled senior staff and directors from any part of the world. The SADC Model Bilateral Investment Treaty Template of 2012 provides examples of these clauses.

Regional Trade Harmonization Treaties Another important source of international law obligations relating to extractive investments in Africa are regional harmonization treaties and regimes that have been codified in the African sub regions. Some of these regimes have supranational character and are directly applicable to investments in member states, even without domestication or implementation [e.g. Organization for the Harmonization of Business law in Africa (OHADA).[19] It is therefore important to check whether the host country is a member or signatory of any regional trade regime and to seek legal opinion on the key provisions that could impact an investment. The main African regimes to watch out for include:

1. North Africa: Union of Arabic Maghreb (UMA) with Algeria, Morocco, Tunisia, Libya and Mauritania
2. West Africa: Economic Community of West African Countries (ECOWAS), West African Economic and Monetary Union and OHADA[20];
3. Central Africa: Economic Community of Central Africa States, Economic Community of the Great Lakes Countries to promote economic integration (ECGL) and Economic and Monetary Community of Central Africa; and
4. East Africa: Southern African Development Community, and Common Market for Eastern and Southern Africa.

---

[18] Umbrella clauses are included in BITs to provide additional protection to investors and to cover directed at covering investment agreements that host countries frequently conclude with foreign investors. See Sinclair (2004), also Yannaca-Small (2008); also Wälde (2005). The clause appeared the *BIT between Germany and Pakistan* in 1959 (Article 7): "Either Party shall observe any other obligation it may have entered into with regard to investments by nationals or companies of the other party." Also in 1967 OECD draft Convention on the Protection of Foreign Property (Article 2) Each Party shall at all times ensure the observance of undertakings given by it in relation to property of nationals of any other Party."

[19] See Olawuyi (2015a, b, c).

[20] See for example, *The Harmonised System: ECOWAS Common External Tariff and other schedules, Act 905* which provides for the harmonized system of customs and import duties in West Africa.

It is pertinent to ascertain the extents to which these regional regimes have been made applicable to the specific country of investment, and how these regimes may impose additional legal requirements such tax, reporting obligations, environmental liability, fair trade, fund repatriation and local participation among others on foreign investments in that country.

Framework Treaties Some treaties do not contain clear, detailed rules or specific rules; they lay down a framework of general principles or requirements for states. The rationale for a framework convention is that as scientific and other evidence grow, or the political will increases to tackle the problem in a more decisive manner, then further specific regulatory protocols are adopted. States to initially commit to engage in cooperative behaviour to manage an emerging environmental problem mostly through a framework convention.

While few specific references to petroleum or mineral exploration and exploitation appear in international framework treaties, other treaties, conventions and declarations on subjects peripheral to the development of natural resources have implications for extractive industry activities. For example, international treaties relating to the management of the environment to prevent oil related pollution especially in marine resources and seas, where most of exploration and mining activities take place, are particularly important. Examples include the Antarctic Treaty of 1959 which provides a legal framework for the management and use of the Antarctica—the earth's coldest, driest, and windiest continent. Article 1 provides that Antarctica shall be used for peaceful purposes only. The treaty prohibits mineral mining, military activities, nuclear explosions and nuclear waste disposal, while supporting scientific research, and activities that protect the continent's ecozone. Being a framework convention, more specific legal obligations about the treaty are fleshed out in protocols to the treaty.[21]

Specialized/Specific Treaties These are designed to address specific subject matters, especially those that relate to the global impacts of petroleum or mining related activities, including business, human rights, environmental and waste management issues. Examples include the 1972 Convention on the Prevention of Marine Pollution by Dumping of Wastes and Other Matters ("the London Dumping Convention of 1972"),[22] and the 1973 International Convention for the Prevention of Pollution

[21] The Protocol on Environmental Protection to the Antarctic Treaty (Environmental Protocol or Madrid Protocol) was agreed in 1991 and came into force in 1998, after it had been ratified by all Antarctic Treaty Consultative Parties (ATCPs). The Protocol prohibits all commercial mining activities in the Antarctica. It also commits the Parties to the "comprehensive protection of the Antarctic environment" to prevent environmental degradation. The Protocol also requires a comprehensive environmental impact assessment (EIA) of all activities in the Antarctica before they are allowed to go ahead.

[22] See generally, the Convention on the Prevention of Marine Pollution by Dumping of Wastes and Other Matters, 29 December 1972, 1046 UNTS 120.

from Ships ("MARPOL Convention of 1973") which were adopted to address the subject of marine pollution.[23]

Umbrella Treaties A good example of this designation of treaty is the United Nations Conference on the Law of the Sea 1982 ("UNCLOS 1982"),[24] which is a convention that consists of a general convention, linked to one or more treaties on specific issues. While lacking a formal provision for adoption of protocols or annexes, it references other agreements, which interpret, amplify, and develop its existing provisions.

5.2.1.2 Customary International Law

Customary international law refers to international obligations arising from established state practice, as opposed to obligations arising from formal international conventions or treaties.[25] According to the order set out in article 38, "international custom as evidence of a general practice accepted as law", is the second most important source of international law, custom simply put is, the way things have always been done by states and parties, then translates to the way things are presently done. Once a rule is identified as a rule of customary law, it confers on it binding status, and it obeyed by states as a binding rule of conduct.[26] According to article 38, there are two elements needed for the establishment/determination of the existence of an international custom. The two elements, which will be considered briefly below are compulsory and must both be present for the creation of customary international law. These elements are:

1. *General Practice*: This amount to consistent state practice over a significant period of time. State practice should be consistent with uniform and constant usage.[27] This can be derived from a number of sources such as domestic law, newspaper reports, and government statements amongst others. This need not be perfectly consistent or conform rigorously. It is not necessary for states to believe that this behaviour constitutes law, what is important however is that there be no objection to it, or challenge by other states. It should also be adopted in the practice of a sufficiently widespread and representative number of states, including those that are not treaty parties. Customs primarily play an evidentiary role. They also provide historical background of the raw material for the emergence of customary

[23] International Convention for the Prevention of Pollution from Ships, 2 November 1973, 1340 UNTS 184.

[24] UNCLOS Convention, United Nations Convention on the Law of the Sea, 1982, 21 ILM (1982), 1261 (adopted on 10 December 1982, and entered into force on November 16 1994).

[25] Talmon (2015).

[26] Ibid.

[27] In *Fisheries Jurisdiction Case* ICJ Rep 1973, at p. 50, a joint opinion by five judges of the court held that 'state practice must be common, consistent, and concordant'. See also *North Sea Continental Shelf case* para 74.

rules. Custom can also develop if a number of states follow a practice consistently because of their authoritative influence over others. An example is the role of maritime powers in establishing customs relation to the law of the sea, and the prominent leadership of the United States of America and the defunct Soviet Union in establishing customary practices on matters related to space law.[28]

2. *Opino Juris*: This translates as "an opinion of law". According to Article 38(1)(b) of the ICJ Statute, the custom to be applied must be "accepted as law". *Opinio Juris* is the conviction or belief that an action was carried out because it was a legal obligation or is accepted by countries as the law.[29] It denotes a subjective obligation, a sense of belief on the part of a state that it is bound to follow a custom in question. It is the second element (along with state practice) that is necessary to establish a legally binding custom. This element exists because of the belief that the practice is required by law, and requires states to act in a certain way under the stated belief that their behaviour is or is becoming law. Note however that whether the practice of a state is due to *a belief* that it is legally obliged to do a particular act is difficult to prove objectively. Therefore, *opinio juris* remains an unsettled and debated notion in international law. It is important to note that specific/special customary norms will prevail over general customary norms unless the general custom is *jus cogens* or a pre-emptory norm of international law. Note also that treaty provisions, as well as United Nations declarations can also provide evidence of a rule of customary law. Customary law is as legally binding as treaty law. However, in order not to be bound by customary law, a state must have objected consistently to the creation of such customary law.[30] If a state only objects after the customary rule has been established, it will most likely be bound by it.[31]

5.2.1.3 General Principles of Law

The third source of international law under article 38(1)(c) are "general principles of law recognized by civilized nations." There is no universally agreed set of principles that comprise these "general principles". They broadly include principles of international law, legal maxims, declarations of principles and sometimes, an

[28] See generally, Bell and McGillivray (2008), pp. 150–151, also Birnie et al. (2009)and Bodansky (2010), pp. 94–102, 192–197.

[29] In *the Maritime Delimitation and Territorial Questions between Qatar and Bahrain Case (Qatar v Bahrain)* ICJ Rep 2001, 40, Qatar and Bahrain accepted that the relevant Articles of the United Nations Law of the Sea Convention of 1982 represented customary international law and should form the basis of the court's ruling.

[30] Charney (1985).

[31] In *Anglo-Norwegian Fisheries Case* ICJ Rep 1951, 116, United Kingdom had argued that there was a customary international law rule that "closing lines' along the mouths of bays can no longer than 10 nautical miles. The ICJ, as well as holding that such a rule did not exist in general customary international law due to inconsistent state practice, also noted that "in any event the ten mile rule would appear to be inapplicable as against Norway as she has always opposed any attempt to apply it to the Norwegian coast."

analysis of principles of domestic law that are recognized by national legal systems. As Alston rightly notes, general principles derive legitimacy from their recognition by states.[32] As such, what is required for the establishment of general principles in international law is an evidence of general recognition by states, even if it is widely violated. Evidence of state practice is not a pre-requisite for establishing the existence of general principles in international law.[33]

Generally, in a broad sense they encompass for example, (a) incontrovertible elements in any legal system such as fair hearing, no one should sit judgement in his own case, restitution for violation of an agreement, good faith or estoppel, equity etc.[34]; (b) principles recognized by international law such as prohibition on the use of force, principles of human rights, freedom of the seas, etc.; (c) or declaratory principles such as those contained in the Stockholm Declaration, and the *Rio Declaration*,[35] as well as other soft law instruments such as the principles of the World Charter for Nature, and the United Nations Environmental Programme's various set of principles. In general, they are said to be norms that reflect the fundamental propositions shared by legal systems around the world. Some treaties also reflect, codify or create general principles of law. The primary effect of these principles is that they serve as a tool for interpretation, application and development of treaties and customary law. They also derive their endorsement, or legality from the endorsement of states. For example, the concept of sustainable development, specifically sustainable resource management can arguably be described as a general principle of international environmental law recognized by civilized nations.

Sustainable Natural Resource Management as a Key Principle of International Law The legal obligation to manage and utilize the world's natural resources in a manner that protects the environment and conserves natural resources has been a cornerstone of international law since the last three decades.[36] The idea of sustainable resource management (SRM) emphasizes the need for a coordinated approach

---

[32] Alston and Samma (1991), p. 102. This point was emphasised in *South West African Cases*, ICJ Report 1966, where Tanaka J held that "the behaviour and consent of States are thus not a necessary prerequisite to the recognition of general principles". See also Hulsroj (1999), p. 245; Brownlie (2003).

[33] See generally, Alston and Samma (1991), p. 102; Hulsroj (1999), p. 245.

[34] In *Rann of Kutch Arbitration* (1968) 50 ILR 2, the tribunal held that "equity forms part of international law, therefore the Parties are free to present and develop their cases with reliance on principles of equity." See also *River Muese Case (Netherlands v Belgium)* PCIJ Ser A/B (1937) no 70, Permanent Court of International Justice, Judge Hudson stating that the Court has some freedom to consider the principles of equity as part of international law. See also Akerhurst (1976), p. 801.

[35] See Bastida and Bustos (2017); also Auty and Mikesell (1988) and Bastida (2008).

[36] See generally, European Union (2003), Communication from the Commission to the Council and the European Parliament - Towards a Thematic Strategy on the Sustainable Use of Natural Resources http://eur-lex.europa.eu/legal-content/EN/TXT/?uri=CELEX:52003DC0572, also Cordonier Segger and Khalfan (2004), Cordonier Segger and Weeramantry, (2004), French (2005, 2010), Gehring and Cordonier Segger (2005), Cordonier Segger et al. (2010) and Jodoin and Cordonier Segger (2015).

that balances social development, economic development and environmental protection in the management and use of petroleum, solid minerals and other natural resources.[37] SRM finds its roots in the concept of sustainable development, which as popularly defined in the 1987 report, Our Common Future, by the World Commission on Environment and Development (known as the Brundtland Commission) is 'development that meets the needs of the present without compromising the ability of future generations to meet their own needs'.[38] In the context of state actions and policy measures, it refers to "state efforts to achieve progress (development) qualified by the condition that such efforts should be possible to maintain over the long term (sustainable)".[39] Sustainable development envisions a symbiotic balance between two pursuits that are often portrayed as mutually exclusive: maintenance of a sound environment, and economic and social development. In essence, SRM refers to the use of natural resources at a rate less than or equal to the natural rate of regeneration or reproduction. It advocates that our use of natural resources today must not diminish their availability and usefulness for future generations.[40] To achieve SRM, countries must promote resource extraction and production methods and techniques that encourage eco-efficiency and the sustainable use of raw materials, energy, water and other resources.[41]

The principle of SRM, which emphasizes the need to protect the environment while pursuing economic growth and development, has become one of the most recognized general principles of international law, and has found its way into several international declarations, treaties and domestic law.[42] SRM has been recog-

[37] Batisda and Bustos (2017), p. 235. See also the 2002 International Law Association (ILA) (2002), pp. 211–216, UN Doc. A/CONF.199/8, 9 August 2002), though non-binding legally, the declaration strongly reaffirms and concretizes the duties of states to incorporate sustainability into the management plans for natural resources and associated activities. It notes that "sustainable development is now widely accepted as a global objective and that the concept has been amply recognized in various international and national legal instruments, including treaty law and jurisprudence at international and national level."

[38] The Brundtland Report (1987).

[39] Bugge and Voigt (2008), p. 88. For detailed and extensive discussions on sustainable development, its evolution, principles and application to national and international laws, see generally, Cordonier Segger and Khalfan (2004), p. 1; Schrijver (2008), p. 1.

[40] Schrijver (1997), Tilton (2003) and Szablowski (2007).

[41] Batisda and Bustos (2017), p. 235, also Vanhulst and Beling (2014).

[42] See Para. 1.2 of the ILA New Delhi Declaration on Sustainable Development stating that "states are under a duty to manage natural resources, including natural resources within their own territory or jurisdiction, in a rational, sustainable and safe way so as to contribute to the development of their peoples, with particular regard for the rights of indigenous peoples, and to the conservation and sustainable use of natural resources and the protection of the environment, including ecosystems. States must take into account the needs of future generations in determining the rate of use of natural resources. All relevant actors (including States, industrial concerns and other components of civil society) are under a duty to avoid wasteful use of natural resources and promote waste minimization policies." See also the 2012 Sofia Guiding Statements on the Judicial Elaboration of the 2002 New Delhi Declaration of Principles of International Law Relating to Sustainable Development, RESOLUTION No. 7/2012 noting in para. 3 that "the sustainable use

nized in many international environmental treaties. Currently, there are over 300 international legal instruments in addition to other soft law declarations, guidelines and General Assembly Resolutions that have identified and recognized the principle of sustainable development.[43] For example, Principle 3 of the Rio Declaration states that the right to development must be fulfilled so as to equitably meet developmental and environmental needs of present and future generations.[44] Principle 2 the Rio Declaration recognizes that States have, the sovereign right to exploit their own resources pursuant to their own environmental and developmental policies. They also have the responsibility to ensure that activities within their jurisdiction or control do not cause damage to the environment of other States or of areas beyond the limits of national jurisdiction. Principle 3 states that the right to development must be fulfilled so as to equitably meet developmental and environmental needs of present and future generations. Furthermore, Principle 27 notes that States shall cooperate in good faith in the further development of international law in the field of sustainable development.[45]

The 1997 UN General Assembly Special Session on Sustainable Development, the 2002 World Summit on Sustainable Development, among other meetings, the 2012 the United Nations Conference on Sustainable Development in Rio de Janeiro, Brazil, called upon countries and stakeholders, to promote and attain the sustainable management and use of the natural and human environment, including the world's natural resources.[46]

More recently in September 2015, UN member States agreed on 'Sustainable Development Goals' (SDGs). Transforming our World: The 2030 Agenda for Sustainable Development.[47] The 17 SDGs and 169 related time-bound targets contain robust ambitions and targets by countries to achieve sustainable management of water and sanitation for all; conserve and sustainably use marine resources for sustainable development; protect, restore and promote sustainable use of terrestrial ecosystems, sustainably manage forests, take urgent actions to combat climate change and its impacts; combat desertification, halt and reverse land degradation and biodiversity loss; and to promote peaceful and inclusive societies for sustainable development, provide access to justice for all and build effective, accountable

---

of all natural resources represents an emerging rule of general customary international law, with particular normative precision identifiable with respect to shared and common natural resources."

[43] For a comprehensive list of these instruments, see the United Nations Sustainable Development Knowledge Platform. See also Schrijver (2017), pp. 99–102, (providing an overview of the ways that sustainable development principles have been incorporated into and operationalized by international treaty regimes and tribunals); French (2005), pp. 177–184: "The Sofia Guiding Statements on sustainable development principles in the decisions of international tribunals," in Cordonier Segger and Weeramantry (2004), pp. 177–184 (highlighting how sustainable development has been incorporated into the juridical activities of international bodies convened under the auspices of UN associated and influenced treaty regimes).

[44] UN *Rio Declaration on Environment and Development* (1992).

[45] UN (1992).

[46] UN (2012).

[47] UN (2015).

and inclusive institutions at all level.[48] These SDGs and their targets guide the UN, member States, other international organizations and stakeholders across all regions of the world towards more sustainable natural resource management globally.[49]

Sustainable development and SRM have also frequently gained recognition in national and international case law,[50] and in national laws and policies.[51] Similarly, a cursory look at how many countries are active parties to environmental treaties and Conventions which are developed based on the idea of sustainable development, show that the concept has gained sufficient recognition by countries. These examples explain how environmental law and sustainable development are intertwined. As the Permanent Court of Arbitration noted in *Arbitration Regarding the Iron Rhine ("Ijzeren Rijn") Railway (Belgium v Netherlands)*:

> environmental law and the law on development stand not as alternatives but as mutually reinforcing, integral concepts, which require that where development may cause significant harm to the environment, there is a duty to prevent, or at least mitigate such harm... This duty, in the opinion of the Tribunal, has now become a principle of general international law. This principle applies not only in autonomous activities but also in activities undertaken in implementation of specific treaties between the Parties.[52]

The rapid ascendancy of the SRM principle in national laws, international treaties, and case law all provide evidence that it has gained recognition as a principle relevant to international law on energy and extractive resources.

---

[48] Ibid.

[49] See Cordonier Segger and Mrema (2016).

[50] See the *US – Shrimp Case*, where the WTO Appellate Body stated that: "...[t]his concept has been generally accepted as integrating economic and social development and environmental protection and the concept is elaborated... so as to put in place development that is sustainable... that meets the needs of the present generation without compromising the ability of future generations to meet their own needs." See also *United States – Import Prohibition of Certain Shrimp and Shrimp Products* (6 November 1998), WTO Doc. WT/DS58/AB/R at note 107. See also *Case Concerning the Gabcíkovo- Nagymaros Project (Hungary/Slovakia)* (1997), I.C.J. Rep., 37 I.L.M. (1998) 162 at 140; *United States – Import Prohibition of Certain Shrimp and Shrimp Products* (20 September 1999), WTO. Doc. WT/DS58/AB/R (Appellate Body Report) and *Permanent Court of Arbitration: In the Arbitration Regarding the Iron Rhine ("Ijzeren Rijn") Railway, between the Kingdom of Belgium and the Kingdom of the Netherlands* (May 24, 2005) at 59, 114, Award of the Arbitral Tribunal; and national cases such as *Vellore Citizens Welfare Forum* v *Union of India* (1996, Supreme Court of India), *Bulankulame* v. *Secretary, Ministry of Industrial Development and Others* (the Eppawela Case) (2000, Supreme Court of Sri Lanka), and *Rajendra Parajuli and Others* v *Shree Distillery Pvt. Ltd. & Others* (Writ No. 3259, 1996, Supreme Court of Nepal).

[51] There is a plenitude of examples across Africa. See for example Kenya's Environmental Management and Coordination Act no. 8 of 1999; also the Nigerian National Environmental Standards and Regulations Enforcement Agency (NESREA) Act (2007); the Niger Delta Development Commission (NDDC) (Establishment etc) (Amendment) Act of 2017 of Nigeria amongst others.

[52] *Arbitration Regarding the Iron Rhine ("Ijzeren Rijn") Railway (Belgium v Netherlands)* (May 24, 2005).

5.2.1.4 Judicial Decisions and Teaching of Highly Qualified Publicists

The fourth source under Article 38 (1) (d) of the ICJ Statute is "…judicial decisions and the teachings of the most highly qualified publicists of the various nations, as subsidiary means for the determination of rules of law". These do not have the same weight as the other sources considered above, as they are classified as subsidiary sources. This source is often considered if there is no treaty on an issue in contention, no customary rule of international law or applicable general principle.[53] Although classified as subsidiary, decisions of the ICJ, tribunals and national courts are important in shaping expectations on legitimacy of claims made under international law. They also provide evidence of what the present law is. Judicial decisions have also articulated several important environmental norms, such as the duty to prevent transboundary harm in the *Trial Smelter* case; 47 the duty to consult and negotiate in the *Lac Lanoux* case between France and Spain; and the Advisory Opinion of the ICJ on the *Legality of the Threat or Use of Nuclear Weapons* in which states' responsibilities not to cause environmental damage beyond their territories or to the global commons was elucidated by the ICJ. There is no rule of binding precedence for judicial decisions in international law, but they serve a persuasive function. According to Article 59 of the ICJ Statute, the decision of the Court has no binding force except between the parties and in respect of that particular case. Despite this, states often make reference to prior decisions of international tribunals and Courts and international tribunals and Courts though not bound by precedence or stare decisis, also frequently refer to their prior decisions in deciding cases before them. Courts will not entirely disregard their earlier pronouncements or decisions even though they might distinguish them.[54]

The second branch of this source is the teachings of highly qualified publicists.[55] These include the reports of international bodies such as the International Law Commission ("ILC"), which provides drafts on topics in international law, which form the fodder for discussion for international treaties or conventions on the subject matter. Writings of eminent scholars and bodies such as the Academic Advisory Group constituted by the International Bar Association's Section on Energy, Environment, Natural Resources and Infrastructure Law (IBA-SEERIL)[56]; environ-

---

[53] Ibid.

[54] See generally, Birnie et al. (2009).

[55] Per Chancellor Kent in 1 Kent, Com. 18: In the absence of higher and more authoritative sanctions, the ordinances of foreign states, the opinions of eminent statesmen, and the writings of distinguished jurists are regarded as of great consideration on questions not settled by conventional law. In cases where the principal jurists agree, the presumption will be very great in favor of the solidity of their maxims, and no civilized nation that does not arrogantly set all ordinary law and justice at defiance will venture to disregard the uniform sense of the established writers on international law. Quoted with approval in *Paquete Habana case* 175 U.S. 677 (1900).

[56] The Academic Advisory Group (AAG), Section on Energy, Environment, Natural Resources and Infrastructure Law (SEERIL), International Bar Association. has published several authoritative texts on extractive industry law including Regulating Energy and Natural Resources (2006), Energy Security (2004) and Public Participation in the Sustainable Development of Mining and Energy Resources (Oxford University Press, 2002) and most recently Zillmanet al. (2018).

mental law academy of the International Union for Conservation of Nature ("IUCN"); World Commission on Environment and development ("WCED"), and the International Law Association ("ILA"), as well as books and articles by leading practitioners that offer authoritative analyses of a wide variety of issues pertaining to international law. In the *Paquete Habana case,* Justice Gray of the United States Supreme Court noted as follows:

> International law is part of our law, and must be ascertained and administered by the courts of justice of appropriate jurisdiction as often as questions of right depending upon it are duly presented for their determination. For this purpose, where there is no treaty and no controlling executive or legislative act or judicial decision, resort must be had to the customs and usages of civilized nations, and, as evidence of these, to the works of jurists and commentators who by years of labor, research, and experience have made themselves peculiarly well acquainted with the subjects of which they treat. Such works are resorted to by judicial tribunals not for the speculations of their authors concerning what the law ought to be, but for trustworthy evidence of what the law really is...[57]

While commenting on the importance of books and articles by eminent scholars and lawyers, Justice Gray, quoted with approval, the seminal book Wheaton, *International Law* (8th ed.), § 15, and stated as follows:

> Wheaton places among the principal sources international law text writers of authority, showing what is the approved usage of nations, or the general opinion respecting their mutual conduct, with the definitions and modifications introduced by general consent. As to these, he forcibly observes: "Without wishing to exaggerate the importance of these writers or to substitute, in any case, their authority for the principles of reason, it may be affirmed that they are generally impartial in their judgment. They are witnesses of the sentiments and usages of civilized nations, and the weight of their testimony increases every time that their authority is invoked by statesmen, and every year that passes without the rules laid down in their works being impugned by the avowal of contrary principles."[58]

Scholarly publications are influential in shaping the development of international law, especially when new, emerging and developing concepts in international law are under consideration. A good example is the *New Zealand v France* case [1995] ICJ Rep 288, in which the dissenting opinion of Judge Weeramantry referred to scholarly and academic opinion on the elements of sustainable development.

5.2.1.5 Other Sources of International Law

Apart from sources recognized in Article 38 of the *ICJ Statute*, soft law instruments also provide normative influences to the development and interpretation of extractive industry law.[59] Soft law instruments are those legal instruments that are not directly enforceable in courts and tribunals but nonetheless have an impact on

---

[57] *Paquete Habana case* 175 U.S. 677 (1900).

[58] Ibid.

[59] For an excellent discussion of the normative effects and importance of UN Resolutions and Declarations see Forsythe (2006), p. 12; also Dupuy (1991), pp. 422–25, 428–31.

international relations and, ultimately, international law.[60] Broadly put, soft law instruments include declarations of international or regional organizations; resolutions of the United Nations Security Council and General Assembly; African Union or decisions of the Conference of Parties to a convention; principles adopted by the United Nations; treaty provisions agreed upon but not yet in force; normative recommendations; declarations of principles; codes of conduct; codes of practice; programmes of action; non-binding conference declarations or statements; and other non-treaty obligations which do not create binding obligations, but often propose guidelines and rules of behavior which are important and useful for responding to emerging challenges.[61]

Soft law is often contrasted with hard law.[62] While some scholars have argued that soft law is not law, some question the importance of soft law generally under international law since they do not create binding obligations.[63] Even though resolutions adopted by the United Nations General Assembly, or General Comments by UN treaty bodies are not international treaties and therefore do not have legally binding force; they have important legal functions and normative effects to the extent that they often elaborate and interpret norms.[64] They provide interpretations and guidelines on how a country should apply a treaty in order to fulfil its international obligations. Soft law instruments play significant roles in the development and evolution of international law. As Higgins rightly noted, the passing of binding decisions by an international body is not the only way in which the development of the law occurs.[65] Legal consequences can also flow from acts which are not, in the formal sense, 'binding'.[66] Not only do soft law instruments provide flexible guidelines on how to tackle emerging concerns, they provide a template on how international policies can be implemented, they also serve as forbearers to binding hard law instruments in the future. We will consider some of the non-binding declarations that have become part and parcel of the body of international law, and by extension, international law on natural resources in subsequent chapters.

Landmark Soft Law Instruments on Sustainable Resource Extraction in Africa" Regional and international bodies have elucidated the importance of sustainable resource management, transparency and human rights in resource extraction in Africa.[67] Of special importance to extractive investments in Africa are declarations and resolutions of the African Union, with respect to sustainable resource management and the protection of human rights in extractive activities.[68]

---

[60] Boyle and Chinkin (2007).

[61] Sloan (1987), p. 93.

[62] Reisman (1988), pp. 371–377.

[63] Guzman and Meyer (2010), p. 171.

[64] Talmon (2005).

[65] Higgins (1994).

[66] Ibid.

[67] For a detailed review of these efforts, see Olawuyi (2015a, b, c).

[68] See Olawuyi (2015a, b, c).

For example the *Resolution on a Human Rights-Based Approach to Natural Resources Governance* (*2012 Resolution*) adopted by the African Commission on Human and Peoples' Rights (African Commission) at its 51st Ordinary Session held from April 18th to May 2nd, 2012 in Banjul, Gambia.[69] The *2012 Resolution* calls on governments and extractive sector investors to ensure that respect for human rights in all matters of natural resources exploration, extraction, toxic waste management, development, and management.[70] The overall aim of the resolution is to ensure the systemic integration of human rights into policies and legislation on resource utilization in Africa. The *2012 Resolution* highlights the interdependence of human rights and development, recalling articles 20, 21, and 24 of the African Charter which protect, respectively: peoples' right to freely determine their political status and pursue their economic and social development according to the policy they have freely chosen, their right to freely dispose of their natural resources, and their right to a satisfactory environment. The *Resolution* also emphasizes and recalls the need to implement previous international law declarations, such as Principles 1 and 22 of the 1992 *Rio Declaration* on human entitlement to healthy and productive life in harmony with nature, and the need to ensure that local communities have a vital role in environmental management and development.[71]

The *2012 Resolution* builds on previous efforts by African governments to introduce human rights concepts into resource governance.[72] There are three key prior important regional efforts in Africa that serve as the foundation for the *2012 Resolution*. They are: the 2009 Economic Community of West African States (ECOWAS) *Directive on the Harmonization of Guiding Principles and Policies in the Mining Sector* (*ECOWAS Directive*),[73] the resolutions of the Pan African Parliament,[74] and the *2009 African Mining Vision,*[75] The *2012 Resolution* brings together these previous efforts to ensure coherence, coordination, and consistency.

First, the 2009 *ECOWAS Directive*, which is applicable to 15 West African countries that are members of the ECOWAS,[76] emphasises the need for stakeholder engagements and consultation in resource development. The *ECOWAS Directive*

---

[69] Resolution on a Human Rights-Based Approach to Natural Resources Governance, African Commission on Human and Peoples' Rights, 51st Sess, (2012).

[70] Ibid.

[71] Ibid. See Report of the United Nations Conference on Environment and Development Annex I: Rio Declaration on Environment and Development, UNGA, 1992, UN Doc A/CONF.151/26 (Vol I) [Rio Declaration].

[72] Olawuyi (2015a, b, c).

[73] Economic Community of West African States, Directive C/DIR.3/05/09, *On the Harmonization of Guiding Principles and Policies in the Mining Sector*. [*ECOWAS Directive*].

[74] See the Recommendations of the Permanent Committee on Agriculture, Rural Economy, Natural Resources and the Environment. See also Pan African Parliament (2011), p. 14.

[75] African Union (2009).

[76] Namely: Benin, Burkina Faso, Cape Verde, Côte d'Ivoire, the Gambia, Ghana, Guinea, Guinea-Bissau, Liberia, Mali, Niger, Nigeria, Senegal, Sierra Leone, and Togo. See Economic Community of West African States, the ECOWAS Commission at a Glance.

sets out guiding principles for harmonizing mining regulatory regimes across member states and is binding on ECOWAS members. It called on member states to respect the free, prior, and informed consents (FPIC) of local communities that face potential negative impacts from mining, hydrocarbon development, or natural resource projects more broadly. The *ECOWAS Directive* urges countries to ensure free, prior, and informed consent when communities will be affected by mineral or hydrocarbon projects. Specifically, it calls on companies to obtain the FPIC of local communities before exploration begins and prior to each subsequent phase of mining and post-mining operations.[77] The *Directive* emphasises the importance of maintaining negotiations and consultations on important decisions, throughout the project lifecycle,[78] *i.e.* from pre-mining, mining operations, closing, and post-closure periods. Furthermore, the directive defines "mineral" to include not only industrial minerals but also petroleum, thereby making the directive applicable to both the solid mineral and oil and gas sectors. The *ECOWAS Directive* therefore arguably remains the most significant FPIC policy document in Africa. It aims to improve transparency and stakeholder consultation and engagement in mineral policy formulation, implementation, and decision-making processes. It also establishes guidelines for countries to mitigate, to the extent possible, the negative impacts of resource development on the environment and the local communities, in line with international best practices. However, the effects and scope of the *ECOWAS Directive* are restricted, as it applies only to 15 of 44 countries on the continent. Furthermore, it provides wide discretion for member states to determine how they will meet the objectives of the *Directive*. The *ECOWAS Directive* also focuses extensively on FPIC and fails to speak to other human rights concerns facing the extractive industry in Africa, such as marginalization of women from decision-making processes, governmental corruption and lack of transparency, forceful acquisition of indigenous lands, and inadequate compensation.

Some of these human rights concerns facing the extractive industry in Africa have been highlighted by the Pan-African Parliament, which has increasingly promoted regional efforts aimed at addressing human rights concerns of resource utilization in Africa. The Pan-African Parliament is the legislative body of the African Union. It has advisory and consultative powers to examine, discuss, or express an opinion on any matter, either on its own initiative or at the request of the Assembly or other policy organs, and make any recommendations it may deem fit relating to, inter alia, matters pertaining to respect of human rights.[79] At a parliamentary workshop held in South Africa in 2011, the Pan-African Parliament expressed concern over the rise of large-scale land acquisitions, also known as "land grabbing", and the impact of domestic and foreign investment on land, water, and related natural resources.[80]

[77] Article 16 (3), ECOWAS Directive. See also Mayer Brown (2010).

[78] Article 16 (4), ECOWAS Directive.

[79] African Union, Pan-African Parliament.

[80] 2011 Parliamentary Resolution, 14. See Greenspan (2014), p. 14.

In order to promote greater participation in decision making processes for resource extraction projects, the Pan-African Parliament called on states to ensure "effective consultations with local communities and various people affected by investment projects and [ensure] that any investment is approved through free, prior, and informed consent of affected communities."[81] The recommendations also call for enhanced land certification and registration systems that take into account pastoralist, women's, and communal rights in advance of investment. The Pan-African Parliament has therefore identified that human rights issues in resource utilization in Africa are exacerbated by pre-existing imbalances in power relations between: men and women, local governments and the governed, elders and young, rich and poor, indigenes and non-indigenes, educated and non-educated.[82] A significant demerit, however, is that the Pan-African Parliament has yet to attain full legislative powers, and as such can only advise and speak on the alarming impacts of the problem of exclusion, without effecting any real legislative or policy change.

Another important soft law instrument is the *2009 African Mining Vision* (the "Vision Report").[83] The Vision Report was put together by the United Nations Economic Commission for Africa (ECA), in conjunction with the African Union (AU), the African Mining Partnership (the intergovernmental forum of African ministers responsible for mining), the African Development Bank (AfDB), the United Nations Conference on Trade and Development (UNCTAD) and the United Nations Industrial Development Organization (UNIDO).[84] The Vision Report highlights human rights challenges that mining projects face, especially inadequate stakeholder engagement and consultation in resource extraction, together with the social and environmental side effects of resource extraction. It catalogues African initiatives aimed at developing a "new social contract for mining" that would balance local benefits with national poverty-alleviation efforts.[85] The Vision Report encourages African countries to develop new legal instruments to facilitate local community participation; multi-stakeholder partnerships of government, private sector, and local communities; and to promote "public participation to secure consent for government and industry actions. In terms of required action plan to improve public participation, importantly, the framework describes public participation not just to include consultation, information sharing, and dispute resolution, but also "participatory decision making."[86]

Similarly, in December 2011, the African Union Conference of Ministers responsible for mineral resources adopted an Action Plan to implement the *African Mining Vision*: the *Addis Ababa Declaration on Sustainable Africa Extractive Industry—*

---

[81] Sixth Ordinary Session of the Pan-African Parliament, para 6(d).

[82] Ibid at preamble (noting that the Pan African parliament is "fully alarmed by the negative impacts on human rights especially on women, including unequal access to land and disruption of access to water").

[83] African Union, Africa Mining Vision (2009).

[84] Ibid 1.

[85] Ibid 12, 39.

[86] Ibid 33.

From Vision to Action (*Addis Ababa Declaration*).[87] The *Addis Ababa Declaration* calls on African states to strengthen transparency, accountability, and access to information, improve public participation and provide capacity building programmes for local communities, civil society, and the legislature in order to provide effective oversight that will create a well-governed mining sector that is inclusive and appreciated by stakeholders. The *Declaration* calls on all African Union member states to "create a mineral sector that is environmentally friendly and is socially responsible through strengthened environmental, health, and safety standards, as well as through the implementation of environmental and social funds."[88]

Though not legally binding., these soft law instruments provide strong normative guidance on how governments, national authorities, investors, and other stakeholders in the African extractive sector can minimize the negative implications of extractive activities and projects on the environment, in particular on the enjoyment of fundamental human rights to life, water, health and cultural values.[89] They emphasise the need for countries to develop project approval frameworks that ensure that resource extraction projects, actions, or measures that do not guarantee or satisfy the requirements of accountability, access to information, participation, equality, and access to justice not be approved, no matter their economic significance or relevance.

5.2.2 Domestic Statutes and Legislation

At national levels, African countries have developed robust bodies of legislation, statutes and regulations that regulate the exploration or exploitation of petroleum and mineral resources. Furthermore, a number domestic legislation, regulations, instruments, directives or even contract, in other broad areas of law have direct relevance to the extractive sector. For example, rules relating to environmental impact assessment, water use, product labelling, corporate reporting, fund repatriation, tax and custom regime; laws on the environment (gas flaring, reclamation, abandonment and clean ups); labour and employment; competition and anti-trust; transportation law; and procurement rules must be comprehensively considered. The use of a broader range of instruments reflects a growth of sources of extractive sector law, in which traditional public law 'command and control' legislation is combined with different regulatory approaches to secure the most effective forms of control of extractive activities to achieve maximum sustainability benefits. It is therefore important to understand not only the specific petroleum or solid mineral law, but to

[87] African Union Commission, African Development Bank & United Nations Economic Commission for Africa, Building a Sustainable Future for Africa's Extractive Industry: From Vision to Action, Action Plan for Implementing the African Mining Vision (2011).

[88] Ibid 32–35.

[89] Ibid.

also understand other national laws and regulations that stipulate requirements for extractive operations to avoid any default.

Although the boundaries of the different types of legislation and rules that are applicable to the extractive industries across Africa vary and are often blurred in practice, the sources of domestic extractive sector legislation can be broken down into three main categories.

5.2.2.1 The Constitution

As we have discussed in Chap. 2, Africa is a dominial jurisdiction. The ownership and control of all minerals, mineral oils and natural gas is vested in the government by national constitutions. The Constitution is the supreme law and it has binding force on all authorities and persons willing to participate in extractive activities in a country. The Constitution sets out the powers of the government to control all mineral and petroleum resources found within the country.[90] The constitution is foundational to the regulation of petroleum and mineral resources, as it sets the tone for all the other legislation, which are all based on the ownership of oil and gas by the government, as well as the government's right to dispose of the resources.

5.2.2.2 Domestic Legislation

Although the boundaries of domestic legislation in the extractive sector are often blurred in practice, the sources of domestic extractive sector legislation can be divided into three main categories:

Primary Legislation Acts of Parliament, also known as statutes, decrees or enactments, typically provide the basic framework for domestic regulation of petroleum and solid mineral operations. They are the basic and most dominant tools through which governments stipulate how petroleum and mineral rights can be acquired in a country. In many African countries, for example Kenya, Ghana, Libya and Nigeria, extractive sector legislation is in the form of a separate petroleum act that governs and regulates oil and gas exploration and production,[91] and a separate mineral and

[90] See the CFRN 1999, supra n 1.

[91] In Kenya, the Petroleum (Exploration and Production) Act Chapter 308 of the Laws of Kenya is the principal oil and gas legislation. It provides a legal framework for the contracting, exploration, development and production of petroleum resources in Kenya. In Nigeria, the Petroleum Act 1969 (the "Petroleum Act") is the principal oil and gas legislation. The preamble to the Petroleum Act provides that its purpose is to provide for the exploration of petroleum from territorial waters and the continental shelf of Nigeria and to vest the ownership of, and all on-shore and off-shore revenue from petroleum resources in the Federal Government. In furtherance of this, section 1 vests the entire ownership and control of all petroleum in, under or upon any lands, which extends to Nigeria's territorial waters, continental shelf and exclusive economic zone in the State.

mining act that regulates the exploration and mining of solid minerals.[92] While in other countries, such as South Africa and Egypt, the legal framework for mining and petroleum industries is codified in a single comprehensive legislation.[93]

Another important general statute, that is a primary source of extractive industry regulation, is the law establishing and empowering the national oil company or a national solid minerals agency, that are often empowered to engage in all commercial activities relating to the petroleum and/or solid minerals industry and to enforce all regulatory measures relating to the general control of the sector. Correspondingly, in many African countries, for example Kenya, Ghana, Libya and Nigeria, there are separate national oil companies,[94] and separate national mining companies respec-

[92] See s. 3 of GHANA's Petroleum Exploration and Production Act of 2016 (Act 919), which vests title to all petroleum existing in its natural state in, under or upon any land in Ghana, rivers, streams, water courses throughout Ghana, the exclusive economic zone and any area covered by the territorial sea or continental shelf, in the President of Ghana on behalf of and in trust for the people of Ghana; also GHANA's Minerals and Mining Act 2006 (Act 703) vests ownership of every mineral in its natural state in, under or upon land in Ghana, rivers, streams, water-courses throughout the country, the exclusive economic zone and an area covered by the territorial sea or continental shelf is the President of Ghana; NAMIBIA's Petroleum (Exploration and Production) Act 1991 (Act 2 of 1991), Minerals (Prospecting and Mining) Act, 1992 and Article 100 of the Constitution vest all natural resources in the State, unless otherwise legally owned.; ZIMBABWE's Mineral Act of 1961, Chapter 21:05 which vests the right and ownership to all minerals, mineral oils and natural gases under the soil, including the rights to search, mine for and dispose of such minerals, mineral oils and natural gases are vested in the President of Zimbabwe on behalf of the state; ZAMBIA's Mines and Minerals Development Act No. 7 of 2008, Mines and Minerals Development (General) Regulations of 2008 and Mines and Minerals Development (Prospecting Mining and Milling of Uranium Ores and other Radioactive Mineral Ores) Regulations 2008 which vest all rights of ownership in, searching for, mining and disposing of, minerals whosesoever located in Zambia in the President on behalf of the Republic; KENYA's Mining Act 2016 which vests ownership of minerals in Kenya in the national government in trust for the people of Kenya; UGANDA's Mining Act 2003 and Mining Regulations 2004 which vest the control and ownership of all minerals in Uganda in the state; TANZANIA's Mining Act, Cap.123 of the Laws of Tanzania (Act No. 14 of 2010); MOZAMBIQUE's Mining Law (Law No. 20/2014, dated 18 August) which vests ownership of all mineral resources found in the soil and subsoil, internal waters, territorial sea, continental shelf and in the exclusive economic zone in the State; and NIGERIA's Minerals and Mining Act 2007 which vests absolute control of all properties and minerals in Nigeria in the State and prohibits unauthorised exploration or exploitation of minerals. All lands in which minerals have been found in commercial quantities shall be acquired by the Federal Government of Nigeria in accordance with the Land Use Act.

[93] The extraction of oil and gas in Egypt is regulated by *Law No. 66/1953, On Mines and Quarries*, and the *Egyptian Mining and Quarries Law 86 of 1956*, which set out the terms and conditions for obtaining relevant concession agreements. In South Africa, the *Mineral and Petroleum Resources Development Act*, 2002 (MPRDA) is the principal legislation for both petroleum and mining activities. It regulates all activities relating to prospecting, mining, exploration, production and related activities of minerals in South Africa. Section 3 of the MPRDA provides that the State serves as the custodian of South Africa's mineral and petroleum resources for the benefit of its people.

[94] See Law No. 20 of 1987 organizing the Egyptian General Petroleum Company (EGPC), also the National Oil Corporation of Kenya was established in April 1981 with a mandate to participate in all aspects of the petroleum industry. The Corporation is wholly owned by the Government of Kenya through a joint ownership by the Ministry of Energy and Petroleum and The National

tively with the powers to oversee the commercialization of petroleum and minerals.[95] In the Nigerian case of *Idoniboye-Obu v. NNPC*, on the issue of the legal status of the national oil company, the Nigerian National Petroleum Corporation (NNPC), the court held that the NNPC is a creation of statute enacted by the National Assembly. Therefore it is a federal government corporation, which performs a central role in the oil and gas industry. Thus, any entrant into the petroleum sector must comply with the relevant directives and guidelines of the national oil company- no more no less.[96] This case highlights the legal characterization and importance of national oil and mining companies that have been established by governments across Africa to oversee, manage and authorize upstream, midstream and downstream extractive investments across Africa.

Apart from direct legislation on mining and petroleum activities, other direct general statutes, such as legislation on the environment, tax,[97] labor, health, company legislation, human rights and extractive sector transparency amongst others,[98]

Treasury. In Nigeria, the *Nigerian National Petroleum Corporation Act of 1977* (the "*NNPC Act*"), establishes NNPC as a body empowered to engage in all commercial activities relating to the petroleum industry and to enforce all regulatory measures relating to the general control of the petroleum sector through its Petroleum Inspectorate department. Also, pursuant to Section 70 of South Africa's MPRDA, authority of the Minister of Mineral Resources, with respect to licensing and permits, is delegated to the Petroleum Agency of South Africa (PASA), which acts as the regulator for petroleum exploration in onshore and offshore South Africa.

[95] See NEITI and C4C, Existing Laws and Policies.

[96] Ibid.

[97] See section 26(B)(2) of South Africa's *Income Tax Act, No 58 of 1962*, which provides for a 28% cap on the tax rate levied for exploration and production activities; and Section 2 of the *Minerals and Petroleum Royalty Act, Act 28 of 2008* which provides that royalties are payable to the state upon the transfer of mineral resources. See also, the *Petroleum Profits Tax Act*, Laws of the Federation of Nigeria 2004, c P13 which provides that profits generated by companies that engage directly or indirectly in petroleum operations are subject to tax. Oil companies are defined in the Act to include companies engaged in crude oil production, petroleum marketing and the servicing companies such as seismic survey, drilling and data collection. In section 2 of the *PPTA*, "petroleum operations" is defined as the winning of, obtaining and transportation of petroleum or chargeable oil in Nigeria by or on behalf of a company for its own account by any drilling, extracting or other like operations or process, not including refining at a refinery, in the course of a business carried on by the company, all incidental operations, as well as any sale or any disposal of chargeable oil by or on behalf of the company. Furthermore, in *Shell Petroleum Development Company of Nigeria (SPDC) v. Federal Board of Inland Revenue (FBIR)* (1996) 8 NWLR pt 466 at 256, the Supreme Court held that by virtue of section 8 of the *PPTA*, any company engaged in petroleum operations is liable to pay tax. On the meaning of "Petroleum Operations" for tax purposes, the court held that "Petroleum Operations" include not only winning or obtaining and transportation of petroleum oil by drilling, mining, etc., but also all activities incidental to such operations excluding refining at a refinery.

[98] See for example *Nigeria Extractive Industries Transparency Initiative (NEITI) Act 2007* which was enacted to ensure due process and transparency in payments made to the Nigerian government by companies operating in the Nigerian Extractive Industry. Pursuant to section 2, the Act is aims to ensure accountability in the revenue receipts of the Federation from companies in the extractive industry and to eliminate all forms of corrupt practices in the determination, payments, receipts and posting of revenue accruing to the Federation. Only a few African countries have enacted transparency legislation. See TANZANIA's *Extractive Industries (Transparency and Accountability) Act, 2015* (TEIA); GHANA's *Petroleum Revenue Management Act 2011* (Act

contain important provisions that are pertinent to successful extractive operations in many African countries. In Nigeria for example, a number of environmental legislation stipulate the need to reduce environmental impacts of extractive activities, and impose comprehensive fines and sanctions for companies that fail to comply.[99] The *Nigeria Liquefied Natural Gas Company (NLNG) Act*, also establishes the NLNG, a limited liability company mainly owned by NNPC, *with a pioneer status and right* to harness Nigeria's vast natural gas resources and produce Liquefied Natural Gas (LNG) and Natural Gas Liquids (NGLs) for export.[100]

Secondary Legislation The second main source of domestic extractive law flows from the first. Beneath the tier of primary legislation on petroleum and mining, there is a wide range of detailed regulations, ministerial guidelines and statutory instruments that are used to flesh out much of the detail of extractive laws. Ministries or departments that oversee extractive permits and licenses often release regulations, rules, notices, warrants and guidelines pursuant to their enabling primarily legislation which provide clarity and further information on how licenses are expected to comply with terms of their licenses in the conduct of extractive operations. Given that they are legal instruments made under powers delegated, directly or indirectly, by an Act of Parliament, their scope and extent are primarily dependent on the powers that have been granted by the primary statute.

A good example is the *Environmental Guidelines and Standards For the Petroleum Industry in Nigeria* ("EGASPIN")[101] a sweeping guideline released by the Department of Petroleum, Resources—the technical arm of the Nigerian Ministry of Petroleum Resources with broad responsibility for implementing government policies, ensuring regulatory compliance, managing the collection of royalties, and licensing of petroleum operations as guided by the Petroleum Act. The EGASPIN specifically deals with the control of pollutants and pollution from the various aspects of petroleum operations and regulates the environmental aspects of petroleum operations. Its primary objective is to regulate environmental management

815); and LIBERIA's *Extractive Industries Transparency Initiative Act, 2009*. See generally Veit and Excell (2015), pp. 65–95.

[99] See for example ss 1–4 of the *Associated Gas Reinjection Act 1979*, Laws of the Federation of Nigeria 2004, c A25, which mandates all oil producing companies to submit detailed programmes for the re-injection or viable utilization of all produced associated gas. It set the limit of October to April 1980 for the oil companies to develop gas utilization projects and 1 January 1984 as end date for gas flaring. The penalty for contravention of this legislation is the forfeiture of the operator's concession. For a detailed overview of these environmental laws, see Olawuyi (2015a, b, c), pp. 57–90. See also Chapter 5 of South Africa's *National Environmental Management Act (NEMA) Act 107 of 1998* which set out environmental standards, approvals, reporting and conditions that must be complied with for extractive operations in South Africa.

[100] The *Nigeria LNG (Fiscal. Incentives Guarantees and Assurances) Act Cap* N87 *2004* (the *NLNG Act*). See also South Africa's *Gas Act 48 of 2001* which provides the legal framework for the storage of gas in South Africa. Also, *Egypt's Law No. 217 of 1980 organising Natural Gas and executive regulations*; and *Law No. 4 of 1988 regarding Oil Pipelines*.

[101] See the Environmental Guidelines and Standards for the Petroleum Industry in Nigeria (1991). [EGASPIN].

practices in the production and discharge of oily wastewaters, sludge and accidental spills of oil from oil and gas production installations. With regards to oil exploration activities, it sets out the procedure of the treatment and control of wastes connected with the oil production process. It also stipulates that to preserve, restore and maintain the chemical, physical and biological integrity of Nigeria's waters, oil and gas installation operators are to ensure that their levels of pollution control are in line with the Best Practicable Control Technology Currently Available ("BPCTCA"). Other wastes such as oil sludge, solid wastes, etc. are to be disposed of by sanitary land filling, encapsulation, solidification, chemical treatment and recycling amongst other treatments It also establishes monitoring requirements for production operations.[102] For gas, it prescribes on flare distances, appropriate burn technology, allowable heat radiation, and noise levels during the gas flaring process amongst other things.[103] It also has provisions for monitoring gaseous point emissions, which include a mandatory frequency sampling of gaseous point emission on a weekly basis or using an approved continued emission monitoring system. Reports on gaseous point emissions are also to be submitted to the Director annually (Table 5.1).[104]

Tertiary Legislation As discussed above, secondary legislation is generally the key means through which ministries and departments exercise statutory powers of providing additional technical information that clarify applicable primary legislation. However, secondary legislation can at times be too technical and may not deal with practical questions or issues that arise on a day to day basis. This has necessitated a rise in guidance documents, circulars, action plans, interpretation aids or statement of regulatory agency practice, all of which provide practical guidelines on the specific activities of a government ministry or department in a non-legalistic language. Such guideline-type documents are often released by regulatory agencies to clarify how the agency has often applied its enabling law or regulations, or in to provide useful information such as how to apply for a license, payment requirements. For example, in many African countries, all bid rounds are guided by clear set of objectives that are released by applicable departments and ministries to set out the general requirements and processes for applying for a license or permit.[105] Tertiary rules and guidance documents supplement formal primary and secondary legislation and generally have persuasive and less binding status. Unless the document contains express intention to carry a binding legal weight, tertiary guidelines provide persuasive principles that serve as normative guidance that cannot be ignored, but are not prescriptive or binding on the agency releasing it.

---

[102] See the EGASPIN, 44–46, 54–59.

[103] Ibid, paras 3.8.8.1 (iii)–(ix).

[104] Another good example of a detailed regulation in Nigeria is the *Oil Prospecting Licenses (Conversion to Oil Mining Leases etc.) Regulations 2004* which clarifies the terms, conditions and procedural requirements for converting an oil prospecting license issued under the Petroleum Act to an oil mining lease.

[105] For example the 2013 Guidelines for Farmout and Operation of Marginal Fields published by the Department of Petroleum Resources, Nigeria (2013) which sets out the broad objectives of the Government's marginal field program. The document notes that "these objectives have been prepared pursuant to the provisions of paragraph-16 of the first schedule of the Petroleum Act 1969, and the Petroleum (Amendment) Act No 23, 1996."

Table 5.1 Network of applicable laws in the extractive industry: the Nigerian example

Source	Instrument
Constitution	1. Constitution of the Federal Republic of Nigeria 1999
Petroleum Laws	2. Petroleum Act of 1969, Cap P10 LFN 2004 3. Nigerian National Petroleum Corporation Act Cap N123 LFN 2004 4. Associated Gas Reinjection Act 2004 and the Associated Gas Re-injection (Amendment) Act 2004; Associated Gas Reinjection Act 1979 5. Petroleum Profits Tax Act Cap P13 LFN 2004 6. Nigerian Oil & Gas Industry Content Development Act 2010 7. Deep Offshore and Inland Basin Production Sharing Contract Act 1999 8. Oil in Navigable Waters Act 1968 9. Oil Pipeline Act 1956 as amended in 1965 10. Hydrocarbon Oil Refineries Act 1965 No. 17 11. Niger Delta Development Commission (Establishment, etc) (Amendment) Act of 2017 12. Petroleum Products and Distribution (Anti-Sabotage) Act of 1975 13. Nigeria LNG (Fiscal Incentives Guarantees and Assurances) Act Cap N87 2004 (the "*NLNG Act*") *Regulations* 14. Deep Water Block Allocation to Companies (Backing Rights) Regulations 2003 15. Environmental Guidelines For the Petroleum Industry in Nigeria (EGASPIN) 16. Oil Prospecting Licenses (Conversion to Oil Mining Leases etc) Regulations 2004 17. Petroleum (Drilling and Production) Regulation 1969 with amendments in 1973, 1979, 1995 and 1996 18. Petroleum Industry Bill, 2012 19. National Domestic Gas Supply and Pricing Regulations of 2008 *Policy* 20. National Energy Policy April 2003 by The Presidency, Energy Commission of Nigeria
Solid Minerals Laws	21. Nigerian Minerals and Mining Act, 2007 22. Nigerian Mining Corporation Act 23. Nigerian Coal Corporation Act 24. Nuclear Safety and Radiation Protection Act 1995 *Regulations* 25. Minerals Oil (Safety) Regulation 1963 *Policy* 26. National Minerals and Metals Policy of January 2008, by the Ministry of Mines and Steel Development

(continued)

Table 5.1 (continued)

Source	Instrument
Environmental Laws	27. National Environmental Standards and Regulations Enforcement Agency Act 2007 28. Environmental Impact Assessment Act 1992 29. National Oil Spills Detection and Response Agency (NOSDRA) Act 2006 30. Water Resources Act of 1993 31. Quarantine Act of 1926 32. Agriculture (Control of Importation) Act of 1964 33. Kainji Lake National Park Act of 1979 34. Criminal Code Act of 1916 35. Land Use Act of 1978 36. River Basins Development Authorities Act of 1987 37. Endangered Species (Control of International Trade and Traffic) Act of 1985 38. Territorial Waters Act of 1967 39. Exclusive Economic Zone Act of 1978 40. National Water Resources Institute Act of 1985 41. Factories Act of 1987 42. Hides and Skins Act of 1942 43. Sea Fisheries Act of 1992 44. Inland Fisheries Act of 1992 45. Animal Disease (Control) Act of 1988 46. Civil Aviation (Repeal and Re-enactment) Act of 2006. 47. Federal National Parks Act of 1991 *Policy* 48. National Policy on the Environment 1988
Transparency	49. Nigeria Extractive Industries Transparency Initiative (NEITI) Act 2007

5.2.2.3 Case Law

In civil law jurisdictions, such as such as Algeria, Angola, Egypt, Gabon, Morocco and Tunisia, most of the formal sources of extractive law are codified and statutory. In such jurisdictions, there is very little judge made law and most of what there is consists of interpretations of comprehensive legislation and statutory provisions that specify all matters capable of being brought before the court, including the applicable procedures and punishments for each infraction.[106] However, in common law jurisdictions such as Ghana, Kenya, Nigeria and Zambia, case law-that is, decisions made by judges in the courts-is an important source of extractive industry law.[107] In addition to reliance on national legislation, countries with a common law tradition of judicial precedence often rely on uncodified case law in interpreting or applying legislation.

[106] See Picker (2012), pp. 1085–1134.

[107] A third, 'mixed' system of law exists in countries such as South Africa and Somalia, constituting a combination of common law and civil law.

Substantive extractive cases come before African courts in the broad ways. First, if there is a dispute about provisions of, or the application of, a primary or secondary legislation, the courts may be called upon to interpret the disputed provision. In such cases, the court's analysis and interpretation of the disputed provision becomes the law. A good example is the case of *The Social and Economic Rights Action Centre and the Centre for Economic and Social Rights (SERAC) v Nigeria,* brought to the African Commission on Human and People's Rights on behalf of the people of Ogoni land.[108] The plaintiffs alleged violations of articles 2, 4, 14, 16, 18(1), 21, and 24 of the *African Charter on Human and People's Rights* resulting from several abuses occasioned the government's stake in oil exploration activities in the area. The key question before the African court was whether environmental degradation and health problems resulting from extractive operations and activities in the Niger Delta area, constituted a violation of the human rights provisions of the African Charter. SERAC alleged that the oil consortium disposed toxic wastes into the environment and local waterways in violation of applicable international environmental standards; that the consortium also neglected and/or failed to maintain its facilities causing numerous avoidable spills in the proximity of villages thus resulting in contamination of water, soil and air, which has had has had serious short and long-term health impacts. Furthermore, SERAC argued that the Nigerian Government has condoned and facilitated these violations by placing the legal and military powers of the State at the disposal of the oil companies. In addition, the Government had neither monitored operations of the oil companies nor required safety measures that are standard procedure within the industry, and also withheld information on the dangers created by oil activities. The Government did not require oil companies or its own agencies to produce basic health and environmental impact studies regarding hazardous operations and materials relating to oil production, despite the obvious health and environmental crisis in Ogoniland, and additionally refused to permit external monitoring of the situation by scientists and environmental organisations. SERAC also alleged that there was no obligation or duty to consult communities before beginning operations, even where the operations posed direct threats to community or individual lands. Furthermore, SERAC alleged that over the years, Nigerian government and security forces attacked, burned and destroyed several Ogoni villages and homes and destroyed and threatened Ogoni food sources through a variety of means, resulting in malnutrition and starvation.

In its decision, the Commission found that the Federal Republic of Nigeria was in violation of Articles 2, 4, 14, 16, 18(1), 21 and 24 of the African Charter on Human and Peoples' Rights; and appealed to the government of the Federal Republic of Nigeria to ensure protection of the environment, health and livelihood of the people of Ogoniland. This case established the principle that African governments are responsible under the African Charter for protecting the rights of local communities that are or could be affected by extractive operations by: stopping all attacks on their communities and leaders and permitting citizens and independent

[108] Social and Economic Rights Action Centre (SERAC) and Another v Nigeria (2001) AHRLR 60 (ACHPR 2001), Communication 155/96).

investigators free access to the territory; conducting an investigation into the human rights violations described above and prosecuting officials of the security forces, and relevant agencies involved in human rights violations; ensuring adequate compensation to victims of the human rights violations, including relief and resettlement assistance to victims of government sponsored raids, and undertaking a comprehensive clean-up of lands and rivers damaged by oil operations; ensuring that appropriate environmental and social impact assessments are prepared for any future oil development and that the safe operation of any further oil development is guaranteed through effective and independent oversight bodies for the petroleum industry; and providing information on health and environmental risks and meaningful access to regulatory and decision-making bodies to communities likely to be affected by oil operations.[109] This decision is evidence of a shift in judicial attitude, from placing a greater premium on revenue from petroleum exploration and exploitation activities, over environmental protection. The case also reflects the influence of regional courts over the interpretation and application of domestic law.

This approach was also endorsed in *Endorois Welfare Coucil v Kenya*. In that case, the African Commission linked property and environmental rights to other human rights when it found that by evicting Kenya's Endorois people from their traditional land for tourism development, the Government of Kenya was in violation of the rights to freedom of religion, property, culture, religion, environment, natural resources and the right to development of the indigenous people of Endrois.[110] As signatories to international and regional treaties on SRM, human rights and environment, African countries have the responsibility to not only respect, protect and fulfil their obligations under regional conventions and treaties, but also to ensure that the objectives of such regional instruments are achieved domestically.

Second, in common law directions such as Ghana, Nigeria, Kenya, Zambia amongst others, previous decisions of courts provide normative influence for the development of obligations and norms in the extractive industry. These include actions in nuisance, negligence, trespass and strict liability that address broad issues that relate to waste management, environmental protection and the live and let's live principle, all of which are relevant to the extractive industry. Notable among these cases are *Gbemre v Shell Petroleum and Development Company Ltd,*[111] a landmark

---

[109] Ibid.

[110] Center for Minority Rights Development (Kenya) and Minority Rights Group International on behalf of the Endorois Welfare Council v Kenya, African Court for Human and Peoples' Rights, Case No. 276/2003.

[111] *Gbemre v Shell Petroleum and Development Company Ltd*, unreported Suit No FH- C/B/CS/53/05. See also important Kenyan cases that discuss the need for public participation and environmental impact assessments, to prevent environmental damage: *In the Matter of the National Land Commission Supreme Court* [2015] eKLR; *John Muraya Mwangi & 495 Others & 6 Others v Minister for State for Provincial Administration & Internal Security & 4 Others* [2014] eKLR; *Mohamed Ali Baadi & Others v The Attorney General & 10 others* Nairobi High Court Petition No. 22 of 2012; and *In the Matter of the Mui Coal Basin Local Community* [2015] eKLR. See also the South African cases of *Agri SA v Minister* for Minerals and Energy 2013 (4) SA 1 (CC); *Minister of Mineral Resources and Others v Sishen Iron ore Company (Pty) Ltd and Another* 2014 (2) SA

environmental case in Nigeria. It is the first known case where a Nigerian Court declared gas flaring as illegal and as a breach of fundamental human rights. The case was instituted by Mr. Gbemre in a representative capacity, for himself and for each and every member of the Iwehereken community in Delta Sate Nigeria against Shell Petroleum and Development Company, Nigeria, the Nigerian National Petroleum Corporation (NNPC) and the Attorney General of the Federation. The plaintiff was asking among others things, for a declaration that the constitutional rights to freedom to life and dignity included by extension, the right to a clean and healthy environment; a declaration that the continued flaring of gas was a violation of these fundamental rights and that the provisions of the Associated Gas Re-Injection Act that allow for gas flaring are inconsistent with the constitutionally guaranteed right to life. In the judgment, passed on 14th November 2005, the Federal High Court of Nigeria sitting in Benin, ordered that gas flaring must stop in the Niger Delta community, as it violates guaranteed constitutional rights to life and dignity. The court ruled that the damaging and wasteful practice of flaring cannot lawfully continue. This ruling made it clear that gas flaring by all oil exploration companies is illegal and constitutes a violation of human rights.[112] A number of other land mark cases have reviewed the grant of licenses and concessions by ministries and departments to determine the scope and extent of arising legal obligations to the contractor.[113]

Third, in addition to judicial decisions, a wide range of quasi-judicial decisions by administrative panels, tribunals and regulatory agencies can provide a source of law. For example, the Minister responsible for the petroleum or solid mineral sector, in their role as ultimate arbiter on questions of participation, licensing and permitting, may set precedents on the manner in which central rules or policies are to be implemented. For example, a Minister may establish an administrative panel to investigate the licensing processes to determine if they have been conducted in line with the stipulated legal processes and procedures. Furthermore, when a ministry or agency issues guidelines or rules that are deemed to be *ultra vires,* unlawful or statutorily incoherent, corporations or stakeholders may approach the court or administrative panel to judicially review the exercise of such statutory powers by the agency.

---

603 (CC) para 63; *Minister of Health and Anor. v New Clicks South Africa (Pty) Ltd and Others* [2006 (2) SA 311 (CC)]; and *Doctors' For Life International v The Speaker National Assembly and Others* (CCT12/05) (2006) ZACC 11.

[112] Other important tort cases are Machine Umudje v Shell B.P. Petroleum Development Company of Nigeria (1975), Adediran & Anor v Interland Transport Ltd; Braide v Adoki.

[113] See for example, *Attorney General of the Federation (AGF) vs. Attorney General of Abia State* (No. 2) (2002) 6 N.W.L.R, part 764, page 542 on the legal implication of Federal Government ownership of petroleum resources in Nigeria; also *Federal Government of Nigeria vs. Zebra Energy* (2002) 18 N.W.L.R. Part 798 page 162 on the legal mode and procedure for revocation of an Oil Prospecting Licence; *South Atlantic Petroleum Company vs. Minister of Petroleum Resources* (unreported suit No. FHC/L/CS/361/2006) on the legal requirement for conversion of Oil Prospecting Licence to Oil Mining Lease; *Shell Petroleum Development Company of Nigeria vs. Federal Board of Inland Revenue* (1996) 8 N.W.L.R. Part 466 page 256 on the computation of adjusted profits for purpose of payment of petroleum profit tax and on the legal authority of Nigerian federal board of inland revenue. to sue for and recover petroleum profits on behalf of the Federal Government of Nigeria.

In such a case, the relevant agency or ministry will have legal obligations to review its actions and process in line with the conclusions of the reviewing tribunal or panel.

5.3 Conclusion

Given the strategic economic importance of extractive resources to the domestic economies of African countries, the extractive industry is a well-regulated industry with a plethora of applicable domestic legislation, and legal instruments at regional and global levels.

Across Africa, the rights and duties of an extractive company towards the public authority emanate from four key sources: the national constitution, domestic legislation, the petroleum or mining license (*'titre minier'*), and the petroleum agreement, which often takes the form of either a *'convention d'établissement'* (establishment agreement) or a production sharing contract. Similarly, other families of law such as international law, environmental law, investment law, labour, human rights and extractive transparency laws and relevant case law, must also be well understood and complied with to avoid legal risks relating to non-compliance.

These networks of laws are supervised by institutions across various tiers of government that have been established to approve, monitor and evaluate extractive sector investments. The next chapter evaluates governance and institutional arrangements across Africa for the regulation and supervision of the extractive industry.

References

African Union (2009) Africa Mining Vision. africaminingvision.org/amv_resources/AMV/Africa_Mining_Vision_English.pdf. Accessed 12 June 2018

African Union Commission, African Development Bank and United Nations Economic Commission for Africa (2011) Building a sustainable future for Africa's extractive industry: from vision to action, action plan for implementing the African Mining Vision

African Union. Pan-African Parliament. au.int/en/organs/pap. Accessed 12 June 2018

Akerhurst M (1976) Equity and general principles of law. Int Comp Law Q 25:801

Alston P, Samma B (1991) The sources of human rights law: custom, Jus Cogens and general principles. Aust Yearb Int Law 12:82

Anglo-Norwegian Fisheries Case ICJ Rep 1951

Associated Gas Reinjection Act 1979, Laws of the Federation of Nigeria 2004

Attorney General of the Federation (AGF) vs. Attorney General of Abia State (No. 2) (2002) 6 N.W.L.R

Auty R, Mikesell R (1988) Sustainable development in mineral economies. Clarendon Press, Oxford

Bastida A (2008) Mining law in the context of development. In: Andrews-Speed P (ed) International competition for resources: the role of law, state and markets. Dundee University Press, Dundee

Bastida A, Bustos L (2017) Towards regimes for sustainable mineral resource management—constitutional reform, law and judicial decisions in *Latin America*. In: Alternative pathways to sus-

tainable development: lessons from Latin America, International Development Policy Series No. 9. Graduate Institute Publications/Brill-Nijhoff, Geneva/Boston
Bell S, McGillivray D (2008) Environmental law, 7th edn. Oxford University Press, Oxford
Birnie P, Boyle A, Redgwell C (2009) International law and the environment, 3rd edn. Oxford University Press, Oxford
Bodansky D (2010) The art and craft of international environmental law. Harvard University Press, Cambridge
Boyle A, Chinkin C (2007) The making of international law. Oxford University Press, Oxford
Brownlie I (2003) Principles of international law, 6th edn. Oxford University Press, Oxford
Bugge HC, Voigt C (2008) Sustainable development in International and National Law. Europa Law Publishing, Amsterdam
Bulankulame v. *Secretary, Ministry of Industrial Development and Others* (the Eppawela Case) (2000 Supreme Court of Sri Lanka)
Case Concerning the Gabcíkovo- Nagymaros Project (Hungary/Slovakia) (1997) I.C.J. Rep., 37 I.L.M
Center for Minority Rights Development (Kenya) and Minority Rights Group International on behalf of the Endorois Welfare Council v Kenya, African Court for Human and Peoples' Rights, Case No. 276/2003
Charney J (1985) The persistent objector rule and the development of customary international law. Br Yearb Int Law 56(1):1–24
Charter of the United Nations (Adopted June 26 1945, entered into force on 24 October 1945)
Cordonier Segger MC, Khalfan A (2004) Sustainable development law: principles, practices and prospects. Oxford University Press, Oxford
Cordonier Segger MC, Mrema E (eds) (2016) Sustainable Development Goals (SDGs): contributions of international law, policy and governance. Series of 10 Legal Issues Papers UNEP/CISDL
Cordonier Segger MC, Weeramantry CD (eds) (2004) Sustainable justice. Martinus Nijhoff, Leiden
Cordonier Segger MC, Gehring M, Newcombe A (eds) (2010) Sustainable development in world investment law. Kluwer Law International, Alphen aan den Rijn
Department of Petroleum Resources, Nigeria (2013) Guidelines for farmout and operation of marginal fields. http://africaoilgasreport.com/wp-content/uploads/2013/12/GUIDANCE-NOTES-FOR-NIGERIAN-2013-MARGINAL-FIELDS-BID-ROUND.pdf. Accessed 12 May 2018
Dominican Republic-Central America Free Trade Agreement (CAFTA-DR) (2006)
Dupuy P (1991) Soft law and the international law of the environment. Mich J Int Law 12:420
Economic Community of West African States. The ECOWAS commission at a glance. afcac.org/en/index.php?option=com_content&view=article&id=57:economic-community-of-west-african-states-ecowas&catid=38:official-recs&Itemid=149. Accessed 12 June 2018
Economic Community of West African States. Directive C/DIR.3/05/09, *On the Harmonization of Guiding Principles and Policies in the Mining Sector*. http://documentation.ecowas.int/download/en/publications/Ecowas%20Directive%20and%20policies%20in%20the%20mining%20sector.pdf. Accessed 12 June 2018. [*ECOWAS Directive*]
Egypt-Germany BIT (2005) http://investmentpolicyhub.unctad.org/IIA/mostRecent/treaty/1347. Accessed 12 Aug 2018
Environmental Guidelines and Standards for the Petroleum Industry in Nigeria (1991) Lagos: Department of Petroleum Resources. Revised in 2002
Environmental Management and Coordination Act no. 8 of 1999 of Kenya
European Union (2003) Communication from the Commission to the Council and the European Parliament - towards a thematic strategy on the sustainable use of natural resources. http://eur-lex.europa.eu/legal-content/EN/TXT/?uri=CELEX:52003DC0572. Accessed 11 June 2018
Extractive Industries (Transparency and Accountability) Act, 2015 of Tanzania
Extractive Industries Transparency Initiative Act, 2009 of Liberia
Federal Government of Nigeria vs. Zebra Energy (2002) 18 N.W.L.R

Fisheries Jurisdiction Case ICJ Rep 1973
Forsythe D (2006) Human rights in international relations. Cambridge University Press, Cambridge
French D (2005) International law and policy of sustainable development. Manchester University Press, Manchester
French D (2010) Global justice and sustainable development. Brill, Leiden
Gas Act No 48 of 2001 of South Africa
Gbemre v Shell Petroleum and Development Company Ltd, unreported Suit No FH- C/B/CS/53/05
Gehring M, Cordonier Segger MC (eds) (2005) Sustainable development in world trade law. Kluwer Law International, Alphen aan den Rijn
Germany-Pakistan BIT 1959
Greenspan E (2014) Free, prior, and informed consent in Africa: an emerging standard for extractive industry projects. Oxfam America, Oxford. https://www.oxfamamerica.org/static/media/files/community-consent-in-africa-jan-2014-oxfam-americaAA.PDF. Accessed 12 June 2018
Guzman A, Meyer T (2010) International soft law. J Leg Anal 2:171–225
Harvard Draft Convention on the International Responsibility of States for Injury to Aliens (1961)
Higgins R (1994) Problems and process: international law and how we use it. Oxford University Press, Oxford
Hulsroj P (1999) Three sources – no river: a hard look at the sources of public international law with particular emphasis on custom and general principles of law. Zeitschrift für öffentliches Recht 54:219
Income Tax Act, No 58 of 1962 of South Africa
International Convention for the Prevention of Pollution from Ships (2 November 1973), 1340 UNTS 184
International Law Association (ILA) New Delhi Declaration on Sustainable Development (2002) International environmental agreements: politics, law and economics 2. UN Doc. A/CONF.199/8, 9 August 2002
International Law Commission (2015) Most-favoured-nation clause final report of the study group on the most-favoured-nation clause. Yearbook of the International Law Commission
Investment Agreement for the Common Market for Eastern and Southern Africa (COMESA) Common Investment Area (2007)
Japan-Lao People's Democratic Republic BIT (2008)
Jodoin S, Cordonier Segger MC (eds) (2015) Sustainable development, international criminal justice & treaty implementation. Cambridge University Press, Cambridge
Land Use Act of Nigeria 1978 Cap L5 Laws of Federation of Nigeria 2004
Law No. 20 of 1987 establishing the Egyptian General Petroleum Company (EGPC)
Law No. 66 of 1953, on Mines and Quarries of Egypt
Law No. 217 of 1980 organising Natural Gas and executive regulations of Egypt
Law No. 4 of 1988 regarding Oil Pipelines of Egypt
Machine Umudje v Shell B.P. Petroleum Development Company of Nigeria (1975) 9-11 S.C. 155 at 166-167
Mayer Brown (2010) Developments: recent legal developments in the mining sector of West Africa states. https://www.mayerbrown.com/publications/recent-legal-developments-in-the-mining-sector-of-west-african-states-01-22-2010/. Accessed 12 Aug 2018
Mexico-United Kingdom BIT (2006)
Mineral Act of Zimbabwe, (Act No. 38) of 1961
Mineral and Petroleum Resources Development Act 2002 of South Africa
Minerals (Prospecting and Mining) Act, 1992 of Namibia
Minerals and Mining Act 2006 (Act 703) of Ghana
Minerals and Mining Act 2007 of Nigeria
Minerals and Petroleum Royalty Act, Act 28 of 2008 of South Africa
Mines and Minerals Development Act No. 7 of 2008 of Namibia
Mining Act 2003 of Uganda (No. 9 of 2003)
Mining Act 2010 of Tanzania (Act No. 14/10)

Mining Act of Kenya Act no.12 of 2016
Mining and Quarries Law 86 of 1956 of Egypt
Mining Law, Law No. 20/2014 of Mozambique
Mining Regulations 2004 of Uganda
National Environmental Management Act (NEMA) Act 107 of 1998 of South Africa
National Environmental Standards and Regulations Enforcement Agency (NESREA) Act of Nigeria (2007)
National Petroleum Corporation Act of 1977 of Nigeria
Niger Delta Development Commission (NDDC) (Establishment etc) (Amendment) Act of 2017 of Nigeria
Nigeria Extractive Industries Transparency Initiative (NEITI) Act 2007
Nikièma SH (2013) Compensation for expropriation. International Institute for Sustainable Development, Winnipeg
North Sea Continental Shelf Case (Federal Republic of Germany/Denmark, Federal Republic of Germany/Netherlands). I.C.J. Reports 1969
OECD (2004) Most-favoured-nation treatment in International Investment Law. OECD Working Papers on International Investment. OECD Publishing, Paris. https://doi.org/10.1787/518757021651. Accessed 11 June 2018
Oil Prospecting Licenses (Conversion to Oil Mining Leases etc.) Regulations 2004 of Nigeria
Olawuyi D (2015a) Increasing relevance of right-based approaches to resource governance in Africa: shifting from regional aspiration to local realization. McGill Int J Sust Dev Law Policy 11(2):113–158
Olawuyi D (2015b) Legal strategies and tools for mitigating legal risks associated with oil and gas investments in Africa. OPEC Energy Rev 39(3):247–265
Olawuyi D (2015c) Principles of Nigerian environmental law. Afe Babalola University Press, Ado-Ekiti
Pan African Parliament (2011) Making investment work for Africa: a parliamentarian response to "land grabs". Pan African Parliament, Midrand. http://www.iisd.org/pdf/2012/land_grabs_africa_en.pdf [*2011 Parliamentary Resolution*]. Accessed 12 June 2018
Paquete Habana case 175 U.S. 677 (1900)
Permanent Court of Arbitration: In the Arbitration Regarding the Iron Rhine ("Ijzeren Rijn") Railway, between the Kingdom of Belgium and the Kingdom of the Netherlands (24 May 2005)
Petroleum (Amendment) Act No 23, 1996 of Nigeria
Petroleum Act 1969 of Nigeria
Petroleum Exploration and Production Act of 2016 (Act 919) of Ghana
Petroleum Profits Tax Act, Laws of the Federation of Nigeria 2004
Petroleum Revenue Management Act 2011 (Act 815) of Ghana
Picker C (2012) International law's mixed heritage: a common/civil law jurisdiction. Vanderbilt J Transnational Law 41:1083–1134
Rajendra Parajuli and Others v *Shree Distillery Pvt. Ltd. & Others* (Writ No. 3259, 1996, Supreme Court of Nepal)
Rann of Kutch Arbitration (1968) 50 ILR 2
Recommendations of the Permanent Committee on Agriculture, Rural Economy, Natural Resources and the Environment. http://www1.uneca.org/Portals/lpi/Documents/Recommendation-Committee-on-agriculture.pdf. Accessed 12 June 2018
Reisman M (1988) Hard look at soft law. Am. Soc Int Law Proc 82:371–377
Report of the United Nations Conference on Environment and Development Annex I: Rio Declaration on Environment and Development. 1992. UNGA, UN Doc A/CONF.151/26 (Vol I) [Rio Declaration]
Resolution on a Human Rights-Based Approach to Natural Resources Governance (2012) African commission on human and peoples' rights, 51st Session. achpr.org/sessions/51st/resolutions/224. Accessed 12 Jun 2018

River Muese Case (Netherlands v Belgium) PCIJ Ser A/B (1937) no 70

SADC Model Bilateral Investment Treaty Template of 2012

Schrijver N (1997) Sovereignty over natural resources. Balancing rights and duties. Cambridge University Press, Cambridge

Schrijver N (2008) The evolution of sustainable development law: inception, meaning and status. Martinus Nijhoff, Leiden

Schrijver N (2017) Advancements in the principles of international law on sustainable development. In: Cordonier Segger MC, Weeramantry CG (eds) Sustainable development principles in the decisions of international courts and tribunals 1992–2012. Routledge, Abingdon

Shell Petroleum Development Company of Nigeria (SPDC) v. Federal Board of Inland Revenue (FBIR) (1996) 8 NWLR pt 466

Shell Petroleum Development Company of Nigeria vs. Federal Board of Inland Revenue (1996) 8 N.W.L.R

Sinclair AC (2004) The origins of the umbrella clause in the international law of investment protection. Arbitr Int 20(4):411–434

Sloan B (1987) General assembly resolutions revisited. Br Yearb Int Law 58:39–150

Social and Economic Rights Action Centre (SERAC) and Another v Nigeria (2001) AHRLR 60 (ACHPR 2001), Communication 155/96

South Atlantic Petroleum Company vs. Minister of Petroleum Resources (unreported suit No. FHC/L/CS/361/2006)

South West African Cases, ICJ Report 1966

Starrett Housing v. Iran, Interlocutory Award No. ITL 32-24-1 (19 December 1983) 4 Iran-United States claims tribunal reports 122

Statute of the International Court of Justice 3 Bevans 1179; 59 Stat. 1031; T.S. 993; 39 AJIL Supp. 215 (1945)

Szablowski D (2007) Transnational law and local struggles. Mining, communities and the World Bank. Hart, Oxford

Talmon S (2005) The security council as world legislature. Am J Int Law 99:175–193

Talmon S (2015) Determining customary international law: the ICJ's methodology between induction, deduction and assertion. Eur J Int Law 26:417–443

The Convention on the Prevention of Marine Pollution by Dumping of Wastes and Other Matters 29 December 1972, 1046 UNTS 120

The Harmonised System: Ecowas Common External Tariff and Other Schedules, Act 905

The Maritime Delimitation and Territorial Questions between Qatar and Bahrain Case (Qatar v Bahrain) ICJ Rep 2001

The Nigeria LNG (Fiscal Incentives Guarantees and Assurances) Act *Cap* N87 *2004*

The United Nations World Commission on Environment and Development Report (the Brundtland Report) (1987) Our common future. Oxford University Press, Oxford

Third Restatement of Foreign Relations Law of the United States (1987)

Tilton J (2003) On borrowed time? Assessing the threat of mineral depletion. Resources for the Future, Washington D.C.

Turkey Model BIT (2009)

United Nations (1992) Rio declaration on environment and development. ILM 31. https://sustainabledevelopment.un.org/content/documents/Agenda21.pdf. Accessed 22 Mar 2018

United Nations (2012) *The future we want*, Resolution of the General Assembly A/RES/66/288 adopted on 27 July 2012. http://www.un.org/disabilities/documents/rio20_outcome_document_complete.pdf. Accessed 22 Mar 2018

United Nations (2015) *Transforming our world: the 2030 Agenda for Sustainable Development*, Resolution adopted by the General Assembly on 25 September 2015. http://www.un.org/es/comun/docs/index.asp?symbol=A/RES/70/1&referer=/spanish/&Lang=E. Accessed 22 Mar 2019

United Nations Convention on the Law of the Sea, 1982, 21 ILM (1982) 1261 (adopted on 10 December 1982, and entered into force on November 16 1994)

United Nations Sustainable Development Knowledge Platform. Major Agreements and Conventions. http://sustainabledevelopment.un.org/index.php?menu=122. Accessed 23 Jan 2018

United States – Import Prohibition of Certain Shrimp and Shrimp Products (6 November 1998) WTO Doc. WT/DS58/AB/R

United States – Import Prohibition of Certain Shrimp and Shrimp Products (20 September 1999) WTO. Doc. WT/DS58/AB/R (Appellate Body Report)

Vanhulst J, Beling A (2014) Buen vivir: emergent discourse within or beyond sustainable development? Ecol Econ 101:54–63

Veit PG, Excell C (2015) Access to information and transparency provisions in petroleum laws in Africa: a comparative analysis. In: Grant JA, Compaoré WRN, Mitchell MI (eds) New approaches to the governance of natural resources. International Political Economy Series. Palgrave Macmillan, London, pp 65–95

Vellore Citizens Welfare Forum v *Union of India* (1996 Supreme Court of India)

Vienna Convention on the Law of Treaties (23 May 1969) 1155 UNTS, 331

Wälde T (2005) The umbrella clause in investment arbitration – a comment on original intentions and recent case. J World Invest Trade 6(2):183

Yannaca-Small K (2008) International investment law: understanding concepts and tracking innovations. OECD, Paris

Yearbook of the International Law Commission 1978, Volume II, Part Two

Zillman D, Roggenkamp M, Paddock L, Godden L (eds) (2018) Innovation in energy law and technology: dynamic solutions for energy transitions. Oxford University Press, Oxford

Chapter 6
Regulatory Institutions

Abstract This chapter surveys and evaluates institutional arrangements across Africa for the regulation and supervision of the extractive sector. It unpacks the key governance approaches adopted for extractive sector supervision in Africa. It then outlines the key institutions, ministries, agencies and government entities that supervise, oversee and approve participation licenses and contracts. It analyses law and policy reasons why many countries today establish national oil and mining companies and the roles such national companies play in extractive sector governance.

6.1 Introduction

The surge in the exploration and production of solid minerals, crude oil, natural gas, liquefied petroleum gas (LPG), gastoliquids (GTLs) and liquefied natural gas (LNG) across the African continent has generated the need for consistent re-evaluation of governmental supervision arrangements to ensure that extractive sector development proceeds in accordance with the visions, priorities and programs of the resource host country. Although Africa is not one homogeneous geographical unit, approaches adopted to approve, oversee and monitor extractive sector investments have been largely similar and consistent. This is mainly the sectorial approach of granting the national oil or mining company, depending on which subsector, the authority and power to oversee, manage, and authorize all upstream, midstream and downstream operations on behalf of the government or supervisory ministry ("Sectorial approach"). With the exception of countries such as Algeria, Equatorial Guinea and South Africa where one ministry oversees both petroleum and mining activities,[1] virtually all surveyed African countries have established separate ministries, agencies and national companies for petroleum and mining sectors

[1] Pursuant to Section 3 of SOUTH AFRICA's Mineral and Petroleum Resources Development Act (MPRDA) Act No 28 of 2002, the Minister for Mineral Resources oversees and supervises South Africa's mineral and petroleum resources. See also ALGERIA, Ministry of Energy and Mining. Also, EQUATORIAL GUINEA, Ministry of Industry, Mines & Energy.

D. S. Olawuyi, *Extractives Industry Law in Africa*,
https://doi.org/10.1007/978-3-319-97664-8_6

respectively. Consequently, any natural or legal person wishing to conduct petroleum or solid minerals operations in must first identify and obtain all applicable licences from the responsible entity and ministry.

However, the multiplicity of regulatory institutions and the historical sedimentation of legislation on extractive sector operations under the Sectorial Approach has led to a situation in which multiple agencies carry out similar or related activities in the same domain.[2] Similarly, inconsistencies and rapid changes in regulatory framework and institutions often make it difficult for investors to develop a consistent mapping of the supervisory mandates of the diverse institutions.[3] While some studies have argued for a more holistic and systemic approach to the regulation of the entire extractive industry under one ministry and institution, others have called for the unbundling of responsibilities, for example the need to establish separate laws and institutions for petroleum and solid minerals in South Africa, as the better governance approach.[4]

Another key debate is whether the national oil or mining company should combine both regulatory and commercial functions, as is the case in a number of African countries such as Nigeria and Ghana; or whether the national oil company should be restricted and reformed to operate as a commercial entity with no regulatory oversight functions, so as to avoid conflict of interest and optimize transparency in their hydrocarbon activities.[5] The Norwegian model of separating policy, regulatory, and commercial functions in the government's administration of petroleum development has been cited as "the canonical model of good bureaucratic design for a hydrocarbons sector".[6] This chapter does not delve into these debates. Rather it develops a profile of key governance approaches and institutions established across Africa to oversee the administration of petroleum and solid mineral resources in the continent.

This chapter surveys and evaluates institutional arrangements across Africa for the regulation and supervision of the extractive sector. After studying this chapter, you should be able to: understand the key governance approaches adopted for extractive sector supervision in Africa; outline the key institutions, ministries, agencies and government entities that supervise, oversee and approve participation licenses and contracts in your country; especially the reasons why many countries today establish national oil and mining companies and the roles such national companies play in extractive sector governance; and appraise the roles of international institutions and agencies in the extractive sector industry.

[2] See Murombo (2015), p. 323; also Roberts (2014), pp. 1–4.

[3] See Olawuyi (2015), pp. 1–20.

[4] See van den Berg and Moster (2018), pp. 257–260.

[5] See Natural Resource Governance Institute (2009).

[6] Thurber et al. (2010), pp. 5–8; see also Al-Kasim (2006).

6.2 Petroleum Sector Institutions

6.2.1 The Ministry of Petroleum Resources[7]

A key element of the sectorial approach to extractive sector governance is the establishment of a separate supervisory ministry with primary oversight over the oil and gas industry. The Ministry in charge of petroleum resources is generally responsible for the formulation and implementation of government policy and general management of the operations of the petroleum industry. The Ministry also regulates and ensures compliance in the conduct of oil and gas sector operations.[8] A supervising cabinet level Minister is often placed in charge of the Ministry and has overarching regulatory powers to implement, enforce and ensure full compliance with the petroleum legislation. The Minister is responsible for granting rights for the exploration, extraction and production of oil and gas through oil exploration licenses, oil prospecting licenses and oil mining leases.[9] The Ministry also oversees the contractual arrangements such as production sharing agreements and service contracts, and carries out other regulatory functions such as consenting to the assignment of licenses and contractual interests in oil and gas assets.

Countries with abundant oil and natural gas resources have segregated the oversight responsibilities of the Ministry into different departments.[10] In Nigeria for example, the Ministry for Petroleum Resources has two technical oversight departments: the Department of Petroleum Resources ("DPR"), and the Department of Gas Resources (DGR). The Ministry exercises its technical, policy and regulatory functions through the DPR.[11] The DPR, supervises and regulates the Nigerian petroleum industry has had a long and interesting history, and over time has transitioned from many government departments.[12] In its regulatory capacity, the DPR is

---

[7] Known as the Ministry of Energy in some African countries, such as Algeria and Zambia, and the Department of Mineral Resources in South Africa. In Zambia for example, the Ministry of Energy and Water Development (MEWD) is the principal government body responsible for energy policy formulation and administration. See Algeria, Ministry of Energy and Mining; Angola, Ministry of Petroleum; Equatorial Guinea, Ministry of Industry, Mines & Energy; Gabon, Ministry of Petroleum and Hydrocarbons; Ghana: Ministry of Energy and Petroleum; Kenya, Ministry of Energy & Petroleum; Libya: Ministry of Oil & Gas; Nigeria, Federal Ministry of Petroleum Resources; South Africa, Department of Mineral Resources.

[8] See generally, Department of Petroleum Resources (2010). [DPR].

[9] See the *Petroleum Act*, s 2 (1). We have highlighted these in our discussions on the *Petroleum Act* above, and will consider them in more detail in Chap. 4.

[10] See South Africa's Hydrocarbons and Energy Planning Branch of the Department of Energy, which is responsible for coal, gas, liquid fuels, energy efficiency, renewable energy and energy planning, including the energy database, Department of Energy.

[11] See Awogbade et al. (2017).

[12] The present day DPR started as a hydrocarbon section of the Ministry of Lagos Affairs, was upgraded to a Petroleum Division within the then Ministry of Mines and Power, and became the DPR in 1970. With the creation of the Nigerian National Oil Company ("NNOC") in 1971 to engage in commercial activities in the petroleum industry, the DPR continued to perform the statu-

responsible for the enforcement of rules and regulations established by the Ministry, and it also performs a dual role in regulating oil and gas alongside each other. The DPR is responsible for the day-to-day monitoring of the petroleum industry and is responsible for supervising all petroleum industry operations carried out under licenses and leases. It is also principally responsible for processing all applications for licenses and leases in the oil and gas industry. The DPR also has primary responsibility for the enforcement of safety and environmental standards in the oil and gas industry. Its functions in this regard include the supervision of all licensed petroleum industry operations to ensure compliance with the applicable laws, regulations and good oil producing practices; enforcing safety and environmental regulations; keeping and updating records on petroleum industry operations, advising government, relevant agencies and the Minister on technical matters and policies, which may impact the administration and control of petroleum; as well as processing all applications for licenses to ensure compliance with laid-down guidelines.[13]

On the other hand, the DGR, which is also a department in the Ministry of Petroleum Resources is responsible for regulating the gas sector, in accordance with the *Gas Supply and Pricing Regulations*,[14] the National Gas Master Plan ("NGMP"), and other policies issued in respect of the gas sector by the Federal Government.[15] The DGR regulates gas exploration, development, gathering, treatment, processing and utilization. It also issues permits and licenses as it relates to the gas industry, supervise gas pricing, and regulate domestic utilization among other things.[16]

6.2.2 The National Oil Company[17]

Another prevailing element of the sectorial approach to extractive sector governance across Africa is the establishment of a national oil and gas company (NOC), that is fully or in the majority owned by the national government. Fifteen of the

---

tory supervision and control duties in the oil industry. In 1975, the DPR was constituted into the Ministry of Petroleum Resources ("MPR"). With the formation of the NNPC by the merger of the MPR and NNOC in 1977, the Petroleum Inspectorate, which was the regulatory arm, was granted semi-autonomous status and became an integral part of the NNPC. With NNPC's commercialization in 1988, the Petroleum Inspectorate was excised from the NNPC, and merged with the new MPR to form its technical arm. See generally, DPR.

[13] See generally, DPR. In addition to being the regulatory arm, the DPR also acts as a technical advisor to the Minister and Ministry of Petroleum Resources. Decisions taken by the Ministry are usually handed down via the DPR. See Odumosu (2007), p. 881.

[14] See the National Domestic Gas Supply and Pricing Regulations 2008 to the Petroleum Act of 1969, Laws of the Federation of Nigeria 2004, c P10, s 1. [Gas Supply and Pricing Regulations].

[15] See the Gas Supply and Pricing Regulations ibid, s 2 (a).

[16] See ibid, s 2, which sets out the functions of the department of gas.

[17] See Algeria, Sonatrach; Angola, Grupo Sonangol; Equatorial Guinea, GEPetrol; Gabon, Societe Nationale des Hydrocarbures du Gabon; Ghana: Ghana National Petroleum Corporation (GNPC); Kenya, National Oil Corporation of Kenya; Libya: National Oil Corporation; Nigeria, Nigerian

world's 20 biggest energy-producing companies are state-owned enterprises controlled by national governments in Africa and the Middle East.[18] Government-controlled NOCs dominate the ranks of the world's largest energy producers. Estimates indicate that NOCs control approximately 90% of the world's oil reserves and 75% of production as well as many of the major oil and gas infrastructure systems.[19] NOCs have therefore emerged as important actors in the international petroleum industry.[20] Thirteen of 14 former and current members of the Organization for Petroleum Exporting Countries (OPEC) have an NOC and so do other non-OPEC member countries in Africa who produce oil and gas.[21]

NOCs are established by governments to oversee, manage, and authorize all upstream, midstream, and downstream petroleum operations in a petroleum producing country.[22] The organizing functions of NOCs include to: serve as the main agent through which the state regulates and controls its oil and gas resources; maximize the value of economic rent and profits from the exploitation of petroleum resources in the country; enhance and spearhead local content contributions to the country; promote the participation of host communities and nationals in the petroleum sector value chain; ensure the steady supply of petroleum products to the local market; facilitate the allocation of petroleum subsidies to nationals through direct employment, development scholarships, modern infrastructure development and supply of petroleum products at reduced rates amongst others; manage internal and external shocks of fluctuating international oil prices; and ultimately create a level playing field through which national industries can compete with international oil and gas companies (IOCs) in the long term exploration of natural resources amongst others.[23] Summarily, NOCs serve as primary institutional vehicles through which resource-rich countries seek to maximize economic efficiency and the generation of social and economic benefits to its people.[24]

However, in some African countries, NOCs take on a whole range and mix of other regulatory, political and international investment roles depending on the policy objectives and priorities of the Government and the Ministry in charge of petroleum resources.[25] Investors must therefore carefully understand the scope of functions and roles assumed by the NOC in the respective jurisdiction. Whatever the mix of commercial and or regulatory role assumed by a NOC, its primary objective

National Petroleum Corporation; South Africa, PetroSA (The Petroleum, Oil and Gas Corporation of South Africa SOC Ltd).

[18] See Marcel (2006). Also Carpenter (2015).

[19] See Tordo et al. (2009).

[20] Victor et al. (2012).

[21] See OPEC; also Cheon et al. (2014).

[22] See Cheon et al. (2014), Ibid; McPherson (2003), pp. 184–203.

[23] Cheon et al. (2014); also Hartley and Medlock (2008), Eller et al. (2011) and Olorunfemi (1991).

[24] See Tordo et al. (2009), also Losman (2010), Cheon et al. (2014), Ibid; also Stevens (2008) and Wolf (2009).

[25] See Heller et al. (2014), pp. 1–2.

is to put an oil-producing country in a position to fully participate in the exploitation of its petroleum resources.[26]

NOCs are often established and organized by a domestic legislation or regulation which spells out their specific mandates, functions and roles.[27] For example, Nigeria's oil company is the NNPC- the largest NOC in Africa, was established in 1977 pursuant to the Nigerian National Petroleum Corporation Act which spells out the powers vested in the corporation.[28] Section 5 (1) of NNPC Act sets out the functions and duties of the NNPC as follows: (a) exploring and prospecting for, working, winning or otherwise acquiring, possessing and disposing of petroleum; (b) refining, treating, processing and generally engaging in the handling of petroleum products and its derivatives; (c) purchasing and marketing petroleum, its products and by-products; (d) providing and operating pipelines, tanker-ships or other facilities for the carriage or conveyance of crude oil, natural gas or their products and derivatives, water and any other liquids or other commodities related to the corporation's operations; (e) constructing, equipping and maintaining tank farms and other facilities for the handling and treatment of petroleum and its products and derivatives; (f) carrying out research in connection with petroleum or anything derived from it and promoting activities for the purpose of turning to account the results of such research; (g) doing anything required for the purpose of giving effect to agreements entered into by the Federal Government with a view to securing participation by the Government or the corporation in activities connected with petroleum; (h) generally engaging in activities that would enhance the petroleum industry in the overall interest of Nigeria; and (i) undertaking such other activities as are necessary or expedient for giving full effect to the provisions of the Act.[29]

Section 6 (1) of the Act grants the NNPC, statutory powers to undertake acts calculated to facilitate its duties including, but not limited to: (a) sue and be sued in its corporate name; (b) hold, manage and alienate movable and immovable property; (c) purchase or otherwise acquire or take over all or any of the assets, businesses, properties, privileges, contracts, rights, obligations and liabilities of any other company, firm or person in furtherance of any business engaged in by the Corporation; (d) enter into contracts or partnerships with any company, firm or person which in the opinion of the Corporation will facilitate the discharge of its duties; (e) establish and maintain subsidiaries for the discharge of such functions as the Corporation may determine; and (f) train managerial, technical and such other staff for the purpose of the running of its operations and for the petroleum industry in general. In its supervisory capacity, the NNPC is charged with managing all

[26] See Victor et al. (2012), Thurber et al. (2011) and Heller and Marcel (2012).

[27] See for example the The Ghana National Petroleum Corporation Law (1983). The Ghana National Petroleum Corporation (the 'GNPC') is responsible for the development, production and disposal of petroleum. The GNPC is required to ensure that Ghana obtains the greatest possible benefits from the development of petroleum resources. The GNPC participates in petroleum operations on behalf of Ghana with contractors under the Petroleum Agreements.

[28] See the World's Biggest Oil and Gas Companies; see also *NNPC Act,* Cap N123 LFN 2004, s 1.

[29] NNPC Act.

governmental interests in the Nigerian petroleum industry. To this end, the NNPC holds all the petroleum assets vested in the government under the *CFRN 1999* and the *Petroleum Act* on behalf of the Government. This includes the Government's interest in various joint ventures with the multinational oil companies.[30]

NOCs across Africa increasingly face challenges that limit their functionality and efficiency. Several studies have therefore highlighted the need to reform the structural make up of NOCs to achieve more independence, accountability, efficiency and transparency.[31] The challenges include lack of flexibility and the presence of civil service culture that impedes entrepreneurial vision; financial mismanagement and corruption; high and disproportionate administrative costs; regulatory overlap which sees NOCs operating as both commercial entities and as regulators; and continued annual budget allocation of funds from the national government which further reduces independence. Reforming African NOCs in line with international best practices can help them develop the efficiency, structural independence, expertise, entrepreneurial vision and financial resources needed to effectively carry out their duties under their enabling law.

6.2.3 Secondary Institutions

In addition to the NOC, a number of supervisory institutions may be established to oversee specific mandates and functions in the oil and gas industry.[32] This includes specialist institutions on pricing, local content development, environmental restoration and infrastructure renewal in oil producing communities, and procurement agencies amongst others. For example, a number of African countries, such as Angola, Ghana, Nigeria and Tanzania have established distinct committees, boards or agencies to oversee local content, job creation through the use of local expertise, goods and services business, financing in the petroleum industry value chain, and value retention at home.[33]

---

[30] See generally the NNPC Act, ibid, s 5 (1)(g).

[31] Heller et al. (2014) stating that while NOCs have contributed heavily to successful efforts to harness benefits from the oil sector and drive broader national development. In other cases, NOCs have become inefficient managers of national resources, obstacles to private investment, drains on public coffers, or sources of patronage and corruption.

[32] See for example Ghana's Petroleum Commission Act of 2011, which establishes the Ghana Petroleum Commission as the regulator of all upstream petroleum activities. The Commission is tasked with promoting local content and local participation programmes, creating the best possible values through prudent and sustainable management of oil & gas resources.

[33] See ss 5–6 of *Ghana's Petroleum (Local Content and Local Participation) Regulations* (2013); also Tanzania's Petroleum (*Local Content*) *Regulations, 2017*. GN. No. 197 which establishes a Local Content Committee to oversee compliance and implementation of the local content regulations. See also the recently established National Petroleum Agency of Angola, Macauhub (2017).

For example, sections 4 and 69(1) of the *Nigerian Content Act* established the Nigerian Content Development and Monitoring Board (the "Board").[34] The Board is generally responsible for developing procedures that will guide, monitor, coordinate and improve the provisos of the *Nigerian Content Act*. The Board is also charged with ensuring measurable, continuous growth of Nigerian content in oil and gas arrangements, projects, operations, activities and transactions in the oil and gas industry.[35] Some of the broad responsibilities of the board include but are not limited to: increasing indigenous participation in the oil and gas industry; building local capacity and competencies; creating linkages to other sectors of the national economy; and boosting industry contributions to the growth of Nigeria's National Gross Domestic Product.

Its functions primarily include capacity building, job and value creation, and infrastructure development. Section 70 of the Act spells out the specific functions of the Board to include implementing the provisions of the Act and its regulations; supervising, coordinating, administering, monitoring and managing the development and implementation of Nigerian content in the oil and gas industry and in the operations of operators and contractors in the oil and gas industry; appraising, reviewing, evaluating and approving Nigerian content plans and reports submitted to it and administering and operating the petroleum e-market place and Joint Qualifications System set up by the *Nigerian Content Act*.

Its functions also extend to assisting local contractors and Nigerian companies to develop their capacity towards attaining the goal of Nigerian content development; making procedures to guide the implementation of the *Nigerian Content Act* and ensure compliance with its provisions; monitoring and coordinating Nigerian content performance of all operators as well as organizing trainings, conferences, and workshops for public education amongst other things.

One key function of the Board, as detailed under section 70 of the *Nigerian Content Act*, is the administration and operation of the Joint Qualifications System ("JQS"). The JQS is the industry databank of available capacities and capabilities in the Nigerian oil and gas industry.[36] It was created pursuant to section 55 of the *Nigerian Content Act,* which states that the Board shall establish, maintain and operate a joint qualification system in consultation with industry stakeholders, which is to be administered in accordance with the Act and regulations made by the minister pursuant to the Act. Section 56 of the *Nigerian Content Act* goes on to say that the JQS is to constitute an industry databank of available capabilities and shall be the sole system for Nigerian content registration and pre-qualification of contractors in the industry. The functions of the JQS include the verification of contractors' capacities and capabilities; evaluation of application of Nigerian content in the operations of oil companies and contractors; building a data base for national skills development pool; and ranking and categorizing of oil service companies based on capabilities and Nigerian content.

[34] The Nigerian Content Act.

[35] Ibid, ss 4–5.

[36] Ibid, s 106.

Furthermore, another function of the Board under section 70 is the administration and operation of a petroleum e-market place. The oil and gas or petroleum e-market place is a virtual platform for the buying and selling of goods and services in the oil and gas industry that allows for speedy and transparent transactions.[37] Section 54 of the *Nigerian Content Act* establishes the oil and gas e-platform and its functions are to provide a virtual platform to facilitate the transactions required for efficient delivery of goods and services in the industry; provide a functional interface to the JQS; provide universal and transparent governance structure, which is to be drawn from industry stakeholders; and track and monitor Nigerian content and preference of operators, project promoters, suppliers and service providers.[38]

In addition, under section 70, the Board is also charged with making procedures to guide the implementation of the *Nigerian Content Act* and ensure compliance with its provisions. Pursuant to this power to make procedures to guide the implementation of the *Nigerian Content Act*, the Board has published a number of guidelines detailing procedures for regulating various aspects of the *Nigerian Content Act*. For instance, the Board has published guidelines with respect to the provision of legal services, the waiver approval procedure, and human capacity guidelines amongst others.[39]

Petroleum Product Pricing Furthermore, in 2000 the Nigerian government set up a Special Committee on the Review of Petroleum Products Supply and Distribution to look into petroleum pricing problems in Nigeria's downstream sector. The objective was to re-position Nigeria's downstream sub-sector for improved efficiency and transparency. One of the recommendations of this body was setting up of a Petroleum Products Pricing Regulatory Agency with sufficient autonomy to superintend the various phases of the proposal embodied in the report especially the liberalization of the downstream sector of the petroleum industry.[40] The *Petroleum Products Pricing Regulatory Agency (Establishment) Act* (the "*PPRA Act*")[41] establishes the Petroleum Products Pricing Agency (the "Agency").[42] Under section 7 of the *PPRA Act*, the functions of the Agency include determining the pricing policy of petroleum products; regulating the supply and distribution of petroleum products; creating an information databank to facilitates the making of informed and realistic decisions on pricing policies; overseeing the implementation of the relevant recommendations and programmes of the Federal Government.

The functions of the Agency also extend to moderating volatility in petroleum products prices, while ensuring reasonable returns to operators; establishing

---

[37] Ibid.

[38] See generally, the Nigerian Content Development and Monitoring Board. [NCDMB].

[39] Ibid.

[40] See generally, The Petroleum Products Pricing Regulatory Agency (Establishment) Act No 8, 2003. [PPRA Act].

[41] Ibid.

[42] PPRA Act, ibid, s 1.

parameters and codes of conduct for all operators in the downstream petroleum sector; maintaining constant surveillance over all key indices relevant to pricing policy and periodically approve benchmark prices for all petroleum products; preventing collusion and restrictive trade practices that are harmful to the oil and gas sector; exercising a mediatory role for stakeholders in the sector; and carrying out activities that are expedient for the full and efficient discharge of its functions.[43]

6.2.4 Joint Development Authorities (JDAs)

JDAs play important roles in overseeing, authorizing and managing petroleum operations on behalf of two or more countries, with overlapping petroleum resources, who agree to jointly explore and exploit the natural resources, and share the resulting revenues, in those joint development areas.[44] A JDA is an organization established by the treaty that regulates the joint development of an overlapping oil and gas producing zone.[45]

A good example is the Nigeria–São Tomé and Príncipe Joint Development Authority established pursuant to a joint development treaty between Nigeria and Sao Tome and Principe's, to oversee and manage their joint development zone in the Bight of Bonny.[46] The Joint Development Zone (the "Zone") is along the maritime Nigeria-Sao Tome and Principe border. The Zone is an area in the region of the Nigeria–São Tomé and Príncipe boundary region that is speculated to be rich in oil and gas reserves. However, because neither country could explore the resources in the zone without interfering with the maritime territory of the other country, the countries agreed in a treaty to create a joint development authority that would assist

[43] Ibid, s 7.

[44] See for example Abu Dhabi-Qatar. 'Agreement on settlement of maritime boundary lines and sovereign rights over islands, signed on 30 March 1969' *United Nations Document* ST/LEG/SER.B/16, 403; Australia-Indonesia (1990), p. 469; Australia-Papua New Guinea (1979), p. 291; Austria-Czechoslovakia (1964), p. 126; Bahrain-Saudi Arabia. 'Agreement concerning the delimitation of the Continental Shelf in the Persian Gulf, done at Riyadh on 22 February 1958.' *United Nations Document* ST/LEG/SER.B/Add. 2, 185; Convention on the Regulation of Antarctic Mineral Resource Activities (CRAMRA), done at Wellington, 2 June 1988 (1988), p. 859; Federal Republic of Germany-Netherlands (1982) , p. 755; France-Spain (1982), p. 445; Iceland-Norway (1982), p. 1222; Iran-Sharjah (1972), Supplement; Kuwait-Saudi Arabia. "Agreement relating to partition of the Neutral Zone, signed on 7 July 1965." *United Nations document* ST/LEG. SER.B/15, p. 760; Norway-United Kingdom (1977); Cmnd. 7043; Saudi Arabia-Sudan. "Agreement relating to the Joint Exploitation of the Natural Resources of the sea-bed and sub-soil of the Red Sea in the common Zone, done at Khartoum on 16 May 1974." *United Nations Document* ST/LEG/SER.B/18, p. 452.

[45] Schofield (2014), pp. 78–98; Woodliffe (1977), p. 338; Barberis and Julio (1975); Brownlie (1979).

[46] See the Treaty between the Federal Republic of Nigeria and the Democratic Republic of São Tomé and Príncipe on the Joint Development of Petroleum and other Resources, in respect of Areas of the Exclusive Economic Zone of the Two States (21 February 2001), art 9.1[*Nigeria-Sao Tome and Principe Treaty*].

both countries in benefitting from the economic potential of the zone.[47] The JDA is established as a person in international law and under the law of each of the States Parties and has the legal capacity as is necessary for the exercise of its powers and the performance of its functions. In particular, the Authority is to have the capacity to contract, to acquire and dispose of movable and immovable property and to institute and be party to legal proceedings.[48] The primary office of the Authority is in Abuja, Nigeria, with a subsidiary office in Sao Tome and Principe.

Under article 9.6 of the treaty, the JDA is responsible for the management of activities relating to exploration for and exploitation of the resources in the Zone, in accordance with the treaty. In particular, the functions of the Authority include the division of the Zone into contract areas, and the negotiation, tendering for and issue and supervision of contracts with respect to such areas; entering into development contracts with contractors, as well as oversight and control over of the activities of contractors; terminating development contracts; collecting and distributing the proceeds or products of the Authority's share of production from development contracts.

The functions of the JDA also include preparing budgets; controlling the movements of vessels, aircraft, structures, equipment and people into, within and out of the Zone; establishing safety and restricted zones, to ensure the safety of navigation, petroleum activities, fishing activities and other development activities and the effective management of the Zone; issuing regulations and giving directions on all matters related to the supervision and control of operations, including on health, safety and environmental issues; regulating marine scientific research; preserving the marine environment; requesting and coordinating action with respect to search and rescue operations, deterrence or suppression of terrorist or other threats to vessels and structures engaged in development activities in the Zone; the prevention or remedying of pollution; as well as the consideration of matters from time to time specifically referred to it.

Given the strategic importance and functions of JDAs, it is important for entrants into the oil and gas markets in Africa to investigate the licensing requirements that may exist in areas of joint development. Financing petroleum agreements with or in one country covered by a JDA without contacting the supervising JDA could be a potential source of complex legal risks.

6.2.5 Department or Ministry of Environment ("MOE")

All African countries have established ministries or departments responsible for implementing Government policy on environmental protection and sustainable development.[49] Ranging from the Federal Ministry of Environment in Nigeria estab-

---

[47] See generally, NSTPJDA.

[48] See Nigeria-Sao Tome and Principe Treaty, Art 9.2.

[49] For a comprehensive list of all environment ministries in Africa, see The World Bank, Africa Environmental Agencies.

lished in 1999, to the Ministry of Environment and Natural Resources in Kenya, the Department of Environmental Affairs in South Africa established in 1994 (as the Department of Environmental Affairs and Tourism); as well as the Ministry for Environmental Affairs in Egypt established in 1997.[50] The broad mandates of MOEs include: formulating and implementing environmental policies; administering environmental impact assessments relating to public and private projects, including solid minerals and oil and gas projects; preparing necessary action plans for environmental protection; stipulating procedures for environmental impact assessment for projects; managing and protecting oceans and coastal resources, promoting conservation and the sustainable use of natural resources; promoting cooperation in environmental science and conservation technology; and cooperating with Federal and State Ministries, Local Government, statutory bodies and research agencies on matters relating to the protection of the environment and the conservation of natural resources.[51]

Typically, headed by a cabinet Minister or Secretary for Environment, MOEs supervise and enforce environmental protection measures on water quality, effluent limitations, air quality, atmospheric protection, ozone protection, noise control as well as the removal and control of hazardous substances. At provincial or state levels, ministries of environment also play similar roles and are supervised by Commissioners for environment.[52]

However, since their respective establishments in the 1990s, African MOEs have been faced with several operational challenges that have stifled environmental protection in extractive operations. Due to differing priorities placed on environmental matters by successive governments, environmental conservation policies and projects have been inconsistently implemented, while legal provisions on environmental protection have not been transparently enforced.[53] This lack of continuity and political commitment often result in mismanagement of environmental restoration funds, and/or the disappearance of programs and structures designed for embedding sustainable resource governance across Government.[54]

6.3 Regulatory Institutions for the Solid Minerals Sector

As earlier noted, virtually all surveyed African countries have established separate ministries, agencies and national companies for petroleum and mining sectors respectively. Any natural or legal person wishing to conduct solid minerals

[50] See Kenya's Ministry of Environment and Natural Resources, About Us, Nigeria's Federal Ministry of Environment, About Us.

[51] Ibid.

[52] See generally, Federal Ministry of Environment of Nigeria, About the MOE-Profile.

[53] See Shelton (2011), Eaton (1997) and Mobogunje (1995).

[54] Nigeria: ICPC Returns N1 Billion Looted Funds to Environment Ministry (Mutum 2015).

operations in must first identify and obtain all applicable licences from the responsible entity and ministry responsible for overseeing the solid minerals sector. This section explores the institutions that primarily regulate the solid minerals sector.

6.3.1 Ministry for Solid Minerals Development

In all surveyed African countries, the administration of the mining industry is vested in a government ministry that is responsible for mining and solid minerals development.[55] The Ministry in charge of solid mineral resources is generally responsible for the formulation and implementation of government policy and general management of the operations of the solid minerals industry. The Ministry also regulates and ensures compliance in the conduct of mining and metallurgical operations. The Minister in charge of the Ministry has overarching regulatory powers to enforce the enabling law and to prescribe guidelines for the consistent implementation of government policy and framework for mineral resources utilization, most especially issues related to sustainable exploration of mineral resources, community engagement, environmental issues and beneficiation from mining. The Minister is responsible for granting licenses and permits for the exploration, extraction and production of solid mineral resources.[56] The Ministry also oversees contractual arrangements such as mineral development agreements and service contracts, and carries out other regulatory functions such as consenting to the assignment of licenses and contractual interests in mineral assets.

Generally, mining ministries operate through departments or directorates that have direct oversight over licensing, regulation, survey and enforcements. In Kenya for example, the *Mining Act 2016* establishes the Directorate of Mines that is responsible for the day-to-day monitoring of the mining industry and is responsible for supervising all mining operations carried out under licenses and permits. It is also principally responsible for processing all applications for licences and leases, and enforcing safety and environmental standards in the mining industry. Its functions in this regard include the supervision of all licensed mining operations to ensure compliance with the applicable laws, regulations and good oil producing practices. The Kenyan law also establishes a Directorate of Geological Survey which is responsible for developing the national geological database and promoting interest in the mining sector.

---

[55] ALGERIA, Ministry of Mining and Energy, which oversees hydrocarbon, energy and mining sectors in Algeria; ANGOLA, Ministry of Geology, Mines and Industry; EQUATORIAL GUINEA, Ministry of Industry, Mines & Energy; GABON, Ministry of Mines, Industry, and Tourism; GHANA: Ministry of Lands and Natural Resources; KENYA, Ministry of Mining; LIBYA, Ministry of Industry, Electricity and Mining; NIGERIA, Federal Ministry of Mines and Steel Development; SOUTH AFRICA, Department of Mineral Resources; ZIMBABWE, Ministry of Mines and Mining Development.

[56] Ibid.

Similarly, in Nigeria, the Ministry of Mines and Steel Development operates through four departments namely the Mines Inspectorate Department, the Mines Environment and Compliance, the Mining Cadastre Office and the Artisanal and Small-Scale Mining Department. The Mines Inspectorate Department ("the Inspectorate") is responsible for operations in exploration, evaluation, mine development and production.[57] Its detailed functions include: general supervision of mining, quarrying and explosives materials to ensure safe mining operations and enhance high production of minerals and revenue generation; supervision and enforcement of compliance with all work programmes, mine health and safety regulations by mineral title holders; carrying out investigations and inspections to ensure that mineral title holders are in compliance with conditions attached to their licences and leases[58]; development of a mining and quarrying operations database; ensuring the preparation of mineral returns by operators as required by the government; granting approval for the import of machinery, equipment and accessories under customs exemptions; and collating records of all mineral production in Nigeria.[59] The Inspectorate also carries out a variety of other regulatory functions under the *Mines and Minerals Act*, ranging from approvals, incident reports, and compliance activities.[60]

The main function of the Mines Environment and Compliance Department is to ensure that companies adopt and maintain environmentally friendly procedures in their operations and adhere to environmental laws and guidelines.[61] It is also responsible for: establishing environmental procedures and requirements that are applicable to mining operations; reviewing all plans, studies and reports, which are required to be prepared by mineral titleholders in fulfilment of their environmental obligations under the *Mines and Minerals Act*[62]; monitoring and enforcing mineral titleholders compliance with environmental requirements and obligations; liaising with relevant agencies of Government with respect to social and environmental issues associated with mine operations, as well as mining closures and reclamation of land; and maintaining and carrying out regular environmental audits to ensure the adoption of, and compliance with sound environmental practices in all mining operations.[63]

The Mining Cadastre Office is primarily responsible for the administration of mineral titles and the maintenance of cadastral registers.[64] It is also in charge of other administrative and regulatory functions, which include: considering applications for mining titles and permits; issuing, suspending and revoking mining titles

---

[57] See Nigeria's MMA 2007, s 16. See also KPMG (2014).

[58] In furtherance of this function, see MMA, ibid, s 20(2).

[59] See generally, ibid, s 17, 25(2) See also KPMG (2017), p. 6.

[60] See generally, MMA, ibid.

[61] MMA, ibid, s 16 (1)(b).

[62] MMA, ibid.

[63] See generally, MMA, ibid, s 18, s 71(1), 90(2), ss 119–120, See also KPMG (2017), pp. 6–7.

[64] MMA, ibid, s 5.

subject to applicable rules and regulations; receiving and processing applications for transfer, renewal, modification, extension and relinquishment of mineral titles; administer the competitive bidding process set out in section 9 of the *Mines and Minerals Act*; maintaining a chronological; record of all applications for mineral titles in order of priority; and maintaining a register and a general registry book.[65]

The fourth department is the Artisanal and Small-scale Mining (ASM) Department, which is directly responsible for regulating and administering the operations of artisans and small-scale miners. The Act, in section 164, defines artisanal mining as mining operations conducted using non-mechanised methods of reconnaissance, exploration, extraction and processing of mineral resources in an area covered by a small-scale mining lease granted by the Minister for exploitation of mineral resources. The ASM department oversees such small scale, low capital intensive, low technology and high labor-intensive ASM operations. Specifically, it responsible for: organizing, supporting and assisting small scale mining operations; providing extension or field services to mining cooperatives on exploration, exploitation, mineral processing, entrepreneurial training, and environment management; and undertaking initiatives, which improve the acquisition of sustainable livelihoods in artisanal and small-scale mining communities.[66]

6.3.2 National Mining Corporations

State-owned national companies that featured prominently in the development of petroleum sector are now becoming increasingly prevalent in the mining sector. National mining corporations (NMCs) explore, exploit, process, market and supply mineral resources on behalf of government. Government-controlled NMCs dominate the ranks of the world's largest solid mineral producers. Large NMCs across the world include the National Copper Corporation (CODELCO) of Chile, a state-owned mining company which currently ranks as the largest copper producing company in the world; and Vale S.A of Brazil, the largest producer of nickel and iron ore in the world, and third largest mining company in the world after BHP and Rio Tinto.[67]

Africa is home to some of the world's largest state-owned mining companies. These include Morocco's OCP Group, one of the leading exporters of phosphate rock, phosphoric acid and phosphate fertilizers in the world[68]; Nigeria's National Iron Ore Mining Company (NIOMCO); Nigeria's Coal Corporation; Tanzania's State Mining Corporation (STAMICO), Democratic Republic of Congo's La Générale des Carrières et des Mines (Gécamines); South Africa's African

[65] Ibid, s 5 (5), s 7, See generally, KPMG (2017), p. 7.

[66] See generally, KPMG (2017), ibid at 7.

[67] Basov (2015).

[68] World Economic Forum, OCP Group.

Exploration Mining and Finance Corporation (AEMFC); Namibia's Epangelo Mining Company; and Kenya's National Mining Corporation amongst others. NMCs have therefore emerged as important actors in the international mining industry.[69]

NMCs are established as the investment arm of the national government, to oversee, manage, and authorize all geophysical exploration, prospecting, refining, grading, producing, cutting, processing, buying, selling, marketing and beneficiation of mineral products.[70] The aim of NMCs include to: serve as the main agent through which the state regulates and controls its solid mineral resources; maximize the value of economic rent and profits from the extraction of solid mineral resources in the country; enhance and spearhead local content contributions to the country; promote the participation of host communities and nationals in the mineral sector value chain; ensure the steady supply of mineral products, such as coal, iron, steel, marble amongst others, to the local market; facilitate the allocation of subsidies to nationals through direct employment, development scholarships, modern infrastructure development and supply of products at reduced rates amongst others; manage internal and external shocks of fluctuating international prices of mineral products; and ultimately create a level playing field through which national industries can compete with international mining companies (IMCs) in the long term exploration of solid mineral resources amongst others.[71] Summarily, MNCs, like NOCs, serve as primary institutional vehicles through which resource-rich countries seek to maximize economic efficiency and the generation of social and economic benefits to its people.[72]

However, in some African countries, NMCs take on a whole range and mix of other regulatory, political and international investment roles depending on the policy objectives and priorities of the Government and the Ministry in charge of solid mineral resources.[73] Investors must therefore carefully understand the scope of functions and roles assumed by the NMC in the respective jurisdiction. Whatever the mix of commercial and or regulatory role assumed by a NMC, its primary objective is to put an oil-producing country in a position to fully participate in the exploitation of its solid minerals.[74]

NMCs are often established and organized by a domestic legislation or regulation which spells out their specific mandates, functions and roles. For example, Kenya's NMC is established pursuant to *National Mining Corporation Regulations, 2016* which provides a framework for the operationalization of the powers vested in the corporation in accordance with sections 22(1), 25(2) and 27 of the Kenyan

[69] On the role of mining corporations in international law and relations, see Dashwood (2012), pp. 73–93.

[70] See for example Section 22 of the Kenya Mining Act 2016.

[71] Ibid, s. 24.

[72] See Tordo et al. (2009), also Cheon et al. (2014); also Stevens (2008) and Wolf (2009).

[73] See Heller et al. (2014), pp. 1–2.

[74] See Victor et al. (2012), pp. 3–31; Thurber et al. (2011); Heller and Marcel (2012).

Mining Act, 2016.[75] In Nigeria, *Nigerian Mining Corporation (NMC) Act* establishes the NMC to oversee all mining activities in Nigeria except for coal and petroleum. Section 4 (1) of the Act sets out the functions and duties of the NMC as follows; (a) exploring and prospecting for, working, mining or otherwise acquiring, processing and disposing of minerals of various kinds occurring in Nigeria other than petroleum or coal; (b) smelting, roasting, assaying, amalgamating, refining, or otherwise treating the minerals aforesaid in any manner calculated to render the products and by-products of any such mineral marketable; (c) purchasing the ores or the products and by-products of any of the minerals aforesaid from any source outside Nigeria in any case where they are not available in Nigeria; (d) marketing the ores, products and by-products of any or all the minerals afore-said; and (e) performing the other functions conferred on it by or pursuant to this Act.[76]

Section 5(1) of the Act grants the NMC, statutory powers to undertake acts aimed to facilitate its duties including, but not limited to: (a) sue and be sued in its corporate name; (b) hold and manage movable and immovable property; (c) construct and maintain such roads, railways, sidings, tramways, bridges, reservoirs, watercourses, buildings, plant and machinery and such other works as may be necessary for, or conducive to, the discharge of the functions conferred on it by or under this Act; (d) purchase or otherwise acquire or take over any asset, business, property, privilege, contract, right, obligation and liability of any company, firm or person in furtherance of the business engaged in by the Corporation; (e) enter into contracts or partnerships with any company, firm or person which in the opinion of the Corporation will facilitate the discharge of its functions under this Act; (f) train managerial and technical staff for the purpose of the running of its operations; (g) undertake such research as may be necessary for the performance of its functions, and for that purpose it may establish such number of research laboratories including mineralogical, chemical, mineral dressing and assay laboratories either independently or in partnership with any company, firm or person, as it may determine in each case (h) exercise such other powers as are necessary or expedient for giving full effect to the provisions of its enabling Act.

Given that the NMC does not oversee coal production, the Nigerian law establishes a separate national company for coal production known as the Nigerian Coal Corporation (NCC).[77] Section 4 of the NCC Act empowers the NCC to work and win coal in all mines of coal in Nigeria; make supplies of coal available in such qualities,

---

[75] See s. 5 of Kenya's National Mining Corporation Regulations, 2016 which authorizes the Kenya Mining Corporation to: (a) engage in mineral prospecting and mining and any other related activities; (b) invest on behalf of the national government; (c) acquire by agreement or hold interests in any undertaking, enterprise or project associated with the exploration, prospecting or mining of minerals; (d) acquire shares or interest in any firm, company or other body of persons, whether corporate or unincorporated, which is engaged in prospecting, mining, refining, grading, producing, cutting, processing, buying, selling or marketing of minerals; (e) carry on its business, operations and activities whether as principal agent, contractor or otherwise, and either alone or in conjunction with any other persons, firms or bodies corporate; and (f) perform any other functions that may be assigned to it by the Cabinet Secretary or under any other law.

[76] Nigerian Mining Corporation (NMC) Act No. 39 of 1972.

[77] Nigerian Coal Corporations Act No. 29 of 1950.

sizes, quantities and prices, to further the public interest; treat, render saleable, supply and sell coal; produce, manufacture, treat, render saleable, supply and sell products of coal, including coke; manufacture, treat, render saleable, supply, distribute and sell gas; bore for, work, win and mine clay, fire-clay, sand and limestone; manufacture, sell and distribute bricks, tiles, cement and glass; plant, maintain or grow plantations of trees and sell the timber and other products therefrom; produce or manufacture any goods which are of a kind required by the Corporation for or in connection with the working and winning of coal or any other of their activities, or which can advantageously be produced or manufactured by the Corporation by reason of their having materials or facilities for the production or manufacture thereof in connection with the working and winning of coal or any other of their activities, and supplying and selling goods so produced or manufactured; advertise the sale of coal or of any goods produced or manufactured by them whether by way of demonstration, exhibition or otherwise; carry out research and experimental work in matters connected with the functions of the Corporation; construct, maintain and improve or subscribe towards the construction, maintenance and improvement of roads; and engage in any other activity, which may be sanctioned by order of the President.

Like the national oil companies, NMCs across Africa increasingly face challenges that limit their functionality and efficiency. The challenges include lack of flexibility and the presence of civil service culture that impedes entrepreneurial vision; financial mismanagement and corruption; high and disproportionate administrative costs; regulatory overlap which sees NMCs operating as both commercial entities and as regulators; and continued annual budget allocation of funds from the national government which further reduces independence. Reforming African NMCs in line with international best practices can help them develop the efficiency, structural independence, expertise, entrepreneurial vision and financial resources needed to effectively carry out their duties under their enabling law.

6.3.3 Secondary Institutions

In addition to the NMC, a number of supervisory institutions may be established to oversee specific mandates and functions in the mining industry. This includes specialist institutions on pricing, local content development, environmental restoration and infrastructure renewal in oil producing communities, and procurement agencies amongst others. For example, a number of African countries, such as Angola, Ghana, Kenya, Nigeria and Tanzania have established committees, boards or agencies to oversee local content, job creation through the use of local expertise, goods and services business, financing in the petroleum industry value chain, and value retention at home.[78]

---

[78] See ss 5–6 of *Ghana's Petroleum (Local Content and Local Participation) Regulations* (2013); also Tanzania's Petroleum (*Local Content*) *Regulations, 2017*. GN. No. 197 which establishes a Local Content Committee to oversee compliance and implementation of the local content regulations.

Furthermore, as earlier noted, a number of African companies have enacted extractive sector transparency initiatives which focus on promoting transparency and accountability in the extractive industry. For example, the Nigerian Extractive Industry Transparency Initiative (NEITI) is established pursuant to the NEITI Act as an autonomous self-accounting body, which reports to the President and the National Assembly.[79] The NEITI is an Agency charged with the responsibility, among other things, for the development of a framework for transparency and accountability in the reporting and disclosure by all extractive industry companies of revenues due to or paid to the Government of Nigeria. The National Stakeholders Working Group (the "NSWG") serves as the governing body of the NEITI. The NSWG oversees the implementation of NEITI, and is responsible for the formulation of policies, programmes and strategies for the effective implementation and discharge of the objectives and functions of NEITI.[80] The NEITI is made up of a secretariat and various departments. The Secretariat, which is under the leadership of the Executive Secretary, is responsible for the day-to-day administration of NEITI. It functions through the Executive Secretary's Office which is supervised by the Director of Legal Affairs. NEITI also has a Communications Department which develops and ensures the implementation of an effective/comprehensive communications strategy, policies and procedures for grass root understanding of NEITI and its functions in Nigeria. NEITI also has a Technical Department which provides qualitative expertise, leadership and technical support to the NEITI's operations. It implements the Secretariat's technical policies and procedures in the extractive sector; develops strategies for data capturing, system strengthening, and revenue management tracking; and a Finance & Administration Department which is responsible for ensuring effective and optimal utilization and allocation of resources to all arms of the NEITI; defining overall operational plan for the management of the day-to-day support related issues; reviewing and updating administrative procedures and accounts manuals; developing and implementing financial information systems, management policies and procedures; as well as coordinating budgetary processes, controls, facilitating and coordinating external audits.[81]

6.4 Conclusion

Although several institutions have been established to supervise and advance the sustainable management and utilization of petroleum and solid minerals resources in Africa, virtually all African countries remain mired in challenges of lack of transparency, corruption, regulatory overlap and lack of institutional coordination, environmental pollution, illegal oil bunkering, illegal mining, unregulated artisanal and

[79] *NEITI Act*, ss 1–2.

[80] *NEITI Act*, ibid, s 5(1)–(3).

[81] See generally, NEITI.

small scale mining, limited technological development, inadequate funding, limited engagement with youths, lack of functional independence, and lack of entrepreneurial vision, that have stifled previous approaches to institutionalization of sustainable resource management in Africa.[82] As a recent law and governance index rightly points out, virtually all African countries remain very weak in terms of effectively supervising and monitoring the transparent management and distribution of extractive resource wealth.[83]

There is an urgent need for African countries to revitalize and reform current institutional arrangements in the extractive sector, to achieve greater transparency and efficiency. Governance arrangements and institutions that are required to effectively and efficiently manage extractive sector will vary for different African countries, and will have to be tailored to suit extant socio-political, cultural and legal systems. While some scholars identify the need for focused and effective supervision as a good reason for splitting solid mineral and petroleum sector governance under diverse ministries and institutions, it appears the countries such as South Africa, Equatorial Guinea and Algeria that have coordinated solid minerals and petroleum sector development under one ministry have achieved stronger performances in the allocation of resource benefits.[84]

Notwithstanding the preferred approach, a critical factor for success is to ensure a clear, consistent and coherent government policy on resource development. A comprehensive national policy on resource management can provide the guiding principles for mainstreaming accountability, transparency and respect for human rights across the entire extractive sector value chain. Adopting a committed, focused and institutionalized governance approach to resource management across the entire value chain will allow for more coordination and regulatory coherence amongst the diverse agencies, departments and ministries operating in the solid minerals and petroleum sectors. It is also important to ensure regulatory independence and autonomy for extractive sector agencies so as to promote their abilities to function transparently and without undue interference from governments.

Advancing transparency and accountability in resource governance will also depend to a great extent on the contractual and fiscal regime put in place by the governments that spells out the terms and conditions of resource development and exploration. Putting in place a robust contractual framework can help African countries get a better deal with respect to advancing local content, environmental safe practice, respect for human rights, transparency and proper reporting of extractive operations. For IOCs operating in Africa, clarifying the expectations of governments, host community stakeholders and other participants can also help reduce legal risks relating to non-compliance.

The next chapter of the book explores contractual and fiscal frameworks adopted in African extractives industries. It examines the key features of, and distinctions

[82] Chuhan-Pole et al. (2011), also Van der Ploeg (2011).

[83] Natural Resource Governance Institute (NRGI) (2017).

[84] Aharoni and Ascher (1998).

between, the different types of fiscal terms and interests granted by African regimes for petroleum, mining and metallurgical operations.

References

Abu Dhabi-Qatar. Agreement on settlement of maritime boundary lines and sovereign rights over islands, signed on 30 March 1969. United Nations Document ST/LEG/SER.B/16

Africa Environmental Agencies. http://web.worldbank.org/archive/website01004/WEB/0__CO-75.HTM. Accessed 21 Sept 2017

Aharoni Y, Ascher W (1998) Restructuring the arrangements between government and state enterprises in the oil and mining sectors. Nat Res Forum 22:201–213

Algeria. Ministry of Energy and Mining. http://www.energy.gov.dz/fr/. Accessed 12 Apr 2018

Algeria. Sonatrach. http://www.sonatrach-dz.com. Accessed 12 Apr 2018

Al-Kasim F (2006) Managing petroleum resources: the 'Norwegian model' in a broad perspective. Oxford Institute for Energy Studies, Oxford

Angola. Grupo Sonangol. http://www.sonangol.co.ao/English/Pages/Home.aspx. Accessed 12 Apr 2018

Angola. Ministry of Geology, Mines and Industry. http://www.mgm.gov.ao. Accessed 2 May 2018

Angola. Ministry of Petroleum. http://www.minpet.gov.ao/. Accessed 12 Apr 2018

Australia-Indonesia (1990) Treaty on the zone of cooperation in an area between the Indonesian province of East Timor and Northern Australia, done over the Zone of cooperation on 11 December 1989. International Legal Materials 29

Australia-Papua New Guinea (1979) Treaty on sovereignty and maritime boundaries in the area between the countries, done at Sydney on 18 December 1978. International Legal Materials 18, 1979

Austria-Czechoslovakia (1964) Agreement concerning the working of common deposits of natural gas and petroleum, signed at Prague on 23 January 1960. United Nations Treaty Series 495

Awogbade S, Bamgbose K, Izekor (2017) Oil and gas regulation in Nigeria: overview. https://uk.practicallaw.thomsonreuters.com/5-523-4794?transitionType=Default&contextData=(sc.Default). Accessed 12 Apr 2018

Bahrain-Saudi Arabia. Agreement concerning the delimitation of the Continental Shelf in the Persian Gulf, done at Riyadh on 22 February 1958. United Nations Document ST/LEG/SER.B/Add. 2

Barberis J, Julio A (1975) Los recursos minerales compartidos entre Estados y el derecho internacional. Derecho de la Integracion VIII(18–19):45–58

Basov V (2015) True giants of mining: World's top 10 iron ore mines. http://www.mining.com/true-giants-of-mining-worlds-top-10-iron-ore-mines/. Accessed 12 June 2018

Brownlie I (1979) Legal status of natural resources in International Law. Some aspects. Recueil des Cours de l'Academie de Droit International 162

Carpenter JW (2015) The World's biggest State-Owned Oil. Investopedia. https://www.investopedia.com/articles/investing/121715/worlds-biggest-state-owned-oil-companies.asp#ixzz5Cj9GpNKz. Accessed 12 June 2018

Cheon A, Lackner M, Urpelainen J (2014) Instruments of political control: national oil companies, oil prices, and petroleum subsidies. J Comp Polit Stud 48:370–402, 1–33

Chuhan-Pole P, Dabalen A, Land BC (2011) Mining in Africa: are local communities better off. World Bank, Washington, DC

Coal Corporations Act No. 29 of 1950 of Nigeria

Convention on the Regulation of Antarctic Mineral Resource Activities (CRAMRA) (1988) International Legal Materials 1988

Dashwood H (2012) Major developments in the global mining industry. In: The rise of corporate social responsibility: mining and the spread of global norms. Cambridge University Press, Cambridge, pp 73–93

Department of Petroleum Resources (DPR) (2010) Our evolution and history. Department of Petroleum Resources of Nigeria. http://www.dprnigeria.com/in_the_news.html

DPR. About DPR. Historical Background. Department of Petroleum Resources of Nigeria. http://www.dprnigeria.com/aboutus.html. Accessed 12 Apr 2018

Eaton JP (1997) The Nigerian tragedy, environmental regulation of transnational corporations, and the human right to a healthy environment. Boston Univ Int Law J 15:261

Eller S, Hadley P, Medlock K (2011) Empirical evidence on the operational efficiency of national oil companies. Empir Econ 40:623–643

Equatorial Guinea. GEPetrol. http://www.equatorialoil.com/gepetrol.html. Accessed 12 Apr 2018

Equatorial Guinea. Ministry of Industry, Mines & Energy. http://mmie.gob.gq/en/. Accessed 12 June 2018

Federal Ministry of Environment of Nigeria. About the MOE-Profile. Federal Ministry of Environment. http://environment.gov.ng/about-moe/profile/. Accessed 12 Apr 2018

Federal Republic of Germany-Netherlands (1982) Supplementary Agreement to the Treaty concerning arrangements for cooperation in the Ems Estaury (Ems-Dollard Treaty) signed at Bennekom on 14 May 1962. United Nations Document ST/LEG/SER.B/15. International Legal Materials 2

France-Spain (1982) Convention sur la delimitation des plateaux continentaux des deux Etats dans le Golfe de Gascogne (Golfe de Biscaye), signee a Paris le 29 Janvier 1974. United Nations Document ST/LEG./SER.B/19. International Legal Materials 2

Gabon. Ministry of Ministry of Mines, Industry, and Tourism. http://www.mines.gouv.ga. Accessed 2 May 2018

Gabon. Ministry of Petroleum and Hydrocarbons. http://www.minesgabon.org. Accessed 12 Apr 2018

Gabon. Societe Nationale des Hydrocarbures du Gabon. http://www.gabonoil.com. Accessed 12 June 2018

Ghana. Ghana National Petroleum Corporation (GNPC). http://www.gnpcghana.com. Accessed 12 Apr 2018

Ghana. Ministry of Energy and Petroleum. http://www.energymin.gov.gh. Accessed 12 Apr 2018

Ghana. Ministry of Lands and Natural Resources. http://www.mlnr.gov.gh/index.php/ministry-of-lands-and-natural-resources/the-ministry. Accessed 12 June 2018

Hartley P, Medlock K (2008) A model of the operation and development of a national oil company. Energy Econ 30:2459–2485

Heller PRP, Marcel V (2012) Institutional design in low-capacity oil hotspots. Revenue Watch Institute, New York

Heller P, Mahdavi P, Schreuder J (2014) Reforming national oil companies: nine recommendations. Natural Resource Governance Institute, New York

Iceland-Norway (1982) Agreement on the Continental Shelf between Iceland and Jan Mayen, done at Oslo on 22 October 1981. International Legal Materials 21

Iran-Sharjah (1972) Memorandum of understanding, signed in November 1971. Middle East Economic Survey XV(28)

Kenya. Ministry of Energy and Petroleum. http://energy.go.ke. Accessed 12 Apr 2018

Kenya. Ministry of Environment and Natural Resources. About Us. http://www.environment.go.ke/?page_id=2. Accessed 21 Sept 2017

Kenya. Ministry of Mining. https://portal.miningcadastre.go.ke/Site/CustomHtml.aspx?PageID=d7f3f61d-4689-4280-a59a-b865f002dd60. Accessed 2 May 2018

Kenya. National Oil Corporation of Kenya. http://nationaloil.co.ke. Accessed 12 June 2018

KPMG (2014) Nigerian mining sector. KPMG, Lagos

KPMG (2017) Nigerian mining sector brief. https://assets.kpmg.com/content/dam/kpmg/ng/pdf/advisory/ng-Nigerian-Mining-Sector.pdf. Accessed 9 June 2018

Kuwait-Saudi Arabia. Agreement relating to partition of the Neutral Zone, signed on 7 July 1965. United Nations Document ST/LEG.SER.B/15
Libya. Ministry *of Industry*, Electricity and *Mining*. http://industry.gov.ly. Accessed 12 June 2018
Libya. National Oil Corporation. http://en.noclibya.com.ly/. Accessed 12 Apr 2018
Losman D (2010) The Rentier state and national oil companies: an economic and political perspective. Middle East J 64:427–445
Macauhub (2017) Angola plans to create a National Petroleum Agency. https://macauhub.com.mo/2017/11/20/ptangola-vai-criar-agencia-nacional-de-petroleos/. Accessed 12 Aug 2018
Marcel V (2006) Oil titans: national oil companies in the Middle East. Brookings Institution Press, Washington, DC
McPherson C (2003) National oil companies: evolution, issues, outlook. In: Davis J (ed) Fiscal policy formulation and implementation in oil-producing countries. International Monetary Fund, Washington, DC, pp 184–203
Mineral and Petroleum Resources Development Act (MPRDA) Act No 28 of 2002 of South Africa
Mining Act 2016 of Kenya
Mobogunje A (1995) The environmental challenges in Sub Saharan Africa. Environment 37(4):4–10
Murombo T (2015) Regulating energy in South Africa: enabling sustainable energy by integrating energy and environmental legislation. J Energy Nat Resour Law 33(4):320–348
Mutum R (2015) ICPC returns N1 billion looted funds to Environment Ministry. AllAfrica. http://allafrica.com/stories/201509160545.html. Accessed 12 June 2018
National Domestic Gas Supply and Pricing Regulations 2008 to the Petroleum Act of 1969, Laws of the Federation of Nigeria 2004
National Mining Corporation Regulations 2016 of Kenya
Natural Resource Governance Institute (2009) Natural resource charter. https://resourcegovernance.org/sites/default/files/NRCJ1193_natural_resource_charter_19.6.14.pdf
Natural Resource Governance Institute (NRGI) (2017) 2017 Resource Governance Index. https://resourcegovernance.org/sites/default/files/documents/2017-resource-governance-index.pdf. Accessed 9 June 2018
NEITI. http://neiti.org.ng/index.php?q=pages/secretariat. Accessed 12 June 2018
Nigeria – Sao Tome & Principe Joint Development Authority (NSTPJDA). http://nstpjda.org/about-us/. Accessed 12 June 2018
Nigeria. Federal Ministry of Petroleum Resources. http://petroleumresources.gov.ng. Accessed 12 June 2018
Nigeria. Mines and Steel Development. http://www.minesandsteel.gov.ng. Accessed 2 May 2018
Nigeria. Nigerian National Petroleum Corporation. http://www.nnpcgroup.com. Accessed 12 June 2018
Nigeria's Federal Ministry of Environment. About Us. http://www.environment.gov.ng/index.php/about-us/about/about-ministry. Accessed 21 Sept 2017
Nigerian Content Development Management Board. Guidelines. http://www.ncdmb.gov.ng/index.php/guidelines. Accessed 12 June 2018
Nigerian Mining Corporation (NMC) Act No. 39 of 1972
Nigeria-Sao Tome and Principe Treaty
NNPC Act
Norway-United Kingdom (1977) Agreement relating to the exploitation of the Frigg Field Reservoir and the transmission of gas therefrom to the United Kingdom, done at London on 10 May 1976. Treaty Series 113
Odumosu I (2007) Transferring Alberta's gas flaring reduction regulatory framework to Nigeria: potentials and limitations. Alberta Law Rev 44:863
Olawuyi D (2015) Principles of Nigerian environmental law. Afe Babalola University Press, Ado-Ekiti
Olorunfemi M (1991) The dynamics of national oil companies. OPEC Rev 16:321–334

Organization of Petroleum Exporting Countries (OPEC). Oil & Energy Ministries/NOCs. http://www.opec.org/opec_web/en/360.htm. Accessed 18 June 2018
Petroleum (Local Content and Local Participation) Regulations (2013) L.I 2204, Ghana
Petroleum (Local Content) Regulations, *2017* of Tanzania
Petroleum Commission Act of 2011 of Ghana
Petroleum Products Pricing Regulatory Agency. http://pppra.gov.ng/. Accessed 12 June 2018
Roberts CWJ (2014) Extractive sector regulations and policy in Africa: old practices and new models for change. North-South Institute, Davie
Saudi Arabia-Sudan. Agreement relating to the Joint Exploitation of the Natural Resources of the sea-bed and sub-soil of the Red Sea in the common Zone, done at Khartoum on 16 May 1974. United Nations Document ST/LEG/SER.B/18, 452
Schofield C (2014) Defining areas for joint development in disputed waters. In: Wu S, Hong N (eds) Recent developments in the South China Sea dispute: the prospect of a joint development regime. Routledge, London, pp 78–98
Shelton D (2011) Problems in environmental protection and human rights: a human right to the environment. GW Law Faculty Publications & Other Works. Paper 1048
South Africa. Department of Energy. http://www.energy.gov.za/files/petroleum_frame.html. Accessed 12 June 2018
South Africa. Department of Mineral Resources. http://www.dmr.gov.za. Accessed 12 June 2018
South Africa. PetroSA. The Petroleum, Oil and Gas Corporation of South Africa SOC Ltd. http://www.petrosa.co.za/Pages/Home.aspx. Accessed 12 Apr 2018
South Africa's Department for Environmental Affairs. Introduction and background. https://www.environment.gov.za/aboutus/department. Accessed 21 Sept 2015
Stevens P (2008) National oil companies and international oil companies in the Middle East: under the shadow of government and the resource nationalism cycle. J World Energy Law Bus 1(1):5–30
The Ghana National Petroleum Corporation Law (1983) P.N.D.C.L. 64. http://en.noclibya.com.ly/. Accessed 12 Apr 2018
Thurber M, Hults D, Heller PRP (2010) The limits of institutional design in oil sector governance: exporting the "Norwegian Model". ISA Annual Convention 2010 New Orleans, LA
Thurber MC, Hults DR, Heller PRP (2011) Exporting the 'Norwegian model': the effect of administrative design on oil sector performance. Energy Policy 29:5366–5378
Tordo S, Tracy BS, Arfaa N (2009) National oil companies and value creation. Washington, DC: World Bank Working Paper No. 218
Treaty between the Federal Republic of Nigeria and the Democratic Republic of São Tomé and Príncipe on the Joint Development of Petroleum and other Resources, in respect of Areas of the Exclusive Economic Zone of the Two States (2001)
Van den Berg HM, Moster H (2018) Challenges to regulating hydraulic fracturing in South Africa: technological innovation and law making for climate change at the cross roads. In: Zillman D, Roggenkamp M, Paddock L, Godden L (eds) Innovation in energy law and technology: dynamic solutions for energy transitions. Oxford University Press, Oxford
Van der Ploeg F (2011) Natural resources: curse or blessing? J Econ Lit 49(2):366–420
Victor DG, Hults DR, Thurber MC (2012) Introduction and overview. In: Oil and governance: state-owned enterprises and the world energy supply. Cambridge University Press, Cambridge, pp 3–31
Wolf C (2009) Does ownership matter? The performance and efficiency of state oil vs. private oil (1987–2006). Energy Policy 37(7):2642–2652
Woodliffe J (1977) International unitization of an offshore gas field. Int Comp Law Q 26:338–353
World Economic Forum. OCP Group. https://www.weforum.org/organizations/ocp-group. Accessed 12 June 2018
Zimbabwe, *Ministry* of Mines and *Mining* Development. http://www.mines.gov.zw. Accessed 2 May 2018

Chapter 7
Legal Rights and Interests in Resource Extraction

Abstract This chapter examines the key features of, and distinctions between, the different types of legal rights and interests granted by African regimes for petroleum, mining and metallurgical operations. After discussing the nature, scope and fundamentals of freehold, leasehold, mineral, surface and royalty interests, the chapter unpacks the characteristics of commonly encountered mining and petroleum interests in African countries. The chapter also examines the common types of petroleum and mineral exploration rights and licenses across Africa; the conditions for the grant of each of those licenses and permits; and the procedures for assignment and revocation of these rights.

7.1 Introduction

In most countries of the world, as we have seen from our discussions on dominial jurisdictions in chapter 4, all extractive resources belong to the State. Corporations or individuals in those countries cannot legally extract and sell any minerals commodity without first obtaining authorization from the government. The right to exploit and explore mineral resources is often granted to third parties via contractual arrangements. Such licenses and contracts stipulate the nature, scope and extent of rights and interests granted to the license holder.

This chapter examines the key features of, and distinctions between, the different types of legal rights and interests granted by African regimes for petroleum, mining and metallurgical operations. To be examined are the nature, scope and fundamentals of freehold, leasehold, mineral, surface and royalty interests. At the end of this chapter, you should be able to: describe the kinds of ownership interests that can be attained in your country; unpack the characteristics of commonly encountered mining and petroleum interests in your country; understand the types of petroleum and mineral exploration rights and licenses that may be granted and the conditions for the grant of each of those licenses and permits; and analyze the procedures for assignment and revocation of these rights.

D. S. Olawuyi, *Extractives Industry Law in Africa*,
https://doi.org/10.1007/978-3-319-97664-8_7

7.2 Types of Legal Interests in Extractive Resources

Legal interests in extractive resources are generally created and conveyed in the same manner as real property interests or interests in land. In this section, we will examine some of the common and frequently encountered legal interests in extractive resources as well as their characteristic features.

7.2.1 Freehold or Fee Interest

In general, a freehold interest, also known as a life estate, is the outright private ownership of land or property for an indefinite or unlimited period.[1] It is an interest in land whose duration is restricted to the life or lives of a particular person or persons holding it. It is also known as an estate in fee or a fee interest, and describes an estate or interests of inheritance, which may be passed from generation to generation. It is an interest in property that is unconditional and represents the broadest ownership interest recognized by law. The owner of a particular parcel of land owns everything under the surface of the land, including minerals, oil and gas, water, the sky, amongst others and has right to drill a well to explore for oil and gas beneath the land, produce oil and gas, and dispose of oil and gas for his or her own benefit. In other words, nothing has been severed from the ownership right, thus all rights remain joined with the property and thus "transfer with" the property. The freehold interest holder has both surface and mineral rights in the land.

Except for parts of Canada and the United States that recognize freehold mineral interest, ownership rights and interests in many parts of the world, and in African countries, are usually segregated from land rights.[2] This means that private owners of land do not own the minerals or resources found in the subsurface, all such natural resources belong to the State.[3] A freehold is the opposite of leasehold, which as we would see below, under which the property reverts to the owner of the land after the lease period, has expired.[4]

In order to be categorized as a freehold, an estate must possess two qualities. First is immobility, which means that the property must be either land, or some interest derived from or affixed to land such as natural and underground resources. Second, ownership of the estate must be for an indeterminate or infinite duration. If

[1] See generally, Lowe (2014), p. 38.

[2] US States where minerals (the mineral estate) are often severed from the surface estate include: Texas, Oklahoma, Pennsylvania, Louisiana, Colorado, New Mexico and others where oil and gas has been produced for decades. Ibid.

[3] Omorogbe and Oniemola (2010), pp. 116–117.

[4] Before the English Law of Property Act 1925, a freehold estate transferable to the owner's heirs and assigns (successors by inheritance or "purchase" [including gift], respectively) was a "fee simple estate." On the other hand, when transfer, by inheritance or otherwise, was limited to lineal descendants ("heirs of the body"/"heirs of the blood") of the first person to whom the estate was given, it was known as a "fee tail estate."

the time of ownership can be fixed and determined, then it is not freehold.[5] A freehold or fee simple estate represents a complete private ownership, which encompasses of both surface and the mineral rights. The holder of a "fee simple", a "fee" or "freehold" interest controls the surface, the subsurface and the air above their property and has the right to sell, lease, gift or bequest these rights individually or entirely to other people. A freehold interest indicates the infinite duration of an interest or estate, as opposed to the ownership rights that it encompasses. Its only similarity to a leasehold interest is that both of them deal with time and duration, one infinite and one of limited duration. In terms of oil and gas and natural resources, an interest in the surface or in the mineral rights may be held in "fee" that is in terms of an immobile resource and for an infinite duration. As Lowe rightly noted, the "fee interest" is used to describe the whole "bundle of sticks" of rights in real property.[6]

7.2.2 Leasehold Interest

A leasehold interest is a temporary right to occupy land or extract resources from it over a defined, finite period of time. A leasehold interest allows a lessee (the petroleum or mining company) to have access to the petroleum and minerals on the property of the lessor (who may be a freehold owner or in most countries, the State) for a specified period of time. An oil and gas lease is the legal document that codifies the terms and conditions under which the lessee may enjoy access to the leased land. Specifically, given that land is owned by the State in dominial regimes, the right holder only acquires the right to use the land or its resources for a given time, on the terms and conditions, stipulated in the lease. The State, as mineral or resource owner (lessor), creates the leasehold interest by signing the oil and gas or mineral lease and transferring exploration rights to the lessee. The lease also stipulates the percentage or share of production proceeds that the lessor receives and how the royalty is received.

In non dominial regimes, a landowner may execute an oil and gas lease under which the lessee is granted the exclusive right to go upon the land to explore for, produce, sever, and remove oil and gas at its sole expense. An oil and gas lease typically has a primary term that consists of a stated number of years and has a secondary term that lasts for so long after the primary term as minerals are produced in paying quantities. In some oil and gas leases, the term may be extended under circumstances other than production in paying quantities, such as certain types of operations, which may include drilling operations or preparations for drilling operations.

In a leasehold interest, the lessor typically retains a possibility of reverter of mineral rights, *i.e.* a right of reversion, as well as a royalty interest in production. A leasehold interest is also known as a working interest or an operating interest because it is the leasehold owner that works or operates the property.[7] Whether a

[5] Lowe (2014), p. 38.

[6] Ibid, 39.

[7] Lowe (2014), p. 43.

leasehold interest will include all the characteristics of a mineral interest depends on the precise wording of the clause granting the interest, as well as the interpretation given to it by the courts.[8]

7.2.3 *Mineral Interest*

A landowner may own the surface land, but not have rights over minerals found in that land. For example, in many dominial regimes, even though individuals are granted rights to own or use the surface land, they do not necessary own or have rights over the minerals or petroleum resources underneath the land.[9] A mineral interest confers on the holder, the right to exploit, mine, and/or produce any or all the minerals lying below the surface of the property. A landowner can create a mineral interest by conveying his land, but reserving or excepting from such conveyance the oil, gas, and other minerals beneath the land. In this case, the surface rights owner can have the full use of the land but not the minerals in it, and will retain surface possession is subject to the rights of the holder of the mineral rights and interests to come in and extract the minerals therein. The severance can either be by reservation in a deed transferring the surface or by a direct grant of the mineral interest. Only the mineral interest holder may execute a lease conveying the petroleum or mineral resources to another party. It is therefore important for an investor to do a thorough check to ascertain that a land or surface right holder actually has clear title to, or ownership rights and interest over, minerals situate on the land. Many countries have land registries that offer mineral title searches through which a party can obtain an ownership abstract or title report that will stipulate the exact holder of the mineral interest.

Whether the mineral interest is severed from the surface by direct grant or reservation, the mineral interest includes an implied easement to use the surface in such ways, and to such an extent, as is reasonably necessary to obtain the minerals under the surface of the property.[10] This means that mineral ownership implies a right to use the surface of the land over which the minerals are found because the mineral ownership will not have value without access to the minerals. To this end, courts in the United States for example, recognize an implied easement burdening the surface of the land and running to the benefit of the minerals on the basis that that it is either the intention of the parties to the severance or that there is public policy in favour of making sure that the property is useful economically. The easement accruing to the mineral owner is only limited by a standard of reasonableness, as well as an obligation to accommodate the uses of the surface owner.

[8] For instance, a lease of "oil, gas and other hydrocarbons", will not include non-hydrocarbons such as sulphur in its ambit. See ibid, 43.

[9] Omorogbe and Oniemola (2010).

[10] An easement is a right either of passage, to cross or use land belonging to another for a specified purpose. It creates a limited property or non-possessory right for a third party to do something on land that does not belong to them. See LeFort et al. (2013).

The five elements of a mineral right are: the right to use as much of the surface as is reasonably necessary to access the minerals; the right to further convey rights; the right to receive bonus consideration; the right to receive delay rentals; and the right to receive royalties.[11] In essence, the mineral interest holder has the right to go upon the land to explore for, produce, sever, and remove oil and gas. The right to profits and the obligation for costs flows from the right to search, develop and produce minerals from the land. It necessarily follows that there is a corresponding tight to profits and the duty to pay costs incurred in use. The mineral interest is a profit sharing and cost-bearing interest.

The right to further convey is the right of the mineral interest holder to transfer the rights they own to search, develop and produce the minerals to another person. The right to lease the mineral interest is often referred to as an executive right. The owner of a mineral interest may separately convey or reserve come but not all of the incidents of mineral ownership. Incidental to the fact that the holder of a mineral interest has the rights to develop or transfer the mineral interest is that the holder has the right to whatever benefits are provided to the lessor under the terms of a lease that transfers the right to develop such as signing bonuses, and shares of production allocated to the lessor such as royalty payments.[12]

7.2.4 Surface Interest

While mineral interest is the legal right over everything below the surface of the property, surface interest is the legal right over everything above-ground. The holder of a surface interest has the right to exploit or use the surface of the land. The surface interest is therefore what remains of the bundle of rights of land ownership after a mineral interest has been severed. Surface interest rights are broader and not just limited to the surface of the land. It includes all the rights that are not included in the mineral interest. The surface interest holder has right to other substances that lie on top of the surface, such as water, gravel, or formations of natural gas etc, which are considered to be part of the land surface.

A severed surface owner's right to use the land is not absolute. Surface interest is subject to the implied easement or right of a mineral interest holder, or an oil and gas lessee, to use as much of the surface when, where and in such ways as are reasonably necessary to search for, develop, and produce minerals. The surface owner's right to the surface is encumbered or burdened by, and subservient to the easement of the mineral interest holder. This means that the mineral interest owner or lessee may not need permission from the surface owner to use the land surface for extractive operations.[13]

[11] Ibid. See also the US case of *Altman v Blake* 712 S.W.2d 117, 118.

[12] See Lowe (2014), pp. 39–42.

[13] Lowe (2014), pp. 43–44.

7.2.5 Royalty Interests

The conveyance of a mineral interest in land does not dispose of all the rights of the land owner.[14] A land owner will most times retain the right to receive a percentage of the profits derived from the petroleum or mineral resources obtained from the land. In such cases, the reasonable costs of the lessee's operations are deducted first and the landowner receives a percentage of the remainder. This is called royalty interest. A royalty is simply a payment to a landowner for the use of the property. In the extractive industry, a royalty interest holder is entitled to share in the resource that may be removed from the land or the proceeds thereof. A royalty interest is therefore a non-possessory real property interest in a share of production, or the value of the proceeds of production, which is free of production and operating expenses, which may be created by grant or by reservation or exception.[15] Oil and gas royalties are typically expressed as fractions or percentages of production, however in the case of minerals, royalties are often expressed as a stipulated sum of money.

Unlike a mineral interest owner, a royalty interest owner is not entitled to go upon the land to explore for and produce oil and gas and has no executive rights (that is, no right to execute oil and gas leases).[16] A royalty is a right to a share of production, not a right to produce. Consequently, a royalty interest owner has no easement for surface use, except to the extent that they need to go onto the property to collect the royalty share of production. Similarly, the royalty interest is not profit sharing or cost bearing: a royalty is a share of production, free from the costs of production. A royalty is paid even if producing is a money-losing venture; and the royalty interest is not responsible for paying any production costs. Furthermore, the royalty interest holder does not have a right to the lease. Royalty interest holders have no right to grant an oil and gas lease because they have no right to search, develop or produce from the land. Similarly, it follows that since royalty interest holders do not have a right to lease, they also have no right to share in lease benefits such as bonuses, delay rentals or royalties. The exception to this is the leasehold benefits, which is given lease benefits because it is created by a lease.[17]

There are different types of royalty interests that have been created and recognized over time. They include a landowner or lessor's royalty interest; an overriding royalty, a non-participating royalty; a term royalty; a perpetual royalty; and a mineral royalty. These are briefly examined below (Table 7.1).

---

[14] See *Hamilton v. Morris* Res., Ltd., 225 S.W.3d 336, 344 (Tex. App.-San Antonio 2007, pet. denied) (first citing Luckel v. White, 819 S.W.2d *459,* 463 (Tex. 1991); then citing Altman v. Blake, 712 S.W.2d 117, 118 (Tex. 1986)). "A conveyance of a mineral estate need not dispose of all interests; individual interests can be held back, or reserved, in the grantor." French v. Chevron U.S.A. Inc., 896 S.W.2d *795,* 797 (Tex. 1995).

[15] See Smith and Weaver (2015).

[16] See Ingram and Conner (2018), Mineral v Royalty Distinction.

[17] Ibid, 46–47.

Table 7.1 Types of royalty interests

1. **Landowner or Lessor's Royalty Interest:** Landowner's royalty interest is the land owner's share or compensation for resource production under the lease. It is an interest in production that is retained when a mineral owner (the "landowner") executes a mineral lease.[a] This interest is effective during the duration of the lease and is determined through a negotiation between the mineral owner and the lessee, typically an oil and gas company or a broker. It is generally a fixed fraction of the gross production, perhaps varying depending on the volume of production.
2. **Overriding Royalty:** An overriding royalty interest is an interest that the lessee carves out of the leasehold interest, most often to compensate workers such as landsmen, lawyers, geologists and other members of the team who helped to structure the drilling venture. It has been rightly described as "[a]n interest in oil and gas produced at the surface, free of the expense of production. In modern times overriding royalty interests usually refers to a non-cost bearing interest carved out of the lessee's working interest under an oil and gas lease."[b]
Overriding royalty interests do not constitute an ownership of minerals under the ground. Instead, overriding royalties constitute ownership of a portion of the revenues generated from oil and gas production, free of the costs of production, and the ownership ends when the oil and gas lease from which it is carved is terminated. Simply put, overriding royalty interest 'overrides' a portion of the working interest. A lessee under an oil and gas lease may create an overriding royalty interest by assigning its interest in an oil and gas lease, but reserving from such assignment a royalty interest.
3. **Non-participating Royalty (NPRI):** This is a royalty paid to an owner who does not share in the right to explore or develop a lease, or receive bonus or rental payments. It is free of the cost of production, and is deducted from the royalty interest. Mineral interest holders who sell their rights often retain a non-participating royalty. The term "non-participating" denotes that the interest owner does not share in the bonus, rentals from a lease, nor the right (or obligation) to make decisions regarding execution or operation of the lease. It is an expense-free real-property mineral interest that does not participate (hence the name) in collecting bonus or delay rentals, leasing, or exploring and developing. As the Texas court explained "It merely entitles its owner to a share of the production proceeds, free of the expenses of exploration and production."[c]
NPRI is useful when a right owner is interested in a steady stream of royalty income, but does not intend to participate in decision making concerning the extraction operations.[d]
4. **Term Royalty:** This is a royalty for a stated duration or term. Such term may be fixed, for example, a 25 year term, and so long thereafter is there is continued production from the property.
5. **Perpetual Royalty:** This is a royalty that may extend forever. It is not limited in time.[e]

[a]See Kulander (2016), pp. 43–44

[b]Kulander (2016), also see Shade and Blackwell (2014)

[c]*Plainsman Trading Co. v. Crews,* 898 S.W.2d 786, 789 (Tex. 1995); *Hamilton v. Morris Res., Ltd.,* 225 S.W.3d 336, 344 (Tex. App.-San Antonio 2007, pet. denied)

[d]See *Laborde Properties, L.P. v. U.S. Shale Energy II, LLC*; No. 04-16-00168-CV, ___ S.W.3d ___ (Tex. App. – San Antonio Oct. 12, 2016). Also Christopher S. Kulander, (n. 18), noting that the NPRI can be reserved or conveyed as a fixed fraction of gross production, commonly 1/16, or it can be dependent upon the lessor's royalty of the existing lease and every lease covering the captioned land thereafter

[e]See generally, Lowe (2014), pp. 44–46. Nature of Perpetual Oil (1935), pp. 363–367

7.3 Conducting Petroleum Operations: The Licensing Regime

As noted in previous chapters, in all African countries, the entire ownership and control of all petroleum resources in lands, territorial waters or the continental shelf is vested absolutely in the government.[18] Consequently, to conduct petroleum operations in African countries, the legal regime typically provides that participants must first obtain a license or permit to explore, exploit and produce petroleum (including natural gas) within the concession area.[19] Across the African continent, the prevailing regulatory model is to have a responsible department or ministry for petroleum resources which develops, oversees and implements the oil and gas development policy; while the national oil company oversees, manages, authorises and grants all licenses, which are required to participate in the exploitation of petroleum deposits.[20] The responsible Ministry also provides regulations for safe working of petroleum operations; prevention of pollution of water courses and the conservation of petroleum resources, among others. In Nigeria, as it is with several African countries, only a company incorporated locally may be granted a licence to conduct petroleum operations.[21] An application must contain prescribed information and be accompanied by the applicable fee.[22]

Access to petroleum resources may be obtained by submitting an application to the relevant ministry for petroleum resources for an oil exploration licence, oil prospecting license, oil mining lease reconnaissance permit, technical co-operation permit, or other application exploration or production permits.[23] Alternatively, the State

---

[18] See the *Petroleum Act*, 1969, s 1. See also Omorogbe (2001), pp. 34–37.

[19] See *Attorney General of the Federation (AGF) vs. Attorney General of Abia State* (No. 2) (2002) 6 N.W.L.R3. Part 764, page 542 stating that only the Nigerian Federal Government alone and not the littoral states can lawfully exercise legislative, exclusive and judicial powers over the maritime belt or territorial waters as well as sovereign rights over the resources therein. See also *Famfa Oil Ltd. vs. A.G.F. & NNPC* (unreported suit No. C. CA/A/173/06) affirming that participation or acquisition of oil mining licenses in Nigeria is subject to the approval and authority of the Federal Government of Nigeria.

[20] See for example the Ministry of Petroleum in Angola and the role of Sociedade Nacional de Combustíveis de Angola (Sonangol) EP which is the holder of concession rights and grants exploration and production licenses. *Angola's Petroleum Law No. 10/04* of *12* November 2004, and Article 3 of the Constitution of Angola, 2010. See also the structure in Nigeria with the Federal Ministry of Petroleum Resources serving as the supervising ministry, while the Nigerian National Petroleum Corporation grants licenses and contracts on behalf of the State. See the Nigerian See *Petroleum Act,* 1969, s 2(1)-(3).

[21] Omorogbe (2001).

[22] See *Petroleum Act,* 1969, s 2(1)-(3).

[23] For example, petroleum rights in Nigeria are acquired via an Oil Prospecting License or Oil Mining Lease as specified by *the Petroleum Act* and explored via contractual arrangements such as Joint Venture Contracts, Production Sharing Contracts and Service Contracts. Production from offshore fields, are mostly governed by Petroleum Sharing Contracts and are regulated by *the Deep Offshore and Inland Basin Production Sharing Contracts Act.*

may announce a licensing round during which exploration and production rights in respect of a block, or blocks, may be acquired following the submission of bids. The Minister may grant a petroleum-prospecting license, to prospect for either crude oil or natural gas; and a petroleum-mining lease, to explore and dispose of either crude oil or natural gas. The petroleum-prospecting license or petroleum-mining lease will clearly state if it is in respect of either crude oil or natural gas, it cannot be for both as under the existing regime. A company may however apply for an additional license in respect of either crude oil or gas.

In summary, any natural or legal person wishing to conduct petroleum operations in Africa must first obtain all applicable licenses from the government (represented by the supervising ministry or national oil company). There are three types of extraction licenses:

7.3.1 The Oil Exploration License ("OEL") or Reconnaissance License

An OEL is a concession which grants the holder an exclusive right to search for oil and gas within the area of the grant. An OEL applies only in respect of a specified area and authorizes the licensee holder to undertake exploration for petroleum in the license area.[24] The intent of an OEL is to encourage exploration in areas with unknown hydrocarbon potential, limited infrastructure and consequently higher investment risk to the operator. OEL encourages oil and gas exploration outside of the known oil and gas reserves to discover resources in virtually unexplored frontier basins. Exploration licensing allows companies to explore these frontier basins with minimal costs to the government. It allows the holder to conduct aerial, terrestrial, and sea surveys of, or any search in any area for purposes of verifying the existence of natural resources therein. This includes all economic, technical, geological, geochemical or geophysical studies carried out to determine the nature and economic value of oil and gas resources in a given area.

An OEL is typically granted to cover an area selected for exploration licensing, typically between 10,000 and 500,000 acres, depending on the exploration ambitions of the country. An OEL cannot be granted in respect of land, which is the subject matter of the grant of an existing mining license or permit. An OEL is typically granted for a short period of time, for example 1 year in Nigeria, renewable for a further 1 year provided the licensee fulfils prescribed conditions.[25] These conditions include the fulfilment of the obligations imposed on them under the terms of the license. In addition, the Minister of petroleum must be satisfied with the work done and the reports submitted by the licensee in pursuance of the license, and the

---

[24] In Nigeria, the OEL confers exclusive rights of surface and subsurface exploration for the production of petroleum in an area not more than 2590 km^2 (1000 square miles) in size.

[25] See Olawuyi (2018), pp. 1–2.

licence holder must have made an application for renewal at least 3 months prior to the date of expiry of the license. An OEL does not confer any right to the grant of a prospecting or mining license on the OEL holder.[26]

The licensing process is typically initiated in one of two ways: companies may submit to the responsible ministry, a proposal to conduct exploratory activity within a specified area, which may then be approved by the Minister.[27] Secondly, the relevant Ministry of Petroleum can, at any time, issue a notice requesting proposals to explore an area designated by the commissioner. Once a request for proposals has been issued, applicants will have a window within which they must notify the Ministry of their intent to submit a proposal, and then a further window to submit the proposal. Submitted proposals must: describe the area proposed to be subject to the license; state the specific minimum work commitment expressed in dollars; describe the amount and form of security to be posted based on the projected cost of the planned exploration work; propose the term of the license (unless already established by law or regulation); and verify that a prospective licensee meets minimum qualifications.[28]

When initial exploration results show that there could be commercial quantities of oil and gas resources in the licensed area, the holder may apply for an oil prospecting license (OPL). Given that an OEL does not grant its holder the right to drill for, win or take away oil and gas from the license area, an OPL is the necessary next step for a company in order to properly evaluate the extent and economic value of deposits believed to be within the license area. An OEL holder does not automatically have the right to obtain a production or mining license for all (or some) of the lands covered by such license. However, based on practice experience with governmental approval processes in frontier African countries, a good record and relationship with the Ministry or national oil company during the term of the OEL is likely to lead the government to be amenable to entering into direct negotiations for the granting of an OPL.

7.3.2 The Oil Prospecting License ("OPL")

An OPL is required to search for and verify the quantity and quality of petroleum resources in an area. It grants the holder the exclusive right to search for oil and gas in a license area by conducting geological, geophysical and geochemical investigations to determine the extent and economic value of the deposit within the license area. This includes necessary drilling, analysis, and detailed studies required to establish the existence of hydrocarbons in commercially exploitable quantities.

---

[26] See generally, *Petroleum Act, supra* n 13, Schedule I, paras 1–4. For more detailed provisions on the rights and powers of the holder of an OEL, as well as their obligations, see generally, the *Petroleum (Drilling and Production) Regulations of 1969 to the Petroleum Act, supra* note 13, ss 10–13 [*Petroleum Drilling and Production Regulations*].

[27] Olawuyi (2018), pp. 1–3.

[28] See for example Mercier and Olawuyi (2014).

After the payment of a significant signature bonus within a prescribed time period, the holder may carry away and dispose of petroleum won during prospecting operations, subject to the fulfilment of obligations imposed on them by the applicable legislation.[29] An OPL may only be transferred or assigned with the permission of the granting Ministry.

The maximum tenure of an OPL, which is determined by the Ministry, including renewal periods, is often limited to a short duration. For example, in Nigeria an OPL is valid for 5 years when granted over land and territorial waters and 7 years when granted in respect of continental shelf and Exclusive Economic Zone areas.[30] The holder of an OPL, on successfully prospecting and discovering oil in commercial quantities, may apply for an oil mining lease in order to enjoy a longer duration and other more comprehensive contractual rights of an OML.[31]

7.3.3 The Oil Mining Lease ("OML")

The OML is the ultimate concession for petroleum exploration and production, giving the holder the right to explore for and dispose of any petroleum discovered within the area covered by the OML. An OML may only be granted only to the holder of an OPL who has satisfied all the conditions imposed on the licence or otherwise imposed on them by law and has discovered oil in commercial quantities.[32] An OML is required in order to carry out all necessary activities aimed at extracting hydrocarbons, including drilling, constructing of tunnels, conducting geological and geophysical works, drilling development wells, installation of collection pipes, assembly of equipment and construction of buildings for exploration, and all other operations necessary for the maintenance of pressure, and for primary or secondary recovery of oil from the ground.[33]

The holder of an OML has the exclusive right to conduct exploration and prospecting operations and to win, get, work, store, carry away, transport, export or otherwise treat petroleum discovered in or under the leased area. The term of an OML often extends up to 20 years and may be renewed with the approval of the

---

[29] See the *Petroleum Profits Tax Act*, Laws of the Federation of Nigeria, 2004, c P13. [PPTA].

[30] See generally, *Petroleum Act,* Schedule I, paras 5–7. For more detailed provisions on the rights and powers of the holder of an OPL, as well as their obligations and restrictions on their activities, see generally, the *Petroleum Drilling and Production Regulations*, ss14, 15 and 16–25.

[31] The conditions and legal requirement for converting an OPL to OML in Nigeria is well detailed in the *Oil Prospecting Licenses (Conversion to Oil Mining Leases etc.) Regulations 2004,* which clarifies the terms and conditions under which an OPL issued under the Petroleum Act may be converted to an OML. See also *South Atlantic Petroleum Company vs. Minister of Petroleum Resources* (unreported suit No. FHC/L/CS/361/2006).

[32] Oil is deemed to have been discovered in commercial quantities by the holder of a prospecting licence if the Minister, is satisfied that the licensee is capable of producing at least 10,000 barrels of crude oil per day from the license area. See *Petroleum Act, ibid,* Schedule I, para 9.

[33] Olawuyi (2018), pp. 1–3.

Minister in accordance with the applicable petroleum law if it is economically reasonable, and subject to the payment of applicable royalties and fulfilment of the lessee's obligations under the OML.[34] In many African countries, the OML is often awarded subject to receiving final executive or parliamentary approval, or alternatively by a decree of the State, further to a proposal by the approving national oil company.

7.4 The Licensing Regime for Solid Minerals

As with oil and gas, ownership of solid minerals in African countries is vested in the government on behalf of its people. Consequently, organizations or individuals cannot legally extract and sell any mineral commodity without first obtaining an authorization from the government.[35] Across the African continent, the dominant governance model is to have a responsible department or ministry for solid minerals or mining which develops and oversees the solid mineral development policy; while the national mining company oversees, manages, and authorises and grants all licenses, which are used to participate in the exploitation of the mineral resources.[36] The right to search for, or exploit mineral resources, is obtained through the grant of permits, licenses and leases namely a technical cooperation permit, reconnaissance permit, an exploration license, a Small Scale Mining Lease, a Mining Lease, a Quarry Lease and a Water Use Permit.[37] This section examines each of these licences and permits.

7.4.1 Technical Cooperation Permit

A technical cooperation permit (TCP) allows the holder to conduct desktop study, and acquire seismic data from other sources including government agencies, in order to ascertain the extent of mineral resources in a given area.[38] A TCP may cover

[34] See generally, *Petroleum Act,* Schedule I, paras 8–13. For more detailed provisions on the rights and powers of the holder of an OML, as well as their obligations and restrictions on their activities, see generally, the *Petroleum Drilling and Production Regulations*, ss14, 15, 16–25.

[35] Part III of the Kenya Mining Act specifies that a person shall not search for, prospect or mine any mineral, mineral deposit or tailings in Kenya unless that person has been granted a permit or license in accordance with the Act. See also s 2 of the Nigerian MMA 2007 which vests control of all properties and minerals in Nigeria in the State and prohibits unauthorised exploration or exploitation of minerals.

[36] See S. 1 of Kenya's Mining Act, 2016 and The Model Production Sharing Contract, 2015. See also *MMA 2007,* ibid, s1 (1).

[37] MMA 2007, ibid, s 46.

[38] See section 77(1) of South Africa's *Mineral and Petroleum Resources Development Act (MPRDA)*.

an area as large as 7.5 million acres.[39] A TCP does not include the right to conduct any prospecting or exploration activities.[40] The holder of a TCP may carry out a technical co-operation study in accordance with a technical co-operation work programme to determine the availability of mineral resources in the area covered by the permit. Furthermore, the holder of a TCP has an exclusive right to apply for, and be granted, an exploration right over the area described in the permit subject to certain terms and conditions. TCPs are often valid for a short period. For example, in South Africa where TCPs are more dominant, they are valid for 1 year and are not renewable or transferable.[41]

7.4.2 Reconnaissance or Prospecting Permit

A reconnaissance permit confers on the holder, the right to search for a specific mineral (or commodity) within the license area, on a non-exclusive basis, by conducting geochemical and photo-geological surveys or other remote sensing techniques.[42] Reconnaissance or prospecting, the first stage of the geological analysis, involves the search for minerals, precious metals or mineral specimens, through digging by hand, physical observation, rock sampling, geological surface analysis, geophysical surveys, geochemical surveys, photo-geological surveys, remote sensing techniques, laboratory testing and assays, or by other non-obstructive surveys or studies of surface geology, to define the extent of a mineral deposit and to determine its economic value.[43] The holder of a reconnaissance permit may carry out operations for, or in connection with, the exploration for minerals or petroleum by geological, geophysical and photo-geological surveys, as well as any remote sensing techniques.[44] It includes intentionally searching for mineral by means of any method—(a) which disturbs the surface or subsurface of the earth, including any portion of the earth that is under the sea or under other water; or (b) in or on any resi-

[39] For example, in South Africa, Falcon Oil and Gas Limited holds a technical cooperation permit covering an area of approximately 7.5 million acres (approximately 30,327 km^2) onshore Karoo Basin, South Africa.

[40] See section 77(1) of South Africa's MPRDA.

[41] See section 78(1) of South Africa's MPRDA.

[42] In Nigeria, in order to qualify for the award of a Reconnaissance Permit, an applicant must be a Nigerian citizen of capacity who has not been convicted of a criminal offence, or incorporated company under the *Companies and Allied Matters Act* ("CAMA") or a mining cooperative. See the Companies and Allied Matters Act of 2018, Laws of the Federation of Nigeria 2004, c 20 [CAMA]. See also MMA 2007, s 47; Kenya's Mining Act, 2016 and The Model Production Sharing Contract, 2015, and Extractives Baraza, Types of Mineral and Hydrocarbon Rights (2018).

[43] See s 1 of Kenya's Mining Act, 2016. See also S. 164, MMA (2007). Prospecting must be differentiated from fossiking. Fossicking in many countries is defined to mean searching for minerals for a purpose other than for commercial gain often to a depth of 2 m by digging by hand; or using hand held instruments.

[44] See s. 1 of Kenya's Mining Act 2016.

due stockpile or residue deposit, in order to establish the existence of any mineral and to determine the extent and economic value thereof; or (c) in the sea or other water on land.[45] The holder of a prospecting permit may search for minerals (or commodities) by conducting geological, geophysical and geochemical investigations to determine the extent and economic value of any deposit within the license area. Drilling, excavation or other sub-surface techniques are permitted under a prospecting permit or license.

It does not however include the right to engage in drilling, excavation or other sub-surface techniques. A reconnaissance permit is often valid for a short period, (for example 1 year in South Africa) and is not renewable or transferable, nor does it grant the holder any exclusive rights.[46] The permit is granted subject to the fulfillment of certain conditions, which include submitting periodical report, conducting activities in an environmentally and socially acceptable manner compensating land users for any damage to their property, and payment of applicable fees.

A reconnaissance permit allows the holder to obtain access into, enter and fly over any land for mining purposes and to search for minerals. The land area over which the permit is granted is often not limited. The reconnaissance permit also allows the holder to obtain and remove surface samples in small quantities. However, the holder is not allowed to conduct other subsurface activities. It has a 1 year duration and can be renewed annually.

7.4.3 Exploration Licence

The exploration licence (EL) allows the holder to, on an exclusive basis, discover, and determine the characteristics and economic value of mineral resources in the designated mining area.[47] Exploration means operations and works carried out to determine the existence, quality, quantity and economic value of minerals for the purpose of commercial exploitation. It includes conducting geological, geophysical, geobotanical and geochemical surveys or drilling; taking samples for the purpose of chemical or other analysis; using appropriate instruments, equipment and techniques; or extracting; or re-processing of existing seismic data, acquiring and processing new seismic data or any other related activity necessary to define the quality, characteristics and economic value of a mineral discovery.[48]

---

[45] S. 1 of South Africa's MPRDA.

[46] MMA 2007, ibid, ss 56–58.

[47] In Nigeria, to qualify for an Exploration Licence, an applicant must be an incorporated company under the CAMA, or a mining cooperative, who hold a Reconnaissance permit, and have fulfilled all the conditions attached to the permit.

[48] See s 1of South Africa's *Mineral and Petroleum Resources Development Act* (*MPRDA*), also s. 164, MMA (2007).

Further to this, the holder of the licence has the exclusive right to erect and maintain plant and machinery on the leased area, carry out all the operations necessary for the exclusive exploration of mineral resources, and to take, remove and export specimen from the land, conduct bulk sampling and trail processing of mineral resources within prescribed limits, sell samples and specimens obtained from exploration or sampling activities, and take timber and use water for watercourses for domestic purposes.[49]

Conversely, the EL also imposes certain obligations on the holders. The holder of an EL is required to conduct their exploration activities in a safe, friendly, skilful, efficient and workmanlike manner. They are also to conduct exploration activities in an environmentally and socially responsible manner, give requisite notice to the occupiers of the land, remediate and restore the land to a safe state after its exploration activities, not divert watercourses or explore in forest reserves, except where permitted, compensate the users and occupiers of the land for any damage resulting from exploration activities, submit periodical reports and pay all applicable fees and charges.[50]

An EL cannot be granted in respect of any land that is subject to an existing exploration licence, mining lease, small-scale mining lease or quarry lease; and the mining area is not to exceed 200 square kilometres (km^2).[51] An exploration right is valid for the period specified in the right, which in most cases is an initial period of 3 years that can be can be renewed for 2 further periods of 2 years, subject to compliance with stipulated work obligations and the satisfaction of all other requirements imposed by the license. The total lifespan of an EL is often limited so as not to exceed 7 years. Subject to certain terms and conditions, the holder of an exploration rights has the exclusive right to apply for, and be granted, a production right in respect of the petroleum and the exploration area in question.[52]

The holder of an EL typically has the exclusive right to apply for one or more Mining Leases, Small Scale Mining Leases, or Quarry Leases in respect of any part (s) of the exploration area, upon due compliance with its exploration obligations. Given that an EL does not include a right to sell or carry away minerals found in the land, the holder of an exploration license who sells any mineral resources discovered will be subject to the payment of royalty as if the mineral resources sold were obtained under a mining lease.[53]

---

[49] MMA 2007, ibid, s 60.

[50] MMA 2007, ibid, s 61.

[51] MMA 2007, ibid, s 59.

[52] MMA 2007, ibid, s 62–63. See also KPMG Report, 8.

[53] See MMA 2007, ibid, s 62–63.

7.4.4 Mining Lease

A mining lease is an exclusive permit granted in respect of an area, which gives the holder the right to mine or extract specified minerals (or commodities) within the lease area. The lease confers upon the holder the right to exclusively use, occupy and carry out mineral exploitation and exploration in the area covered by the lease. The leaseholder also has the right to market, sell, export or dispose of the mineral products obtained from the mining activity. They also have the right to use the wood and water found in the mining area.[54]

The mining lease imposes obligations on a leaseholder which include commencing exploration activities within the stipulated time frame, carrying out mining operations in a skilful manner, maintaining their operations and the lease area in a safe manner in compliance with applicable health and safety regulations, compliance with social obligations, and all the requirements from an environmental impact assessment, compensate owners or lawful occupiers for the revocation of their rights to use the land, submit information and periodical reports as required by the Act and pay all applicable fees, rents, charges and royalty. The leaseholder is not to commence operations until they have met all the stipulated preconditions set out in the Act[55] Mining leases are granted for a longer period, for example 25 years in Nigeria (renewable for another 20 years), and 30 years in South Africa (renewable for another 30 years). Renewal is only possible if the holder has complied with their minimum work obligations and met all the requirements imposed on them.[56]

7.4.5 Small Scale Mining Lease

Artisanal and Small-scale Mining (ASM) broadly refers to subsistence mining. ASM operations involve low level of technology, low capital expenditure and high labor-intensive methods of mineral processing and extraction.[57] ASM is often practiced by individuals, groups or communities typically working with limited technology or capital.[58] ASM, or subsistence mining, occurs in approximately 80 countries worldwide, especially in developing countries in Africa, Asia, Oceania, and Central and South America.[59]

---

[54] MMA 2007, ibid, s 68.

[55] MMA 2007, ibid, s 70–71.

[56] MMA 2007, ibid, s 66. See generally, KPMG Report, 8.

[57] Mining Facts (2012), also Extractive Hub (2017).

[58] See Ledwaba and Nhlengetwa (2016), pp. 26–30.

[59] According to the World Bank, there are approximately 100 million artisanal miners globally, accounting for 80% of global sapphire, 20% of gold mining and up to 20% of diamond mining. See World Bank Group (2013).

While ASM is recognized and regulated in countries such as Nigeria, Ghana, and Kenya, ASM is unrecognized, and even considered illegal, in several other African countries.[60] For example, ASM in South Africa takes place outside of the formal structures created to regulate the mining sector, leading to confusion as to its legality in South Africa.[61] ASM in South Africa has been categorized as a "mining activity employing less than 50 people, and has annual turnover of less than R10 million with fixed and moveable assets of less than R15 million."[62]

In countries where ASM is recognized in formal legal structures, a small scale mining lease or permit is often granted to artisanal or small scale miners to extract specified minerals within an area.[63] The holder of a Small Scale Mining Lease is granted the exclusive right to carry out small-scale mining operations within the title area. The mining area, *i.e.* the area for which a mining right or permit is granted; is often designated in the permit.[64] The leaseholder is under obligation to carry out rehabilitation and reclamation activities in the lease area on the completion of their mining activities.[65]

7.4.6 Quarry Lease

The quarry lease grants its holder an exclusive right to operate a pit or a quarry, *i.e.*—an excavation site and place (not being a mine) in which persons work at the removal of rock, earth, clay, sand, soil, gravel, limestone, or such other common mineral substances by means of explosives or any rock-crushing or treatment plant.[66] A quarry lease is granted for the purpose of extracting quarriable minerals from a quarry. Quarriable minerals include asbestos, china clay, fullers earth, gyp-

---

[60] See Small-Scale Gold Mining Law, 1989 (PNDCL 218) and the Precious Minerals Marketing Corporation Law, 1989 (PNDCL 219), which set up the Precious Minerals Marketing Corporation (PMMC) to promote the development of small-scale gold and diamond mining in Ghana and to purchase the output of such mining, either directly or through licensed buyers. See also s 91 of Nigeria's MMA 2007, ibid, s 49; and Second Schedule to Kenya's Mining Act, 2016.

[61] Ledwaba (2017); also Chamber of Mines (2016).

[62] See Minerals and Mining Policy of Republic of South Africa, 1998; also Scott et al. (1998), pp. 2–3; Mutemeri and Petersen (2002), pp. 287–288.

[63] MMA 2007, ibid, s 49.

[64] In Kenya, the proposed prospecting area for ASM may not exceed 25 contiguous blocks; or (b) in the case of mining operations, the proposed mining area may not exceed two contiguous blocks. An activity may still be designated as ASM if (a) in the case of mining operations, the actual or estimated annual extraction of minerals or material bearing minerals does not exceed 25,000 m^3; or, (b) the prospecting or mining operations do not employ specialised prospecting, mechanized mining technologies ,chemicals including mercury and cyanide or explosives; or, (c) the proposed prospecting or mining operations, do not involve an investment or expenditure which exceed such amount as may be prescribed by the Cabinet Secretary. See Second Schedule to Kenya's Mining Act, 2016.

[65] *MMA 2007,* ibid, s 90. See also KPMG Report, 8.

[66] See definition of quarry of in Section 2 of Fiji's Quarries Act, Cap 147.

sum, marble, limestone, mica, pipe clay, slate, sand stone, and gravel amongst others.

A quarry lease gives the holder the right to carry out quarrying operations and also dispose of the quarriable minerals. The leaseholder also has the right to make all excavations, erect, construct and maintain, housing, machinery, engines, workshops, electrical transmission lines, communication equipment and other structures that are necessary for their operations.[67] Like other mining licenses and leases, a quarry lease is only awarded to individuals who have technical capacity and have an established record of conducting mining or quarrying operations. It may also be granted to an exploration license holder, mining cooperatives, or to construction companies that need to extract minerals for the construction of roads, railway lines, dams and other engineering works or structures of public interest.[68]

To obtain a quarry lease, an applicant must demonstrate the need for exclusive use. The applicant must also provide a robust description of proposed approaches and methods for the protection, reclamation and rehabilitation of the site, including, where applicable, details pertaining to environmental protection, backfilling, contouring, benching, sloping, grading, fencing, screening, construction of berms and revegetation. On approval, the lease applicant is required to submit a legal survey of the proposed site. The applicant is also required to pay an annual land rental fee and reclamation security. The term of quarry leases can vary up from country to country. A quarry lease may also be renewed for a term equal to the original term or such lesser term as the Minister may decide. The Minister may cancel or suspend a quarry lease if the holder of the lease does not comply with the terms and conditions of the lease.

7.4.7 *Water Use Permit*

In several African countries, a water use permit (WUP) is required, from relevant water ministries or departments, for any activity or operations requiring the use of water beyond average daily rate.[69] For example, withdrawing of water for agriculture, industrial, mining, power generation, recreation, urban and water supply activities.[70] A WUP typically allows water to be withdrawn from groundwater or surface

[67] MMA 2007, ibid, s 76–78. See also KPMG Report, 8.

[68] MMA 2007, ibid, s 51.

[69] In many parts of the United States, a water use permit is required before persons may withdraw water in quantities that average 100,000 gallons per day (equivalent to 70 gallons per minute for 24 h straight) or 1 million gallons per year from groundwater or surface water. See for example, Department of Natural Resources Minnesota.

[70] See for example Section 151(1) of South Africa's National Water Act (Act 36 of 1998) which provides that no person may use water otherwise than as permitted under this Act, fail or refuse to give data or information, or give false or misleading data or information when required to give information under the Act, fail to register an existing lawful water use when required by a responsible authority to do so and fail to register a dam with a safety risk. The Act gives the Department

water for any reasonable-beneficial public uses—such as public supply (drinking water), mining, agricultural and landscape irrigation, commercial use, and power generation. Due to the water intensive nature of extractive operations, anyone proposing mining or petroleum exploration operations must have a WUP prior to beginning work and may be required to mitigate the impacts of such intensive water use. The aim of a WUP is to ensure that the amount of water requested for mining operations is reasonable, and that water use does not have unreasonable impacts on the available water resource for neighboring users.

A WUP authorizes the holder to consume or use particular quantities of ground or surface water for mining operations. It gives the holder the exclusive right to obtain and convey water and/or occupy the land for the purposes of conveying water.[71] A WUP is usually granted to holders of an exploration license, a mining lease, a quarry lease and a small-scale mining lease. The WUP is usually for the same duration as the applicable mining permit or license pursuant to which it was granted.

A WUP applicant is typically required to submit a water conservation plan, providing measures to reduce water use and preserve water resources for other beneficial uses. Furthermore, the applicant must demonstrate use of the lowest acceptable quality source of water.

Generally, several of the foregoing permits and licenses are usually awarded following a competitive bidding process. Corporations wishing to obtain concessionary rights in an African jurisdiction can obtain such permits by direct proposals or applications to relevant ministries or departments, or by participating in public invitations for bid or licensing rounds that are often conducted in accordance with the direct provisions of the applicable underlying legislative framework applicable to extractive industry operations in the country. Public (open) tenders are typically published regularly in local newspapers. In case of classified tenders, an invitation to tender is issued to pre-selected contractors and vendors. The eligibility and cooperate structure vary from one jurisdiction to another and often change. Some countries, such as Liberia and Nigeria, often stipulate that to be eligible to bid for a petroleum or mining licence, the interested company must be registered and operating as a local company. In addition, some specify that at least 51% of the beneficial interest of the company must be owned by one or more citizens; that the company must have upstream extraction experience; and that the company's memorandum and articles of incorporation must authorize the company to conduct oil and gas exploration and production activities.[72] These legal requirements and steps must be carefully ascertained to avoid legal risks.

---

of Water and Sanitation the power to grant water use permits and to mandate the registration of water use. Registration of water use is compulsory to all mining or exploration operations in South Africa.

[71] MMA 2007, ibid, s 52. See also KPMG Report, 8.

[72] See for example Mercier and Olawuyi (2014).

7.5 Conclusion

The strategic importance of extractive resources to the domestic economies of African countries means that petroleum or mining activities may not be lawfully undertaken without obtaining applicable licenses and permits from national authorities. Securing relevant mineral title or permit is therefore a critical component of successfully investing in Africa's promising extractive sector. This chapter has explored the types of concessionary rights, interests, permits and licenses that are applicable to prospective extractive sector investors across Africa. These licenses stipulate the nature, scope and extent of rights and interests granted to the license holder.

In most cases, foreign companies may participate in the licensing process by satisfying incorporation, registration and procurement formalities in accordance with several locally applicable corporate, business organizations, foreign investment, energy, trade, currency importation, competition, tax and finance laws. To avoid legal risks relating to non-compliance, it is very pertinent for a foreign company going to invest in Africa to carefully examine and negotiate fiscal and contractual terms that maximize profits and reduce risks exposure.

The next chapter provides a comprehensive overview of the nature and characteristics of the different fiscal arrangements that have been developed across Africa to govern relationships across the various sectors of the extractive industry value chain.

References

Altman v Blake 712 S.W.2d 117 [US]

Attorney General of the Federation (AGF) vs. Attorney General of Abia State (No. 2) (2002) 6 N.W.L

Chamber of Mines (2016) Illegal and artisanal mining. Fact sheet 2016. Johannesburg, South Africa

Companies and Allied Matters Act, 2018 Laws of the Federation of Nigeria, c 20, 2004

Department of Natural Resources Minnesota. http://www.dnr.state.mn.us/waters/watermgmt_section/appropriations/permits.html

Extractive Hub (2017) Artisanal and small scale mining. https://extractiveshub.org/servefile/getFile/id/4197. Accessed 12 June 2018

Falcon Oil and Gas Limited. http://www.falconoilandgas.com/karoo-basin-sa. Accessed 12 June 2018

Famfa Oil Ltd. vs. A.G.F. & NNPC (unreported suit No. C. CA/A/173/06)

French v. Chevron U.S.A. Inc., 896 S.W.2d *795*, 797 (Tex. 1995)

Hamilton v. Morris Res., Ltd., 225 S.W.3d 336, 344 (Tex. App.-San Antonio 2007, pet. denied)

Ingram S, Conner T (2018) Mineral v Royalty Distinction. http://www.landman.org/docs/educational-material-(pdf)/mineral-v-royalty-aapl-paper.pdf. Accessed 12 June 2018

Kulander CS (2016) Fixed vs. Floating non-participating oil & gas royalty in texas: and the battles rage on. Texas A&M Law Rev 4(1)

Laborde Properties, L.P. v. U.S. Shale Energy II, LLC; No. 04-16-00168-CV, ___ S.W.3d ___ (Tex. App. – San Antonio Oct. 12, 2016)

Ledwaba P (2017) The status of artisanal and small-scale mining sector in South Africa: tracking progress. J South Afr Inst Mining Metall 117(1)

Ledwaba P, Nhlengetwa K (2016) When policy is not enough: prospects and challenges of artisanal and small-scale mining in South Africa. Afe Babalola Univ J Sustain Dev Law Policy 7(1)

LeFort D, Mauel JG, Stiffler JC (2013) Legal characteristics of U.S. Oil and gas interests and the U.S. Oil and Gas Lease. https://www.pillsburylaw.com/images/content/6/5/v2/650/Article201308ShaleInvestmentEnglishtranslation.pdf. Accessed 12 June 2018

Lowe J (2014) Oil and gas law in a Nutshell, 5th edn. West Academic Publishing

Mercier T, Olawuyi D (2014) Liberia Basin Bid Round 2014. Norton Rose Fulbright. http://www.nortonrosefulbright.com/files/liberia-basin-bid-round-2014-pdf-115kb-119874.pdf

Mineral and Petroleum Resources Development Act of South Africa

Mineral Rights. Basic Information about Mineral, Surface and Oil and Gas Rights. http//www.geology.com

Minerals and Mining Policy of Republic of South Africa, 1998

Mining Facts (2012) What is artisanal and small-scale mining? http://www.miningfacts.org/communities/what-is-artisanal-and-small-scale-mining/. Accessed 12 May 2018

Mutemeri N, Petersen FW (2002) Small-scale mining in South Africa: past, present and future. Nat Resour Forum 26:287–288

National Water Act (Act 36 of 1998) of South Africa

Olawuyi D (2018) Regulation of upstream oil & gas. Gulf Legal Advisor. Lexis Nexis

Omorogbe Y (2001) Oil and gas law in Nigeria. Malthouse Press Limited, Lagos, Nigeria

Omorogbe Y, Oniemola P (2010) Property rights in oil and gas under dominial regimes. In: McHarg A, Barton B, Bradbrook A, Godden L (eds) Property and the law in energy and natural resources. Oxford University Press, Oxford

Petroleum Act, 1969 of Nigeria

Petroleum Law No. 10/04 of 12 November 2004 of Angola

Plainsman Trading Co. v. Crews, 898 S.W.2d 786, 789 (Tex. 1995)

Precious Minerals Marketing Corporation Law, 1989 (PNDCL 219) of Ghana.

Scott R, Rockey N, Hudson R (1998) The status of small-scale mining in South Africa – a preliminary study. Minerals and Energy Policy Centre and The Marketing Shop 1998

Shade J, Blackwell R (2014) Primer on the Texas law of oil and gas, 5th edn. Lexis Nexis

Small-Scale Gold Mining Law, 1989 (PNDCL 218) of Ghana

Smith E, Weaver J (2015) Texas law of oil and gas, 2nd edn. Lexis Nexis, pp 2–61

The Model Production Sharing Contract, 2015

Types of Mineral and Hydrocarbon Rights (2018) Extractives Baraza. http://extractives-baraza.com/resources/types-of-mineral-and-hydrocarbon-rights. Accessed 12 June 2018

World Bank Group (2013) Artisanal and Small-Scale Mining. WBG. http://www.worldbank.org/en/topic/extractiveindustries/brief/artisanal-and-small-scale-mining. Accessed 12 June 2018

Chapter 8
Contractual and Fiscal Arrangements in the Extractive Industry

Abstract This chapter provides a detailed analysis and discussion of the prevalent fiscal and commercial arrangements in extractives industries in Africa. It examines the meaning, nature and essential characteristics of the ten main contractual arrangements relied upon by state-owned enterprises and governments in Africa to convey titles and interests: concessions; production sharing contracts; joint ventures; service contracts; farm out contracts; pooling agreements; unitization contracts; long term oil and gas supply contracts; mining development agreements and impact benefit agreements. It also discusses the legal effects of the agreements on the conduct of extractive operations.

8.1 Introduction

This chapter provides a detailed analysis and discussion of the prevalent fiscal and commercial arrangements in extractives industries in Africa. Although there are several fiscal arrangements that have been developed to govern relationships across the various sectors of the extractive industry value chain, the ten main arrangements relied on by state-owned enterprises and governments to convey titles and interests are: concessions; production sharing contracts; joint ventures; service contracts; farm out contracts; pooling agreements; unitization contracts; long term oil and gas supply contracts; mining development agreements and impact benefit agreements. This chapter provides a detailed analysis of the meaning, nature and cope of these arrangements.

After studying this chapter, you would be able to highlight the existing fiscal regimes and tools utilized to govern extractive operations in your country; the unique commercial and technical features that set extractive contracts apart from general commercial contracts; appreciate the types of commercial arrangements and contracts utilized in the extractive industry; understand the similarities and differences in the various types of arrangements and contracts; and understand the legal effects of the agreements on the conduct of extractive operations.

D. S. Olawuyi, *Extractives Industry Law in Africa*,
https://doi.org/10.1007/978-3-319-97664-8_8

8.2 Fiscal Arrangements in the Extractive Sector

The fiscal regime of a country refers to the set of instruments, laws, regulations, agreements or tools (taxes, royalties, dividends, etc.) that stipulate how revenues from oil and mining activities are to be shared between the state and extractive companies (ECs).[1] It defines the relationship between the State and the EC, allocates risks and benefits between the parties, and stipulates how the capital and operational investments in the extractive operations will be allocated inter se. The procedure or formulae for allocating the risks and benefits resulting from resource extraction are often spelt out in the country's legal framework, as well as in the commercial agreement entered into between the state and the EC.

The fiscal regime of a country is the most fundamental reflection of the overall energy policy and priority of the country. When designing a fiscal system, a government aims to maximise revenue from its extractive resources, while at the same time providing sufficient incentives to foreign investors to participate. Selecting the right fiscal regime is fundamental for a State to maximize the social, economic and environmental benefits of its geological affluence. For an EC, negotiating the right fiscal regime will go a long way in determining the profitability, sustainability, viability and business strategy of an extractive investment. For the EC, the ultimate aim is to balance the risks and investments during the resource exploration phase with returns gained during the production phase. Several factors, such as the level of participation by the state (whether solely as regulator or as commercial participant as well); nature of returns expected (whether purely royalties or focus on share of oil, gas or minerals produced); pick up location of proceeds (well head or export point); timing of the revenues (early on or later stages of production); risks tolerance or neutrality of the state (share of risk borne by the state and the EC); expectations on transparency, local content and accountability; and willingness to attract ECs (through reduced tax obligations and other fiscal incentives) are all factors that governments consider when selecting the preferred fiscal regime and arrangement in a country.

As Table 8.1 below shows, fiscal regimes adopted across Africa can be broadly categorized into two: concessionary and contractual.[2] Each of these are discussed below.

[1] Radon (2005), p. 63; Natural Resources Governance Institute (2015), pp. 1–2.

[2] Kaiser (2004), p. 2; Radon (2005), p. 63.

Table 8.1 Fiscal regimes in select African Countries

Country	Petroleum resource	Fiscal system	Regime fiscal term specified	Summary
Algeria	Oil	Concession	Quasi-Progressive	*Production based royalties (daily production):* Base of royalties is equal to the quantities of extracted hydrocarbons multiplied by the monthly average of the base prices. Effective royalty rate is subject to negotiation between the State and the EC, however, the law sets minimum legal rates for each production bracket
Hydrocarbon Law Act No. 2005-07, as amended by Law n°13-01 dated 20 February 2013			Production based royalties; production based tax on revenues, Tax is levied on extraordinary income	*Production based tax on revenues:* based on cumulative production since beginning of production phase.
				Additional tax on earnings: This consist of annual revenues minus deductions, depreciations, royalties, and operating expenses
Angola	Oil	PSC	Progressive. Rate of return based profit oil sharing	*Petroleum production tax:* levied on quantities of crude oil and natural gas measured at the wellhead at a rate of 20%. May be reduced to a minimum of 10% if: (a) petroleum exploitation is conducted in marginal fields, (b) in offshore depths greater than 750 m, (c) in fields designated by the government as difficult to reach
1997 Model production sharing contract				
Law on Taxation of Petroleum Activities Law No. 13/04				*Petroleum income tax:* Levied on national concessionaires that do not enter into associations with any entity, business associations, JVs, RSCs
Presidential Legislative Decree 3/12				*Income Tax*: Oil income is calculated based on standard income tax rate (about 35%)
Cameroon	Oil	Concessions and PSC	Progressive: R-factor based profit oil split (PSCs); Additional petroleum tax (Concessions)	In PSCs: profit oil is shared between the investor and the government in a manner set out by the contract. Profit oil split is based on R-factor
Law No. 99/013 Petroleum Code of 1999				Concession agreements have additional petroleum tax. Other corporate taxes include land royalties and flat fees

(continued)

Table 8.1 (continued)

Country	Petroleum resource	Fiscal system	Regime fiscal term specified	Summary
Gabon *Law no. 011/2014 dated 28 August 2014*	Oil and Gas	PSC	Quasi-Progressive/Progressive. Production based or ROR production sharing (contract based)	*Exploration period:* The exploration period must not exceed 8 years (initial 6-year period with the possibility of renewal). The production period is currently fixed at 10 years (15 years for gas) with the possibility of two renewal periods of up to 5 years each
2015 Model Production Sharing Agreement				• *State participation:* Minimum State participation of 20% which is carried through exploration by the contractor
				• There is an option for Gabon Oil Company to purchase up to 15% in the PSC at market rates
				• *Royalties:* Non-deductible royalty based on cumulative production between 9% and 15%
				• Cost recovery ceiling of 75% on gross revenue
				• Cumulative production first tranche profit share floor of 50%
				• *Taxation:* Corporate tax is payable. Contractor is subject to the following fees/taxes: (a) Bonuses: Surface area rental of 5000 CFA Francs per hectare (b) 35% Corporate tax—not included in State's share of profit oil
Ghana *Petroleum (Exploration and Production Law), 2016 Act 919*	Oil	Concession and PSCs	Progressive: Additional oil entitlement on revenues based on RoR	In PSCs: All costs associated with petroleum operations are capitalized until production. At that time, costs may be recovered and used as deductions against petroleum revenue
				RoR Formulae: Contractor is entitled to a share of revenue based on rate of return
Revenue Administration Act, 2016 (Act 915)				*Taxation:* Taxes are payable in accordance with laws of Ghana. Unless specifically provided in a PA, Contractors are required to pay tax on chargeable income at a rate of 35%

Kenya	Oil	PSC	Progressive: profit petroleum sharing based on R Factor	Profit oil will be shared between government and contractors (specific rates set out in individual contracts)
Petroleum (Exploration and Production) Act of 2012, Chapter 308				Profit Oil shall be exclusive of all taxes payable by the contractor. Taxes are payable under PSCs with applicable rates negotiable at time of signing PSCs
2015 Model production sharing contract				Capital expenditure is recoverable at a rate of 20% per annum based on amortization
				State participation: A minimum government participation of 10% is required. The government may elect to participate in petroleum operations and acquire an interest up to the percentage specified in contract
Libya	Oil	PSC	Progressive: profit petroleum sharing based on R Factor	The production share (M factor) was set at 65%; NOC would take 65% of gross production and pay 50% of capital costs and 65% of operating costs
1988 Exploration and Production Sharing Agreement II (EPSA II)				

(continued)

Table 8.1 (continued)

Country	Petroleum resource	Fiscal system	Regime fiscal term specified	Summary
Mozambique	Oil and Gas	Concession and PSC	Progressive: R-factor based profit oil sharing	*Profit oil split:* R-factor based split of profit oil between government and contractor
Petroleum Law, Law 3/2001				*Production tax:* Tax is payable at the rate of 10% (Crude Oil) and 8% (Natural Gas) on value of petroleum produced (including petroleum lost)
Decree No. 19/2004 of June 2004 Petroleum Production Tax Royalty Regulation				*Royalties:* Fixed Rate Royalties are payable as follows
Fixed rate royalties				Crude oil: 8% (for offshore based on depth: Shallow: 7% to >500 m: 3%)
Fiscal Law on Petroleum Activities 12/2007				Natural gas: 5% (for offshore based on depth: Shallow: 4% to >500 m: 2%)
Mining and Petroleum Incentives 13/2007				*Fiscal benefits:* Contractors enjoy exemptions from customs duties, Value Added Tax (VAT), and Specific Consumption Tax for a period of 5 years following the approval of a development plan
Regulations on Petroleum Operations (Decree 24/2004)				

Namibia	Oil and Gas	Concession	Progressive: return based additional profits tax	Petroleum income tax: 35% on income and dividends
Petroleum Laws Amendment Act 1998 Act 24				Additional profits tax: calculated based on net cash
Petroleum (Taxation) Act of 1991				
Petroleum (Exploration and Production) Act 1991 (Act 2 of 1991)				
Petroleum (Exploration and Production) Act, 1991: Regulations				
Petroleum (Exploration and Production) Amendment Act 1993 (Act 2 of 1993)				

(continued)

Table 8.1 (continued)

Country	Petroleum resource	Fiscal system	Regime fiscal term specified	Summary
Nigeria Petroleum Act of 1969	Oil and Gas	Concession and PSC JV between IOC and Nigerian Government Risk service contracts	Quasi-Progressive	Profit oil sharing: if price of oil exceeds $20/barrel, the PSC can be amended to increase government share of profit oil
Deep Offshore and Inland Basin Production Sharing Contracts Decree No. 9: 1999				Royalty regime graduated
				Up to 200 mL 16.67%; 201–500 m: 12%; 501–800 m: 8%; 801–1000 m: 4%; 1000 m+: 0%; Inland basin: 10%
Petroleum Profits Tax Act 2004				*Taxation:* Corporate Income Tax is payable. 50% flat rate on chargeable profits for the duration of the PSCs
				Chargeable profits are assessable profits less Capital allowances; assessable profits are adjusted profits less loss incurred by a company during a previous accounting period; adjusted profits are the value of profits less outgoings and expenses
				Price based taxes: the realizable price as defined in the PSC established by the Corporation or the Holder in accordance with the provisions of the PSC, in respect of crude oil, etc. shall be used to determine the amount payable on royalty and petroleum profit tax in respect of crude oil produced and lifted pursuant to the Production Sharing Contract
				Petroleum profits tax: on chargeable profits
				First 5 years (new companies): 65.75%
				First 5 years (existing companies): 85%
				Subsequent years (all companies): 85%
				Royalty rates for Joint Venture: Onshore Production (20%); production in territorial waters less than 100 m (18.5%); offshore production beyond 100 m (16.67%)

Senegal	Oil and Gas	PSC	Progressive: R-factor based additional petroleum tax, R-factor based profit oil split	Additional petroleum tax: Contractor is liable for additional petroleum tax calculated based on R-factor (overall profitability of the project) as specified in PSC
Petroleum Code No. 98-05 1998				Royalties: The amount of royalty and the rules relating to basis and recovery of costs are as stipulated in the PSCs. The typical ceilings are: onshore Liquid Hydrocarbons: (2–10%); Offshore Liquid Hydrocarbons (2–8%); Gaseous Hydrocarbons, both Onshore and Offshore (2–6%)
				Profit oil split: based on R-factor defined as Net cumulative revenue (after corporation tax) over Cumulated Investment (research, evaluation and development)
				Cost recovery: The amount of revenue available for cost recovery is negotiable but typical ceilings are 60% onshore, 70% in shallow water and 75% in deep water

8.2.1 *Concessions*

The most widely used fiscal arrangement in African extractives industries is the concession. Algeria, Cameroon, Egypt, Ghana, Mozambique, Namibia, Nigeria, South Africa, as well as several other frontier extractives jurisdictions in Africa all adopt the concessionary system. A concession is a right granted by the government to an EC to explore, develop, sell and export oil or minerals in a specific geographic area.[3] Also known as a lease, permit or license, a concession is generally "a government grant for specific privileges; the voluntary yielding to a demand for the sake of a settlement; a rebate or abatement."[4] In extractives context, a concession is a right granted by the State to an EC to search for, win, explore, extract, and mine extractive resources from a defined area of land over a stipulated time period, and if petroleum or mineral was discovered, to produce, market and transport the resource. The oil exploration license (OEL), oil prospecting license (OPL), oil mining lease (OML) and the mineral mining lease that, have been discussed in Chap. 7 of this book, are all examples and forms of concessions.

The central organizing principle of the concessionary system is that the State or land owner (in freehold countries) transfers its ownership of resources in the subsurface to a commercial entity, often a partnership of companies. The companies obtain the exclusive right to extract hydrocarbons from a defined area for a defined period. The private investor that has obtained a concession is known as a concession holder and the geographic area for which the concession is granted as the concession area. Typically, the concession holder receives all of the petroleum or minerals produced from the field, but must pay a royalty and a tax to the State or land owner.[5]

Concessions have a long history in the extractive industry.

8.2.1.1 Traditional Concession

Traditional concession refers to the earliest forms of licensing arrangements between governments and ECs. In the early days of extractive resource production, many States had little or no experience with fiscal design. Therefore, some of the earliest forms of concessions granted broad, lopsided and uninformed rights to ECs to explore extractive resources.[6] First, in terms of the scope of rights, the early forms of concessions failed to clearly distinguish between petroleum and minerals, so they often granted ECs the exclusive right to explore, produce and market all natural

[3] Ibid.

[4] Black's Law Dictionary (2004). Also, Williston (1959).

[5] Radon (2005).

[6] See Likosky (2009), pp. 2–3 stating that the earliest forms of concession contained terms that were "highly uneven, at times teetering on the verge of the unconscionable. Companies paid small sums to the host government for the rights over its natural resources." Also see Omorogbe (1997), pp. 58–60.

resources in the country and to dispose them as the EC deemed fit.[7] Similarly, the earliest forms of concessions often required ECs to pay comparably small sums to the host government for the rights over its natural resources. Typically, the compensation was tied to volume produced and not the value of the resource itself.[8] Consequently, given that ECs themselves determined the volume of production, this meant that ECs had great control over the amount of production and profits declared. This also placed ECs under considerable pressure to maintain production even when pricing and other factors were not favourable.

Furthermore, the scope of the traditional concession was broad, particularly with respect to duration and geography.[9] In many cases it extended over very large tracts of land of the host country and at times over the whole country.[10] Similarly, the duration was very long, usually between 40 years and 75 years. For example, The Kuwait Oil Concession of 1934 was to run for 75 years.[11] The very large scope of the traditional concession unduly tied host countries to ECs even when certain exploration and production parameters are not being met. They also gave ECs great freedom to determine the nature, scope and extent of exploration and production.[12]

The broad and lopsided scope of traditional concessions meant that the interests of the host government and ECs were not adequately balanced in extractive operations and resource management.

---

[7] The following is an example of this type of clause found in a concession agreement in Oman in 1937:

> In consideration of the payments described in Article (...) the Sultan hereby grants to the company for the remainder of the period of this Agreement the exclusive right to explore, search for, drill for, produce, win, refine, transport, sell, export and otherwise deal with or dispose of the substances and to do all things necessary for all or any of the above purposes within the Leased Area.

[8] For example, the Oil Concession of 1934 between the State of Kuwait and the Kuwait Oil Company Limited (United Kingdom) states:

> "(d) For the purpose of this Agreement and to define the exact product to which the Royalty stated above refers, it is agreed that the Royalty is payable on each English ton of 2.40 lb. of net crude petroleum won and saved by the Company from within the State of Kuwait-that is after deducting water sand and other foreign substances and the oil required for the customary operations of the Company's installations in the Sheikh's territories" (Oil Concession of 1934: Article 3(d)).

Similarly, the Iraqi Agreement with the Khanaquin Oil Company in 1926 called for payment of four gold shillings per ton of crude oil produced and saved; while the 1949 Agreement between Saudi Arabia and Getty Oil provided for a royalty of US $00.55 per barrel.

[9] For example, in 1948, The Kingdom of Saudi Arabia granted an oil concession covering the entire kingdom to the Standard Oil Company of California. The King received for the concession an immediate payment of $1,500,000 in gold.

[10] In Nigeria for example, the early concession granted to Shell in 1938 was in respect of the entire mainland of Nigeria. See Omorogbe (1997).

[11] Article 1 of the Oil Concession of 1934 between the State of Kuwait and the Kuwait Oil Company Limited (United Kingdom).

[12] Likosky (2009).

Consequently, these forms of arrangements began to collapse in several African and Middle East countries with the end of colonization and in the second half of the twentieth Century. The New International Economic Order and the creation of the Organization of the Petroleum Exporting Countries (OPEC) resulted in concerted efforts to rebalance the relationships between host countries and ECs, to get rid of the inequitable and lopsided features of the traditional concession.

8.2.1.2 Modern Concessions

Modern concessions are the new generations of concessions that reflect the shift from an inequitable relationship between host governments and ECs to a partnership based relationship.[13] Many African countries, including Nigeria and South Africa, have retained the authority of the State to grant a right to an EC the exclusive right to explore for petroleum and solid minerals, and if discovered, to produce market and transport the commodities, and in turn for the EC to pay specified costs and taxes to the State. Consequently, like traditional concessions, modern concessions grant the EC the right to express rights to explore and produce natural resources over in a stipulated land tract, over a period of time. Concessions are now often called by less politically charged names such as permits, licence or lease.

Furthermore, most of the terms that characterized the old concession have now evolved. The scope of grant is now narrowly defined to include either crude oil, natural gas, or minerals. Similarly, the geographical area of coverage has also been reduced. In Nigeria for instance, the maximum area that can be granted under the oil mining lease is 500 square miles.[14] While the duration is normally for an initial period of 20 years. Furthermore, financial obligations are greatly increased, with ECs now liable to pay to rents, royalties, and a higher tax rates. ECs also increasingly have local content obligations to develop the capacity of nationals, local employees and industries to take part in extractive activities. Some of the essential characteristics of modern concessions are discussed below.

Bidding There are two ways in which modern concessions are allocated and concluded: bilateral negotiation and competitive bidding. When a contract is negotiated bilaterally, the firm, usually a multinational, approaches a country's government in order to obtain a concession for exploration, development, and export of a mineral deposit. The second, which host governments are increasingly adopting, is the granting of concessions through a competitive bidding process. The impetus behind the adoption of a competitive bidding process is manifold. By publicly inviting bids from all interested stakeholders, a competitive bidding process can foster greater transparency and decrease risks of corruption in the award of concessions. A competitive bid process also allows national authorities to anticipate, plan and prioritize the benefits that they would like to realize when awarding concessions. When gov-

[13] Likosky (2009); also Blinn et al. (2009).

[14] Etteh (1973).

ernment objectives are clearly outlined in the form of pre-qualification criteria, it allows national authorities to select the most aligning bid that can allow the effective realization of such objectives.[15] Therefore increasingly, concessions are normally obtained by investors through a competitive bidding process in which the host government drafts and publishes an agreement—usually a model contract is relied upon by many states—which contains usual and financial terms attached to the concessions. However, South Africa adopts a first-come-first-serve system for the granting of petroleum and mineral rights, as stipulated in Section 9 of the Mineral and Petroleum Resources Development Act 28 of 2002. Under this system, concessions are not allocated based on competitive bidding or negotiation. Rather, any prospective investor may apply to acquire exploration or production rights. Rights are granted on the basis of who applied first and met the legislative requirements for acquiring such rights. Where more than one application is received on the same day, preference is given to historically disadvantaged applicants.

Taxes, Royalties and Fees Host governments charge a fixed fee for every license that is given to a private investor. All private investors must pay this fee to obtain the license to explore regardless of whether any natural resources or minerals are discovered. On the other hand, if the private investor finds oil the governments earns royalties in the form of 'gross revenue and/or a profit tax based on net income; both of which are based on the quantity of production and the price at which the production is sold'.[16] Additionally, revenues of private investors are subject to income tax. Therefore, the government earns its revenue through taxes, royalties and fees.

Model Contracts The use of model contracts is an important feature of modern concessions. When inviting competitive bids, countries increasingly develop model contracts that contain draft clauses and terms that will form the basis of an acceptable agreement by both sides.

The benefits of a model contract are threefold. First, they serve as a starting point for negotiations and significantly decrease the time it takes to finalize a deal. Second, model contracts allow host governments to incorporate standard terms that reflect national policy directions and objectives, and to provide private investors limited leeway in altering such fundamental terms. Third, model contracts assist both sides to focus their time, efforts and resources on key commercial and contractual terms rather than on fundamental or standard provisions.

Title to Resources and Equipment Modern concessions have a fixed duration and acreage.[17] They are renewable upon expiration, subject to approvals from the relevant governmental agency. More often than not, concessions give a private investor the exclusive right to operate in the allotted acreage. The ownership of petroleum remains with the host government and title only passes to investor after production

---

[15] Extractive Industries Transparency Initiative (2014), p. 21.

[16] Radon (2005).

[17] Consider the example of Nigeria whereby there is a fixed limit on the acreage for all oil concessions. This limit is fixed at 500 square miles.

reaches the wellhead. Some would argue that title passes to the investor too early in comparison with other arrangements, for instance a production sharing agreement which is discussed below. Similarly, the ownership of equipment that is fixed to the ground, costs of which are entirely borne by the investor, 'generally passes to the state at the expiry or termination of the concession and the investor is typically responsible for abandonment and site restoration.'[18]

Generally, there tends to be a separate license for the main phases in the petroleum or mining process. The license takes the shape of an agreement that governs the relationship between the private investor and the government. The purpose of a license is to 'secure an exclusive right to undertake a particular activity on a particular piece of land'.[19] The different types of licenses that exist include exploration or reconnaissance, prospecting and production licenses which cover different phases of the petroleum exploration or mining project.[20]

Revocation and Termination Modern concessions have clearly specified events or conditions that could result in termination or revocation. A concession, lease or license may be revoked by the host government if the licensee fails to comply with any of the terms and conditions of its grant. The enabling petroleum typically stipulates the events of default that could result in a revocation. Under Nigeria's Petroleum Act for example, a concessionary right may be revoked if the licensee or lessee becomes controlled directly or indirectly by a citizen or subject of, or a company incorporated in, any country which is a country other than the licensee's or lessee's country of origin; or a country the laws of which do not permit citizens of Nigeria or Nigerian companies to acquire, hold and operate petroleum concessions on conditions which in the opinion of the Minister are reasonably comparable with the conditions upon which concessions are granted to subjects of that country.[21]

Furthermore, under paragraph 24 of schedule 1 of the Act, the Minister may revoke any OPL or OML if the licensee or lessee: (a) is not conducting operations continuously and in a vigorous and business-like manner in accordance with good oil field practice, or (b) has failed to comply with any provisions of the law or any regulation, or fails to fulfil its obligations under the special conditions of the license/lease; or (c) fails to pay due rent or royalties, whether or not they have been demanded by the Minister within the period specified by or in pursuance of the law, or (d) has failed to furnish such reports on its operations as the Head of the Petroleum Inspectorate may lawfully require. If the government decides to revoke any license, it shall inform the licensee or lessee of the grounds on which the revocation is contemplated and shall invite the licensee or the lessee to make any explanation if it so

[18] Tordo et al. (2010).

[19] Ibid 13.

[20] Ibid 12.

[21] Paragraph 23(1) of Schedule 1 of the Petroleum Act. See also Section 90 of South Africa's MPRDA which holds that the Minister has the authority to suspend or cancel exploration and production rights. Section 47 establishes the conditions upon which such suspension or cancellation may occur.

desires. If the government is satisfied with the explanation, it may invite the licensee or lessee to rectify the matter complained of within a specified period. If the licensee or lessee makes no sufficient explanation, or does not rectify the matter complained of within the specified period, the Minister may revoke the license or lease. The notice of revocation is then published in the Federal Government official gazette, but the licensee or lessee remains liable for all obligations incurred before the effective date of such revocation.

Similarly, the holder of a concessionary right may at any time, terminate the license or lease by giving to the Minister not less than 3 months' notice in writing to that effect.[22] The holder is also entitled at any time on giving 3 months notice in writing to the Minister to surrender the license or lease in respect of any particular part of the licensed or leased area.[23]

8.2.2 Contractual Regime

Host countries, in search for a new type of relationship that is based on mutual benefit and partnerships with foreign companies, have increasingly adopted petroleum contracts as the basis for allocating rights and interests in resource extraction. In some countries, such as Angola, Ghana, Kenya, and Nigeria amogst others, these contracts exist alongside the concessions that have been granted, to provide a basis for public- private partnerships between the host government and the private investor or EC.[24] Unlike concessions (licenses and permits) that tend to be concise and in template form, a contractual framework allows the government and EC to share control over the financing, exploration, production and marketing of natural resources based on clear, comprehensive and negotiated terms and conditions. These contractual agreements range from the production sharing contract, joint venture contract, service contract, to participation agreements such as farmout contracts, polling and unitization contracts, all of which are treated in this section in sequence.

8.2.2.1 Production Sharing Contract (PSC)

A production sharing contract is one of the most commonly utilized agreement in the petroleum sector.[25] It is an agreement between a host country and a contractor or EC, whereby the EC bears all exploration risks, production and development costs

[22] Para 18 of Schedule 1 of the Petroleum Act.

[23] Para 19 of Schedule 1 of the Petroleum Act.

[24] Likosky (2006) Chapter 2.

[25] Indonesia was first country to utilize employ production-sharing agreements in 1966, also they have remained one of its most consistent users. In direct response to some of the criticisms of concessions, the Indonesian government refused to grant new concessions in the 1960s. Instead

in return for its stipulated share of (profit from) *production* resulting from this effort.[26] A PSC is defined in Nigeria's *Deep Offshore and Inland Basin Production Sharing Contracts Act, 1999* as: "any agreement or arrangements made between the Corporation or the Holder and any other petroleum exploration and production company or companies for the purpose of exploration and production of oil in the Deep Offshore and Inland Basins."[27] Under a PSC arrangement two parties agree on oil exploration and production, with one party (the EC) bearing all the risks of exploration, operation and management. Whenever oil is discovered in commercial quantity, the EC is entitled to recoup its investment first, while the remainder of the oil is shared with the other entity in predetermined proportions. All costs and risks of exploration and production are typically borne by the EC, and in case the exploration and production were successful and a commercial discovery is achieved, the contractor (EC) is entitled to be repaid out of a proportion of the oil produced known as (cost recovery) and further, as a reimbursement for the development achieved, the contractor is entitled to a share in the remaining oil known as (profit oil).[28] The state, however, remains the owner of the petroleum produced subject only to the contractor's entitlement to its share of production. The government or its NOC usually has the option to participate in different aspects of the exploration and development process.

A standard PSC will contain detailed terms on the sharing formula of taking, term and duration of the contract, place of delivery, provisions on default, termination, dispute resolution clause and the governing law. This contract-based regime grants a private investor or a consortium the right to explore for and produce hydrocarbons barred by geographic area and a certain time limit; making a production sharing agreement similar to a concession. Like concessions, under PSCs the investor runs oil operations at its own risks.

However, a PSC differs from the concession in four main respects. First, unlike a concession which grants the EC ownership rights over the resource found in the defined area, in exchange for payment of royalties and tax to the host country, a PSC

PSAs were developed and considered acceptable because they allowed governments to retain national ownership of produced resources, with the foreign investor accepting all of the E&P risks and expense. If hydrocarbons were produced from the field, the foreign investor would be compensated for its efforts with a share of the production. The idea was based upon the agricultural practice of sharecropping. The earliest PSAs were approved in 1960. However, the first significant contract was signed in 1966 with a US consortium known as IIAPCO. These first generation PSAs allowed for up to 40% of exploration and operation costs to be recovered each year. See Barnes (1995). See also Machmud (1993), p. 179; Fabrikant (1975), p. 3030.

[26] Bindemann (1999), pp. 1–15.

[27] PSC was first introduced in Nigeria in 1973 *(Ashland PSC)* in a contract between the Nigerian National Oil Corporation and Ashland Oil Nigeria Company. The first round of PSCs were executed in 1993 and the second round in 2000. The PSC has undergone a few amendments both formal and informal. Section 18 of the Petroleum Act defines "deep offshore" as any water depth beyond 200 m and "Inland Basin" means any of the following Basins, namely, Anambra, Benin, Chad, Gongola, Sokoto and such other basins as may be determined , from time to time by the Minister."

[28] Bindemann (1999).

does not grant the EC ownership rights over the resource. Under a PSC, the State retains legal title to the unproduced natural resources and only transfers title to the IOC's share of the oil once it has been produced based on pre-determined sharing formulae. This addresses the important issue of ownership of oil reserves which has made this contract form to be politically acceptable in most African countries. Second, unlike a concession which grants the EC rights over the total production in the defined area minus fees and royalties, the PSC does not grant the EC rights over total production. Rather the EC's interest in the produced oil is limited to recouping of sunk costs plus the profit oil. Typically, the EC assumes all risks and costs associated with exploration and production. The investor, however, recoups its costs upon production as the production sharing formula is applied only after the EC has neutralized its expenses. Many countries, however, place a limit on the percentage of production that can be used for cost recovery.[29] Third, the title of production, unlike concessions, transfers to the investor at the delivery or export point. Similarly, the EC only gets title over its share of production instead of the entire production as is the case in a concession. Fourth, unlike in concession where a concession holder pays a bundle of taxes and royalties to government for oil produced, under a PSC the EC pays the government in form of a portion of produced oil.

PSCs provide a good opportunity for host governments wanting to avoid the commercial and financial risks of oil production to achieve this goal, as the costs of exploration are solely borne by the investor. The EC usually bears all of the operational and financial risks at the exploration stage if no oil is discovered. PSCs also serve as a tool for technology and expertise transfer aimed at eventual independence of the host government in the extraction and production of its natural resources. Consequently, in most PSCs, ownership of the equipment and facilities used in petroleum operations is vested in the State. An IOC is entitled to deduct capital expenditures on equipment, labour and technology in the form of cost oil. From an EC or IOC perspective, not having ownership rights over equipment and facilities could reduce legal risks relating to having to transfer oil and gas assets at the end of commercial production activities, while for the State, it provides an opportunity and incentive that could enable the country to benefit from technology and expertise transfer that can aid the country to eventually take over resource production by itself. As can be seen from the *Agreement on the Exploration, Development and Production Sharing for the Shakh Deniz Prospective Area in the Azerbaijan Sector of the Caspian Sea*:

> Whereas, Contractor has the technical knowledge and experience, the administrative and managerial expertise, and financial resources to efficiently develop and produce the Petroleum resources of the Contract Area, and desires to contract with SOCAR for that purpose.[30]

This acknowledges the expertise transfer objective of PSCs. To realize the goal of eventual independence is the goal of the host government, production sharing

[29] The recovery cost from production is set at thirty percent in Gabon, see Bindemann (1999).

[30] Final Consolidated Version 3/30/96: Preamble.

agreements provide a great degree of oversight over decision-making. Host government assumes control after the investor has recouped its costs and made a profit but the host but the oversight mechanism by the host government is included in the agreement from the very start of the project. A PSC defines the processes and procedures for sharing and recouping production costs, bonuses, royalty oil and profit oil amongst others.

Defining Features of a PSC As Table 8.1 below shows, the terms, conditions and sharing formula adopted in PSCs vary from country to country. However, a number of common and organizing principles have emerged. A few of these common features are presented below:

1. *Bonuses:* The two main financial incentives offered to a state or landowner under a PSC are the bonus and the royalty. A bonus is a fixed fee which is paid by the EC to the host government at differing stages of the project. The Nigerian PSC for example contains signature and production bonuses. Signature bonuses are paid immediately after the completion of negotiations and signing of the PSC while production bonuses are paid when the production from a specific contract area reaches a particular threshold.
2. *Royalty Oil:* Whenever oil or gas production begins, the host government is entitled to part of the total production. A royalty is agreed upon as a percentage of the lease, minus what was reasonably used in the EC's production costs. Royalty oil is the quantum of oil allocated to the host state that will generate proceeds equal to the actual royalty payable each month and the rent payable each year.[31] While most PSCs levy fixed royalties, some countries such as Algeria, Egypt, Ethiopia, Gabon, and Nigeria incorporate sliding scales. Royalties paid are based on production and correlate with water depth, they essentially decrease as water depth increases.
3. *Cost Recovery Oil:* After royalty oil is deducted, the EC is entitled to a pre-specified share of production for cost recovery. Cost recovery oil is allocated to the EC in such quantum as to enable it to recover all its operating costs incurred on petroleum operations as per reviewed or approved work programme and budget.[32] Cost recovery oil is therefore the oil allocated to EC for the cost borne by the company to make the commercial discovery.[33] Most PSCs provide for operating costs to be recovered in the year of expenditure while capital costs may be recoverable in equal installments over a period of years in accordance with the method prescribed under the law.[34] In addition, certain qualifying oper-

[31] Royalty payable for deep offshore and inland basin PSCs are determined in accordance with section 5 of the Deep Offshore Decree 1999, while royalty for onshore and shallow water PSCs are determined by the provisions of the Petroleum (Drilling and Production) Amendment Regulations 1969. Section 7, Deep Offshore and Inland Basins Production Sharing Contracts Decree 1999.

[32] See Section 8, Deep Offshore Decree, ibid.

[33] Muscolino et al. (1993).

[34] Most of the earlier PSCs in Nigeria did not place any cap on cost recovery, however the 2005 Model PSC has capped cost recovery at 80%.

ating or capital costs incurred by the EC are taken into account and are deducted when calculating chargeable profits for the purposes of petroleum profits tax. Cost oil allowances vary from zero (in Libya) to 100% in countries such as Liberia, Algeria and Nigeria.[35]

4. *Tax Oil:* The next portion is tax oil and is the quantum allocated to the host country so as to allow it pay on behalf of itself and the IOC, the petroleum income tax due each month in accordance with applicable tax legislation. In many domainial regimes, a *tax* is often imposed on the removal of extractive resources. For example, section 3 of Nigeria's Deep Offshore Decree provides that Petroleum Profits Tax applicable to deep offshore contract areas shall be 50% flat rate of chargeable profits for the duration of the contract and the tax shall be split between the NNPC and the IOC in the same ratio as the split of profit oil between the parties.[36]
5. *Profit Oil:* Once the allocation of royalty oil, cost oil and tax oil has been made, the remainder of the production, so called profit oil, is then shared between government and IOC at a stipulated share. The profit oil is often shared based on a pre-agreed production formulae contained in the PSC (for example 65% for the government and 35% for the EC).
6. *Production Sharing Formulae:* Sharing of produced oil is the defining feature of PSCs. The production sharing formula is specified in the agreement and, often times, also in the host government's legislation. There are four main production sharing formulae that have been developed by host governments to increase their profit share.[37] These are briefly outlined below.

   a. Daily Rate of Production (DRoP): According to this formula, the government share increases with the increase in production, calculated on a daily basis. The downside of formula is that field size is a poor indicator in determining profitability. Moreover, the formula does not effectively account for fluctuations in oil prices and costs.
   b. Cumulative Production from Project: Government share is dependent on the total production of oil so it increases with an increase in the total production. This formula does not accurately help calculate the rate of return of the private investor.
   c. R-Factor: R-factor is the ratio of the private investor's cumulative revenues and costs. The R-factor is calculated by dividing total contractor receipts over total contractor expenditures. As the R-factor increases, so does the government's profit share with it. The R-factor is an improvement from

[35] The main implications of PSCs with no cost recovery are twofold. First, total profit oil increases meaning that the IOC and the host government each obtain more crude in terms of volume from their respective shares in profit oil. Second, the IOC has to recover its costs out of its share of profit oil. See Bindemann (1999).

[36] See Section 12, Deep Offshore Decree. Also, section 4 of the decree also provides statutory backing for Investment Tax Credit/Investment Tax Allowance provisions which were originally present in the 1993 PSC model clause.

[37] International Monetary Fund (2012).

DRoP in that it measures profitability. However, the R-factor does not recognize the time value of money.

d. Rate of Return (RoR): If exploration costs are borne by the private investor, which is often the case, the RoR is a form of rent tax. The cumulative RoR of the private investor is calculated and a benchmark rate is set. If the RoR does not meet the benchmark rate, no tax is levied on the profits of the private investor. And the profit share of the government is set by reference to the cumulative RoR of the investor.

7. *Term and Relinquishment:* Given time scale of oil field developments, PSCs are often signed for a period of 25–30 years or more in some cases (10 years exploration and 20 years production), with a provision for relinquishment of as much as 50% of the initial contract.[38]
8. *Management Committee:* PSCs often establish a management committee which is similar to the operating committee under joint ventures. The committee consists of representatives of both the host government and the IOC. The management committee is often entrusted with advisory functions to review and approve the annual work programmes and budgets in respect of exploration operations; review proposals for surrender or relinquishment of any part of the contract area by the Contractor; review and approve proposals for an appraisal programme; and also to evaluate the declaration of a discovery as a commercial discovery.
9. *Minimum Work Obligations:* A PSC typically mandates the IOC to prepare a work programme and budget which is to be approved by the management committee. The IOC also provides the necessary funds for the execution of the work programme. Failure to meet specific work obligations in the work programme will result in early termination of the license and withdrawal of all rights and privileges pertaining thereto
10. *National Interest Provisions:* PSCs typically contain clauses that protect the national economic interest by providing for technology transfer, training of local employees and preference for local suppliers as is required by the applicable local content legislation.
11. *Miscellaneous Provisions:* These include general provisions on the termination of the contract; the governing law of the contract; economic stabilization of the contract so as to protect the interests of the IOC; dispute resolution clause on binding and enforceable arbitration should a dispute arise between the parties.

Although PSCs have been increasingly adopted in African countries, its complex nature, coupled with the various legal challenges that have arisen from the fluctuations in the price of oil over the years, has resulted in a gradual shift to other forms of petroleum agreements. Also, some countries have developed hybrid forms of PSCs that include the characteristics of concessions and PSAs. For example, PSCs can take the form of an Exploration and Production Sharing Agreement (EPSA)

[38] See Section 2 Deep Offshore Decree; section 10 Petroleum Act and section 12(1) Petroleum Act respectively.

where there are no proven reserves, or a Development and Production Sharing Agreement (DPSA) where there are proven reserves to be exploited.[39] Under a typical EPSA/DPSA, the contractor is given the exclusive right to explore for, appraise, develop and produce crude oil and non-associated gas from a designated contract formation in a designated contract area. This right includes the right to store, transport and export for sale, or otherwise deal with or dispose of, the crude oil and non-associated gas. Nevertheless, the state reserves ownership rights over the produced oil and gas unto itself.

8.2.2.2 Joint Venture Contracts (JVCs)

Joint venture, as the name suggests, is a partnership-based arrangement between two parties (usually state-owned enterprise and the EC) with a view to jointly run an extractive venture. Unlike concessions and PSCs, it provides a structured means for shared decision-making, with both parties in the joint venture jointly undertaking coordination and control responsibilities which have already been laid out in a joint venture agreement. Normally, a jointly owned company is created to fulfill the mandate of a joint venture agreement.[40] However, depending on the applicable company law legislation in a jurisdiction, joint ventures may also be run on the basis of contracts alone, without the creation of a spate legal entity (this is known as unincorporated joint ventures). Unincorporated joint ventures offer great flexibility for IOCs that may be willing to avoid complex commercial registration or incorporation requirements in the host country.

The most important document for joint ventures is the contract between the co-venturers. The essential features of JVCs are:

a. *Shared risks and costs:* As the involved parties run a jointly owned enterprise, the risks and costs are shared between the host government and the private investor. Profits, taxes and royalties are also shared by the involved parties.
b. *Shared decision-making and efficient technology transfer:* The jointly owned operation provides a corporate structure that makes shared decision-making and technology transfer easier and efficient. The state-owned enterprise can rely extensively on the expertise of the private investor which usually carries a wealth of experience. The target of a joint venture is the eventual independence of the state-owned enterprise. As such, the extent of technology transfer boils down to the strength of the host government; and
c. *Joint liability:* The natural disadvantage of sharing is that risks and costs must also be shared. Therefore, the host government is a direct party to the natural

---

[39] EPSAs and DPSAs are commonly utilized in Qatar to limit the scope of operations of an IOC to either exploration and/or development.

[40] Nigerian National Oil Company used JV's for a considerable amount of time as an exception in the petroleum industry. But this has changed, Nigeria has shelved JVs in favor of PSAs. See Omorogbe (1997), Radon (2005). Tordo et al. (2010).

resource extraction process and can also be held liable for environmental damage.[41]

Furthermore, a JVC will typically specify the percentages held in the enterprise by the various contracting parties.[42]

8.2.2.3 Joint Operating Agreement

The most common form of agreement utilized in the oil and gas sector to codify or operationalize unincorporated JV relationships is the joint operating agreement (JOA).[43] A JOA defines the terms and conditions of a joint venture between two parties (typically the EC and the host state) and it generally takes effect when joint application becomes successful by the award of licence by the host government.[44] A JOA is therefore a contract between two or more parties establishing, coordinating and setting out the terms of a joint venture between them under which petroleum exploration, development and production operations will be conducted.[45] Taylor and Winsor rightly opine that "it is the constitution by which the joint venture is governed and performs essentially the same role as a partnership agreement or the memorandum and articles of association of a company."[46] The general principle underpinning the JOA, consistent with its joint venture nature, is that participants agree to collaborate to jointly develop and explore extractive resources.[47]

The JOA defines the rights, duties, legal relationship and other obligations of the partners to the joint venture in respect to a particular licensed area. Under the JOA, one of the parties will be appointed as the operator and the other party is invariably the non-operator serves as the financial partner. The JOA stipulates the functions of the operator. The operator is also to comply with the provisions of the law and any other regulations and directions given. The operator owes a fiduciary duty to the joint venture partners. The functional tasks of a JOA are to: (1) define the initial operations in which all parties having development rights will participate; (2) provide a mechanism for conducting subsequent operations to develop and maintain

[41] See Wilkinson (1997), pp. 1–10.

[42] For example, the Azerbaijan JVC provides that: "The Rights and Obligations under this Agreement of each of the Contracting Parties shall be held in the following respective percentage of Participating Interests as of the date this Agreement is executed…".

[43] Per Wigley (1978) "I know of no other agreement in use in the petroleum industry, or any other industry for that matter, that can be compared to the Operating Agreement from the standpoint of frequency of use and the multitude of complicated situations and eventualities it is required to anticipate in its provisions."

[44] See Al-Emadi (2010).

[45] Taylor and Tyne (1992).

[46] Ibid.

[47] On the legal nature of JOAs, see *Ithaca Energy (UK) Ltd v North Sea Energy (UK) Ltd* (Rev 1) [2012] EWHC 1823; also *Euroil Ltd v Cameroon Offshore Petroleum Sarl* [2014] EWHC 12 (Comm) (06 January 2014).

production from the contract area; (3) provide for the day-to-day management of the properties by a designated "operator"; (4) define the rights and duties between the "operator" and the "non-operators". The JOA will typically make reference to: the operator's right to conduct joint operations, prepare work programmes, budgets and authorisations for expenditure; represent the parties before the government and third parties; oversee staffing and contracting for materials and services; and responsibility to deliver a certain standard of performance; and (5) define the rights and duties between the parties to the agreement and those who are not parties to the agreement.[48] Under the JOAs, host states and IOCs own both the equipment and facilities of the project, as well as the oil and gas productions. The power to control and manage the activities of a joint venture are typically vested in the joint operating committee (OpCom). The OpCom comprises a nominated representative of each party, with each representative having a single voting right weighted in accordance with the participating percentage interest of the party which it represents. Parties must carefully consider which decisions of the OpCom require unanimous approval and the approval threshold for other decisions.

Generally, to conclude a JVC or JOA, especially with a third party, the Minister of Petroleum Resources has to review and approve the agreements in line with the provision of the applicable petroleum legislation so as to ensure that the right is being transferred to a reputable company with sufficient knowledge, experience and financial resources to work the license.

8.2.2.4 Service Agreement

Of all the contractual arrangements discussed so far, service agreements afford the greatest deal of control over resources to the host government. In a service agreement, the host government basically hires a contractor to perform the services stipulated in the agreement. Remuneration of the contractor is, therefore, not based on production but on the service performed. Usually, a fixed sum is paid to the contractor without any share of the produced resource or profits for the contractor. The State retains ownership of the hydrocarbon at all times and pays the IOC a "Service Fee" for the work done.

Service contracts are often seen as the ideal choice in circumstances characterized by nationalism. The host government maintains significant control over the resources at all stages while the contractor provides technological and managerial expertise. The provision of a service as a mere contractor and no route to acquisition of a stake in the enterprise can have a detrimental effect on the productivity of contractor. This can serve as a disincentive for the contract towards efficiency and innovation. This structure is less suitable for exploration work and is generally undesirable from an IOC standpoint.

---

[48] Pierce (2007), pp. 1–2.

Types of Service Agreements The three main types of service contracts are the pure service contract, risk service contract, and the technical assistance contract. These are discussed below.

Pure Service Contract This is the most straightforward form of a service agreement. The host government hires a contractor to perform a specific service for which it is then remunerated accordingly. In other words, all risks are borne by the state, and the contractor performs its stipulated services and is paid a flat fee for this service. The contractor may acquire an interest in the extracted resource. Pure service agreements are extremely rare in Africa, and have been mainly utilized in Middle East countries such as Saudi Arabia, Kuwait, Qatar, Bahrain, and United Arab Emirates.[49]

Risk Service Agreement (RSA) RSAs are arrangements under which the contractor bears the risks associated with exploration.[50] The contractor would undertake exploration for the host government and would only get remunerated if it discovered commercially exploitable resources. If commercially exploitable resources are identified, then the company receives cash remuneration for its efforts in addition to a possible stake in the subsequent enterprise. Additionally, the contractor would either also receive a stake in the subsequent enterprise or a discounted price on a given percentage of the total production. If resources are not found, then the company bears all the financial risks from out of its own pocket. RSAs are extensively used in South America and Latin America, especially Brazil, Argentina and Colombia, where a part of their appeal appears to lie in the fact that sovereignty over the resource at all times is assumed.

Technical Assistance Agreement This type of a service agreement is much narrower than the other types of service contracts. The contractor receives a fixed compensation for the services provided and does not acquire any interest in the resources explored or extracted. On the face of it they are remarkably similar to the pure service contract. The main difference is that the company is engaged to provide technical services, without any interest in the oil at any time. Its remuneration consists solely of a specified and pre-agreed fee.

---

[49] These are more prevalent in OPEC countries where technological know-how is of a requisite level and capital is in abundance. Barrows gives an example of the Saudi Arabian service contract with Aramco. Under the contract, Aramco carries on production operations for a net fee of 15 cents per barrel after tax. Often the service contract is accompanied with a legally unconnected but parallel purchase contract for part of the oil being produced from the contract area as is the case in Saudi Arabia. See Tordo et al. (2010), p. 10.

[50] Smith (1992); also Neto (1985), p. 114.

8.2.2.5 Assignment and Participation Agreements

A holder of a concession or petroleum contract may, for financial, technological or other operational reasons, choose to assign or transfer rights in the applicable license or contract. Assignment arrangements are used to accomplish transfers of lease rights. Furthermore, a right holder may choose to combine their operations with those of other rights holders in adjoining lands to enhance efficiency and eliminate waste. This section explores a few of the commonly used agreements for the transfer or combination of oil and gas rights and interests.

Farmout Agreement (FOA) A farm out is an agreement that codifies the terms and conditions under which the holder of an oil and gas lease (the "farmor") assigns all or part of the working interest in that lease to another party (the "farmee"), to explore, prospect, win, work and carry away any petroleum encountered the area during the validity of the lease.[51] A FAO is generally created whenever a license holder is unable or unwilling to drill on a lease, but is willing to assign an interest therein to another who will assume the drilling obligations and save the lease from expiring or defaulting. The character and purpose of the FOA has been rightly summarized as: "a contract to assign oil and gas lease rights in certain acreage upon the completion of drilling obligations and the performance of any other covenants and conditions therein contained. It is an executory contract."[52] The primary characteristic of a FOA is the obligation of the farmee to drill one or more wells on the assigned acreage, to a certain depth, at a specified location, and within a certain time frame, at the farmee's own risk and expense. The farmor may use a FOA to preserve or maintain a lease by securing production close to the end of the lease primary term, to comply with an implied covenant to offset or to obtain an interest in production without additional cost. A farmout helps create a win-win scenario for involved parties as it isolates the farmor from risks associated with the exploration and production—the farmee bears all risks and costs—while providing the farmee with an eventual right to an interest in the concession area, if commercially exploitable oil is discovered. Therefore, when the risks in the petroleum industry reach high levels for whatever reasons, farmouts become prevalent. Generally, the greater the risks a farmee takes, the greater the area in which the farmee will earn an interest.

Entities that hold oil and gas leases generally enter into farmout agreements for the following reasons.[53]

a. *Concession Preservation:* Oil concessions are accompanied with a primary term which is the longest period of time a concession owner can hold the land for, without undertaking any drilling. Once this period lapses without any drilling

---

[51] See Holland and Hart LLP (2014). Also, para. 17 of Schedule 1 to the Nigerian Petroleum Act defines a farm out as "an agreement between the holder of an oil mining lease and a third party which permits the third party to explore, prospect, win, work and carry away any petroleum encountered in a specified area during the validity of the lease".

[52] Martin and Kramer (2009); Martin and Kramer (2012), p. 500.

[53] For detailed and extensive discussion, see Lowe (1987).

being done, the concession automatically expires at the end of the primary term. Often times companies are unable to act within the primary term due to credit shortages, lack of personnel and managerial inefficiencies among other issues.[54] Farmout agreements are entered into to preserve the concessionary right and avoid this automatic expiration. Furthermore, entering a farmout arrangement can serve to ensure the flow of fresh capital into the concession area to ensure continuity of operations.

b. *Risk Sharing:* Owing to the high cost of drilling wells, many companies rely on farmout agreements to share the burden and risks of such costs. While joint venture is probably a better option for companies as they remain in control of the operations, if the farmee is an experienced and reliable operator, companies tend to choose farmout agreements instead. Essentially, a farmout agreement entered into for the purposes of risk sharing adopts the structure of a joint venture agreement specifying that drilling is an obligation of the farmee and not an option. The farmee's interest is in this case conditional upon drilling.
c. *Exploration and Evaluation:* FOAs have also become helpful to companies, who are unable to obtain leases or license, but are interested in obtaining geological information in order to evaluate other prospects in the same concession area. A farmee can gain access the farmout area in order to conduct exploratory searches and studies.
d. *Obtaining Reserves:* If a company's primary business is not exploration and production but it requires oil or gas resources for its operations, then it may farm out its leases to drilling companies (farmee) who will drill on the land and make oil and gas available to the farmor for tis primary business. The types of businesses that rely on such a model are pipeline and refinery operators who are under obligations to deliver oil or gas to their consumers. In such cases, the farmee will maintain and drill lands covered by the lease and then sell the production back to the farmor, such that the farmor is able to access a steady flow of petroleum products that will is required for their primary business.
e. *Drilling an 'Obligation' Well:* A concession holder has the obligation to protect the interests of the host government i.e. to ensure development or exploration in the concession area. In many African countries, drilling a well to ensure continuous production is therefore an implied obligation. When a lease or license holder is for technical or financial reason unable to meet this obligation, it may farm out the acreage to a farmer to meet its legal obligation under the lease. Governments may also mandate a farmout or directly recover and farmout a non-producing acreage to a third party.

For example, in Nigeria, as a result of the increasing number of oil and gas acreages that remained unproduced after discovery, or in a few cases partially appraised after the initial oil and gas discovery was made, the government enacted the *Petroleum Amendment Decree*. The aim of the Amendment Decree was to open the unproduced fields to exploration and production activities. Consequently, the Guidelines for the Farmout and Operation of Marginal Fields (the "Farmout

[54] Lowe (1987).

Guidelines") were released pursuant to the provisions of paragraph 16 of the first schedule of the *Petroleum Act 1969*, and the Petroleum Amendment Decree. The Guidelines stipulate that all marginal fields, as defined under the Act, are to be farmed out to other companies, mostly indigenous Nigerian companies, who have the capacity to produce them.[55]

The Guidelines set out in general terms some of the provisions of a Farmout Agreement, which would govern the relationships between the parties to include indemnity provisions, a definition of the Farmout area, terms and conditions governing deep-drilling, abandonment, the operation of straddle fields, the responsibility for community development, the operator of the project and the rules governing the appointment and change of operators, non-performance, encumbrance, termination and renewal.

Under the Guidelines, the Farmee has all the rights of an OML leaseholder in respect of the Farm-out Area containing the field(s) once the farm-out is concluded and all the rights interests and duties of the previous leaseholder shall be transferred to the new leaseholder. The Farmee shall also have the right/obligation to deal directly with the government as the new leaseholder. All the rights, interests, obligations and liabilities of the farmor in respect of farm-out Area containing the field(s) shall automatically transfer to the farmee and the farmor shall be relieved of these rights and obligations from the date of the execution of the Farm-out Agreement. In addition, operations carried out by the farmee in the Farm-out Area containing the field(s) shall be in accordance with the existing safety and environmental regulations applied in the industry.[56] A farmout agreement generally contains provisions relating to the duty imposed on the farmee; the earning factor for the farmee; interest earned by the farmee; number of wells to be drilled; specifications of each well to be drilled; and the form of the agreement.

To formalize a farmout arrangement, the approval of the host state government is typically required. In Nigeria for example, without the prior consent of the Minister, the holder of an oil prospecting licence or an oil mining lease shall not assign his licence or lease, or any right, power or interest therein or thereunder.[57] The Farmor must seek and obtain the consent of the Minister, pay the prescribed fee after which

[55] Under the Guidelines, a marginal field is defined as any field that has reserves booked and reported annually to the Department of Petroleum Resources and has remained un-produced for a period of over 10 years. See the Office of the Presidential Adviser on Petroleum and Energy, "Guidelines for the Farmout and Operation of Marginal Fields (August 2001). [the Marginal Fields Guidelines].

[56] See generally, the Marginal Fields Guidelines.

[57] Para 14 of Schedule 1 to the Petroleum Act. See also section 11 of South Africa's MPRDA, which prohibits the cession, transferal, letting, subletting, alienation, assignment or disposal of a mining or prospecting right, interest therein, or a controlling interest in respect to a company or closed corporation, in the absence of ministerial consent. See also the South African case of *Mogale Alloys (Pty) Ltd v Nuco Chrome Bophuthatswana (Pty Ltd) and Others (2011)* holding that where a majority shareholder of a petroleum or mining company disposes of shares in such a way that would result in a change of control, ministerial consent is required for such share transfer.

the Minister's consent for the assignment may be given on payment of such other fee or such premium, or both, and upon such terms, as the Minister may decide. The Minister shall not give his consent to an assignment unless he is satisfied that (a) the proposed assignee is of good reputation, or is a member of a group of companies of good reputation; (b) there is likely to be available to the proposed assignee sufficient technical knowledge and experience and sufficient financial resources to effectually carry out a programme satisfactorily to the Minister in respect of operations under the licence or lease which is to be assigned; and (c) the proposed assignee is in all other respects acceptable to the Federal Government.[58]

Pooling Agreement A pooling agreement, as the name suggests, is an agreement concluded by two or more holders of petroleum permits or rights to join together portions of their respective tracts for the purpose of combining costs, resources and efforts by interest owners in such a pooled unit. Its main use is to join or pool together interests in multiple tracts of land to form a single drilling unit, mostly for the purpose of having sufficient acreage to receive a well drilling permit under the relevant spacing laws and regulations, and for the purpose of sharing production costs, resources and efforts by interest owners in such a pooled unit.[59]

Pooling may be voluntary or compulsory. Voluntary pooling arrangement occurs when a right holder combines or pools its tract with another right holder for the purpose of sharing in the costs and revenues associated with drilling and producing a well from that spacing unit.[60] Compulsory pooling is common in countries with spacing requirements or legislation.[61] In those countries, pooling is a mandated by spacing requirements or laws that provide minimum acreage or distance requirements for drilling units, so that exploration, drilling and production can proceed in

[58] Para 16, ibid. Also Para 17 of Schedule 1 to the Petroleum Act provides that "The holder of an oil mining lease may, with the consent of and on such terms and conditions as may be approved by the President, farm out any marginal field which lies within the leased area". The President shall not give his consent to a farm-out or cause the farm-out of a marginal field unless he is satisfied that: (a) it is in the public interest so to do, and, in addition, in the case of a non- producing marginal field, that the marginal field has been left unattended for an unreasonable time, not being less than 10 years; and (b) the parties to the farm-out are in all respects acceptable to the Federal Government.

[59] Kramer and Martin (2006), pp. 1–3.

[60] See for example, Section 1509.26 of the Ohio Revised Code, which provides that "The owners of adjoining tracts may agree to pool the tracts to form a drilling unit that conforms to the minimum acreage and distance requirements…".

[61] See Section 4.021 of Alberta's *Oil and Gas Conservation Rules* (*OGCR*) which provides that no well shall be produced unless there is common ownership throughout the drilling spacing unit (DSU). This means that if there are separate tracts within a DSU with different ownership, all owners within the DSU must have an arrangement to share in the costs and revenues associated with drilling and producing a well from that spacing unit. Applications for compulsory pooling are made under section 80 of the *Oil and Gas Conservation Act* (*OGCA*) RSA 2000, c O-6 ["*OGCA*"], specifically, s 79 (1), and generally, ss 79(1)-(6).

the most efficient and economical manner.[62] The name spacing unit refers to how pooling agreements or pooling clauses came about. Spacing rules were introduced to prevent the economic and physical waste that was a result of drilling unnecessary wells.[63]

Historically, in the early years of oil and gas exploration, the quest by license holders to avoid the implications of the rule of capture caused energy companies to drill large numbers of wells into subsurface oil and gas pools and produce these wells as rapidly as possible in order to extract the pool's reserves before adjoin license holders could. The race to finish approach resulted in waste and also impaired the ultimate recovery of oil and gas from many pools. Oil and gas conservation legislation was introduced to curtain the effect of the rule of capture. Consequently, most oil and gas producing jurisdictions introduced spacing requirements or oil and gas conservation legislation to prevent wasteful practices. A spacing unit was conceptualized as the smallest area within which no more than one well could be produced from any particular pool. Typically, only one well could be drilled within a spacing unit. The oil and gas conservation legislation typically contains provisions for forced or compulsory pooling in situations where voluntary pooling cannot be accomplished in a reasonable period of time, or if a tract owner is missing and untraceable, or if there is a dispute about the ownership of a tract. In such cases, the right holder wishing to drill a well may apply to the regulator for a compulsory pooling order.[64]

In a broader sense, pooling prevents indiscriminate drilling which adversely affects the resources beneath the surface. Furthermore, the combination of small tracts of land significantly lowers exploratory, development and production costs associated with the drilling of wells as few wells can undertake the same task. In effect, also prompting the royalty owners to undertake production in an equitable manner from the wells.[65]

While pooling is mainly undertaken to comply with spacing rules, it could also be helpful for consolidating royalty payments. In a typical lease, the lessor is entitled to a royalty payment usually based on the production from the leased land only. In the aftermath of a pooling agreement, this changes. The lessor now obtains royalty payments on a pro rate basis regardless of whether the production occurred in the land or from other tracts of land included in the agreement.[66]

---

[62] Handlan and Sykes (1984); also Anderson (1985), pp. 91–92; also see Ballem (2008), pp. 251–255; Neave (1970), pp. 231–232. See also section 1509.27 of the Ohio Revised Code which provides that "If a tract of land is of insufficient size or shape to meet the requirements for drilling a well thereon…and the owner of the tract who also is the owner of the mineral interest has been unable to form a drilling unit under agreement…, on a just and equitable basis, such an owner may make application to the division of oil and gas resources management for a mandatory pooling order."

[63] Handlan, ibid; also Bankes (1997). For more on unit operations under Alberta's conservation regime, see Onuma (2015), p. 5. See also the *OGCA*, s 80(1).

[64] Ibid, also Alberta Energy Regulator, Compulsory Pooling.

[65] Ibid.

[66] Martin and Kramer (2012).

In the African extractive industry context, pooling is not a widely known concept given that in a majority of African countries, the state holds title to all lands.[67] Nonetheless, companies may decide to enter into pooling agreements to achieve greater profitability and efficiency.

Unitization Agreement Unitization is the consolidation or unified development and operation of a producing reservoir, which extends across two or more concession areas, so that exploration, drilling and production can proceed in the most efficient and economical manner by one operator.[68] Whereas pooling agreements combine ownership interests in multiple tracts of land to form a single spacing unit, a unitization agreement is a more complex combination of leases to form an area-wide joint operation. Unitization occurs when separately owned tracts and interests in a tract either desire, or are forced to combine their tracts and interests to provide for a larger area of joint operation, often called a "unit," by a single operator. It normally occurs for the purpose of engaging in joint enhanced recovery operations so as to substantially increase the recovery of oil and gas.[69] With unit operations, a reservoir can be exploited as one unit, thus avoiding duplication of efforts, and eliminating fears of adverse drainage, and sharing costs of conservation programmes.[70]

Analogous to pooling, unitization can also be voluntary or mandatory.[71] A unit can be established if the right holders or owners of the land overlying the pool voluntarily ask for it. However, a regulator may force a unitization if it finds it reasonably necessary to increase substantially the ultimate recovery of oil and gas, and the value of the estimated additional recovery of oil or gas exceeds the estimated additional cost incident to conducting the operation.

In Nigeria for example, under the Farmout Guidelines, marginal fields, are subject to compulsory unitization in their development where the fields straddle or overlap each other.[72] A unitization agreement will be a condition precedent for the approval of a Farm-out arrangement between the parties.[73] Furthermore, under the *Drilling and Production Regulations*, unitization or unit development may be compulsorily ordered under two circumstances. Firstly, where the individual tracts form part of one geological reservoir and the field is capable to being developed as a unit according to good oil field practice, or secondly, where in the interests of the

[67] In other parts of the world, usually, a pooling agreement gives lessee the option to join up other leases in the vicinity of the lessee's own lease to form a single unit. In terms of numbers, the optimal size of an oil field is 40 acres in accordance with spacing regulations while that of gas is 640 acres. Larger areas normally require governmental approval. See M Wigley, Review of Pooling and Unitization. Also Williams and Meyers (1957).

[68] Kramer and Martin (2006), pp. 1–3, also Andrews Kurth Kenyon LLP (2014).

[69] See Anderson (1985), pp. 91–92. See also Ballem (2008), pp. 251–255.

[70] See Neave (1970), pp. 231–232.

[71] Neave (1970), pp. 231–232.

[72] See the Marginal Field Guidelines, 8, 20–21.

[73] See the Marginal Fields Guidelines, ibid.

Nigerian government, unitization is essential to secure the maximum ultimate recovery of petroleum. The main purpose behind unitization is to achieve efficiency by maximizing the production capacity of a reservoir. In essence, unitization is pooling at a much larger scale. Unitization make is possible for a single operator to conduct joint operations without regard to individual surface property rights.[74]

After unitization, competition between the various energy company-lessees holding leases of the unitized zone within the unit area is eliminated. A unit operator is appointed and the unit operator develops the unitized zone within the unit area on behalf of the unit working interest owners in a more efficient and cost-effective manner. Similarly, by eliminating competition between the working interest owners within the unit area, the unit operator can achieve significant cost savings for the unit working interest owners in the design and construction of gathering and processing facilities.[75] Also, unitization provides the unit operator with greater flexibility to account for production to the right holders or land owners with one set of tanks as opposed to multiple tanks for multiple parcels, leases and owners.

The large-scale nature of unitization agreements makes it easier and more effective for operators to undertake operations that would otherwise be economically unfeasible. For instance, operations like cycling, pressure maintenance and secondary recovery operations make unitization a requisite for efficient undertaking of these tasks. Where pooling is mainly a result of spacing rules, unitization is commonly relied upon to add vast quantities to producible reserves of a field that would be irrecoverable without unitization.[76]

A unitization agreement specifies the unitization area, more commonly 'unit area'; expenditure commitments of each party; appointment of the unit operator; authority and duties of the unit operator and conduct of unit operations; decommissioning; and provisions on assignment and withdrawal.

8.2.2.6 Oil Sale Agreement

Oil sale agreements govern the sale of oil and gas and other petroleum products by the host government. The sale and purchase of oil and gas are very common transactions in the oil and gas industry and happen almost on daily basis. There are various ways in which governments sell their oil. First, is the term sale under which a national oil company negotiates a contract with a buyer for a stipulated period, typically a year or more, and the agreement stipulates how the price will be determined against a specified spot benchmark. Second, is the spot sale, a quick sale in which buyers and sellers of individual cargoes find each other and transact a one-time deal for immediate delivery of a specific quantity of oil at a specific location. Therefore, in a spot market, prices for specific grades of crude oil are determined and purchased 'on the spot' at current market rates. Third, is the hybrid sale in

[74] Handlan and Sykes (1984).

[75] See generally, Freehold Owners Association (FHOA), Unit Agreements.

[76] M Wigley, Review of Pooling and Unitization.

which some producers sell half of their production on term contracts and half in the spot market, or they allow term contract holders to sell the oil in the spot market. Fourth, oil sales can proceed through oil tenders. Tenders can take many shapes but typically involve a seller (or buyer) that invites counterparties to bid for a defined volume and quality of oil. Fifth is the retroactive pricing sale under which seller determines what a buyer needs to pay after all product prices are known. A fifth method is the use of long term delivery or offtake agreements under which a producing country enters into a take or pay supply arrangement with another country or party.[77] Long term delivery agreements provide the seller with an assured revenue stream that ensures an adequate return on the significant project capital investment and risks for energy infrastructure such as pipelines and refineries. Such agreements place the risk of deteriorating market conditions on the buyer by requiring it to always be responsible for the payment for a minimum purchase commitment (sometimes referred to as volume risk shifting), leaving the seller with only the market price risks to manage, which in some cases may be hedged. This creates a secure outlet for their crude and alter the price depending on the refining margins.

Finally, oil sales can proceed through outsourcing. Some producers prefer to hire international trading firms or their oil company partners to sell their oil as they lack the skills or the commitment to invest in skills'.[78]

Generally, the two most common types of oil sales in Africa are term sales and spot sales.[79] By extension, the two most common types of oil sale agreements are term and spot contracts. These are discussed below.

Term Contracts A term contract is generally an agreement between two parties in which one party agrees to supply a commodity on a continuing basis to a second party for a price or for other considerations.[80] In the oil and gas context, it is a contractual relationship, between oil producer or supplier and a buyer seeking to purchase and lift crude oil for a specified period of time. Term contracts account for two-thirds of all exported oil because they guarantee security of supply for refineries and of demand producers.[81] They help guarantee a fixed source of sale and revenue for an oil exporting country over a fixed period of time. The contract includes the formula price which specifies how the price will be set. The contracts also outline the volume to be sold and the place of delivery. Term contracts also ensure operational predictability because refineries can predict the quality, volume and delivery dates, and produces know how much of which crude stream to produce.[82]

---

[77] This is also known as a take or pay contract. A take-or-pay contract is essentially an agreement whereby the buyer agrees to either: (1) take, and pay the contract price for, a minimum contract quantity of commodity each year (the " TOP Quantity"); or (2) pay the applicable contract price for such TOP Quantity if it is not taken during the applicable year. See Rogers and White (2013).

[78] van Schaik (2012).

[79] Ibid.

[80] United States Energy Information Administration (EIA).

[81] van Schaik (2012).

[82] van Schaik (2012). Also see for example, General Conditions for Sale and Purchase of Nigerian Crude.

A key component of a term contract is the price formula. This is usually calculated with two inputs: a price differential and an established international benchmark.

Spot Contracts A spot contract is a contract for the sale and purchase of oil and for immediate settlement or payment at current market rates. A spot transaction is "a one-time open market transaction for immediate delivery of a specific quantity of product at a specific location where the commodity is purchased 'on the spot' at current market rates."[83] A spot contract differs from a futures contract under which contract terms are agreed now but delivery and payment occur at a future date.[84] Spot cargoes are typically loaded a month or more after the date of the deal because of the time it takes to book space on pipelines, secure a vessel and sail it to the port for its cargo.

8.2.2.7 Mining Development Agreement (MDA)

As the name suggests, MDAs are contracts solely utilized in mining projects. A MDA governs the relationship between the government and a licence holder for the exploration of a certain area of land for minerals or for the mining of minerals in a certain area in exchange for royalties, taxes and other obligations.[85] Mining agreements are not used in all countries, especially those countries that have well established concessionary frameworks. In such countries, parties rely on the applicable licenses, leases and applicable law to govern their respective rights and obligations.[86] As the contractual regime is more commonly followed by countries that do not have a very elaborate minerals law, the MDA, unlike concessions, is typically a lengthy document. Under an MDA the host government (as the owner of the minerals, for example, coal, iron ore) grants an investor a licence to undertake mining, development and (sometimes) exploration activities, and to provide the requisite finance and technical skills in order to explore for (and hopefully produce) minerals in accordance with the local mining legislation.[87]

MDAs can be contrasted with PSAs, that are predominantly used in the petroleum sector. Unlike in PSAs, under which the contractual relationship is based on sharing proceeds of production on a predetermined basis by government and the EC, in a MDA, the investor is granted an exclusive right to exploit within a defined

[83] United States Energy Information Administration (EIA)b.

[84] While many countries that export oil use a mixture of both term and spot contracts. There are others that solely export oil through term contracts, for instance Saudi Arabia and Nigeria. Then there are others that solely trade their oil in the spot market. For instance, Congo and Norway sell all their oil on the spot market. See General Conditions for Sale and Purchase of Nigerian Crude Oil.

[85] See the Negotiations Portal for Host Country Governments.

[86] South Africa is one African nation that does not rely on mining agreements as it has a strong and robust legislative framework on mining. See Extractive Industries Transparency Initiative (2014).

[87] Allen and Overy LLP (2013).

contract area and to process, market and sell any minerals it successfully extracts. In return, it pays the state for this right, by way of taxes, royalties and rentals.[88]

A typical MDA will clarify the EC's obligations during various stages of development, its rights to extract minerals and what minerals are covered, and rights and obligations with respect to infrastructure. It may also protect the EC from unlawful confiscation of its property, and will usually provide for special dispute resolution procedures. In many cases, the MDA references and incorporates general law, but can also provide for departures from general law. For instance, in the tax sections, the MDA may modify certain of the income tax rules or it may provide for different tariffs'.[89]

Key Provisions in an MDA Some of the key provisions in a typical MDA are discussed below.

a. *Operations and Control:* Typically, the Contractor or EC will be solely in charge of operations under the contract. However, in some jurisdictions, an operator may be appointed to assist the Contractor to carry out mining operations (e.g. Guinea). In such cases, the MDA clarifies the operator's role, responsibilities, appointment, remuneration, and dismissal. Similarly, a committee may be appointed to assist with carrying out certain administrative tasks relating to the operations.[90]
b. *Work Program and Budgets:* This provision typically requires the private investor to submit minimum expenditure commitments on an annual basis. The scope and extent of the work, as well as expenditure obligations vary across jurisdictions. In some, the Contractor must use commercially reasonable efforts to incur capital expenditure.[91] In other countries, the private investor is required to submit detailed work and expenditure programs to the host government.[92] Where the Contractor fails to meet its minimum expenditure commitments, it is typically required to compensate the State to account for the deficit.[93]
c. *Relinquishment:* This provision obliges the private investor to surrender certain parts of the contract area back to the host government. Generally, the parts surrendered are unexplored. Relinquishment helps promote efficiency as it encourages the investor to focus on areas of interest. As such, the private investor usually has the prerogative in deciding which areas it desires to relinquish.
d. *Commercial Discovery/Feasibility Study:* The requirement to submit a feasibility study to the host government must be met before the investor can undertake any production. A feasibility study highlights the strengths and weaknesses of the proposed venture. The investor is generally obliged to prepare and submit a feasibility study to the State for its review and/or approval. Following approval, the

[88] Ibid.

[89] Ibid, 35.

[90] Ibid.

[91] Liberia follows this approach. Allen and Overy LLP (2013), p. 6.

[92] Sierra Leone is an example for this approach. Ibid.

[93] Guinea follows this approach. Ibid.

investor is able to construct and operate the mines necessary to extract the minerals. The investor may also be obliged to prepare and produce geological and other types of maps and reports and make these available to the host government.[94]

e. *Taxation:* The IOC is typically required to pay various taxes in accordance with the applicable laws of the jurisdiction, including income tax, export tax and duties, import duties, and stamp duty. The tax provisions of a MDA are some of the most heavily negotiated, as stability of the tax regime and keeping costs as low as possible are very important to the investor.[95]
f. *Royalty:* Royalties are payable by the EC to the host government, as compensation for its exploitation of the host government's minerals, and are typically agreed upon as a percentage of the volume or value of extracted minerals from the contract area, and may vary between different minerals.

8.2.2.8 Impact and Benefit Agreements (IBAs)

Most MDAs, and even petroleum contracts, include or reference an IBA, also known as a community development agreement. An IBA is a consultation-based agreement between the EC and local communities that addresses both how the community can take advantage of development gains offered by the project, and how environmental, human rights or social impacts from the project may be mitigated and managed. The primary purpose of an IBA is threrefore twofold. First, is to address the adverse effects of mining activity on the local communities and their environment. Second, to ensure that the people in communities, where the project is situated, receive benefits of mineral resource development, including pro-poor initiatives and other long term development strategies which may be beyond the immediate scope of impacts of a project.[96] IBAs have become recognized as effective mechanisms for addressing exclusions of poor and marginalized communities in natural resource development.[97] IBAs comprehensively address social, environmental, and economic impacts of the project and project closure.[98]

IBAs are not specific to the extractive industry and are in fact also used in the agriculture sector.[99] Their predominant use, however, occurs in the mining industry. IBAs are signed in addition to the main agreement that governs the right and obligations directly related to the mining project. The main agreement is usually between the host government and the private investor. IBAs, on the other hand, are between the local community and the private investor. They are entered into to ensure com-

[94] The submission of geological maps and reports to the government is mandatory in Liberia. See Allen and Overy LLP (2013), p. 7.

[95] Ibid.

[96] World Bank Sustainable Energy – Oil, Gas and Mining Unit (2012), p. 5.

[97] Brickner (2016).

[98] For a model template Mining Development Agreement, see generally, Mann et al. (2013).

[99] Brickner (2016).

munity development by creating a relationship between the extractive industry and local communities.[100] IBAs aim to ensure that a reasonable share of mining benefits, profits and technology goes to those communities close to and disproportionately affected by a mining operation.[101] The parties to IBAs are therefore the local communities that are affected by mining operations and the mining industry. IBAs have become so prevalent that in some countries they are required by law.[102]

Typically, IBAs contain provisions that address advancing employment skills training, provision of educational and business opportunities for members of the community, and compensatory payments for those members of the community who have incurred losses due to mining development in the area. Additionally, IBAs also include environmental obligations and monitoring mechanisms despite being separate from the routine environmental assessment process.[103]

Elements of a Successful IBA A study conducted by the World Bank on IBAs identified the following 'building blocks' of successful IBAs.[104]

a. *Responsibilities and Obligations:* A comprehensive IBA must clarify and define roles, responsibilities and expected behaviors of all signatories to the agreement. It is also important to consider and clarify the role of government and civil society to avoid overlaps.
b. *Timeframe and Process:* Schedules for continued monitoring and evaluation must be included in the agreement. Additionally, the potential to withdraw may also be included in order to help stakeholders manage risk and expectations.
c. *Stakeholder Engagement:* An effective way to administer IBAs is to ensure transparent multi-stakeholder consultation and engagement.[105] Engaging with indigenous and local communities during project design and implementation provides a practical opportunity for national authorities and IOCs to ensure free prior informed consent (FPIC) in the allocation and utilization of benefits and

---

[100] Agreements focused on community development take many names other than IBA. Some examples are community development agreements, participation agreements, landowner agreements, benefits sharing agreements, etc. Community Development Agreement is the term adopted by the World Bank. For a complete list of different names given to IBAs and also a complete guide towards negotiating an effective and sustainable IBA, see World Bank Sustainable Energy – Oil, Gas and Mining Unit (2012).

[101] Sustainable Development Division of Economic Commission of Africa (2004).

[102] Papua New Guinea, South Africa and Chile are examples of countries where IBAs are required by the law. See World Bank Sustainable Energy – Oil, Gas and Mining Unit (2012).

[103] Ghana provides a good example of a working IBA. Newmont Mining Corporation, in line with its commitment to economic and social development of impacted communities, developed formal agreements with local communities covering responsibilities of different stakeholders. The main focus was on employment and development of a new foundation titled the Newmont Ahafo Development Foundation (NADeF). Leaders from ten Ghanaian villages entered into IBAs with Newmont. Ibid.

[104] For the complete list see ibid 11.

[105] On the importance of multi-stakeholder approach, see Odumosu-Ayanu (2014).

resources from indigenous and local communities.[106] FPIC in international law requires national authorities and private actors to respect, protect and fulfil the rights of concerned stakeholders to meaningfully participate in decisions directly impacting their lands, territories, and resources.[107] FPIC underlines the idea that consent must be freely given, obtained prior to implementation of activities, and be founded upon an understanding of the full range of benefits and challenges resulting from the activity or decision in question.[108]

Indigenous communities have the right to fully understand the local content benefits that could result from an extractive activity, and to be able to take part in and influence decision making processes relating to the delivery of such benefits. 'Free' in the FPIC framework implies no coercion, intimidation or manipulation in decision making processes.[109] 'Prior' implies that consent and approval of host communities must be sought in advance of any authorization or commencement of extractive operations.[110] 'Informed' refers to the full disclosure of the intent, scope and benefits of the projects and policies, most especially local content benefits. 'Consent' means that the final decision of host communities with respect to in-community value must be collectively granted or withheld.[111]

A comprehensive stakeholder engagement process provides a chance for national authorities, host communities, and IOCs to cooperatively clarify and agree upon what constitutes success in terms of local content and "in-nation" value addition in accordance with FPIC tenets. Local content objectives must be specific, measurable and achievable to avoid ambiguities and misalignments. Rather than adopting a rigid and inflexible approach, a collaborative approach that clarifies the expectations of governments and host communities, while providing the IOC with the flexibility to develop its local content plans and procurement procedures, could achieve greater results. Also, rather than stipulating blanket and unrealistic timeframes, countries, in defining LCRs and ambitions, must recognize that the timeframe for delivering in-country value may vary from project to project and could be affected by political and other factors not envisaged during contract negotiation.

---

[106] Ruggie (2011), paras 8–12; Special Rapporteur on the Rights of Indigenous Peoples (2009), *Promotion and Protection of All Human Rights, Civil, Political, Economic, Social and Cultural Rights, Including the Right to Development: Report of the Special Rapporteur on the Situation of Human Rights and Fundamental Freedoms of Indigenous People*, 48, UN Doc. A/HRC/12/34 (15 July 2009).

[107] See Articles 1, 12, 20, 27, 30 and 32 of the United Nations Declaration on the Rights of Indigenous Peoples (UNDRIP) A/RES/61/295, adopted 2 October 2007; also Art. 1 of both the International Covenant on Civil and Political Rights and the International Covenant on Economic, Social and Cultural Rights (ICESCR); Articles 6, 7, 16, 16 and 22 of the International Labour Organization's *Convention on Indigenous and Tribal Peoples in Independent Countries - 169/1989.*

[108] Anaya (2005), p. 7; Colchester and MacKay (2004), pp. 8–14.

[109] See United Nations Environment Program (2012), p. 7.

[110] Ibid.

[111] Ibid.

d. *Grievance and Dispute Resolution Mechanisms:* A comprehensive IBA should include dispute resolution mechanisms that allow local communities to seek a review or evaluation of a project if agreed conditions are not clearly met in operations activities. Adequate funding must be earmarked for these mechanisms as this is a key component of a successful IBA.[112]

8.3 Conclusion

With greater reliance on extractive resources, as dominant source of income by African countries, comes a continuous search for fiscal systems that could allow resource rich African countries to maximize economic benefits from the geological affluence situated in the continent. This search for higher extractive benefits has resulted in a move away from traditional concessions, and the evolution to the modern concessions which contain more balanced, equitable and flexible terms for host states. The need for greater cooperation and partnership-based relationships in resource development also saw the emergence of contract based models such as the production sharing contract, joint venture, and service contracts, all of which give the host governments more control over their resources, reduce their risk and more often than not, increase their rewards. Furthermore, other agreements such as technical assistance agreements, conservation agreements, such as pooling and unitization agreements, oil sale contracts and mining development agreements have provided greater flexibility for host countries to attune their relationships with ECs in line with modern realities and government priorities.

In addition, over the last decade, African countries have increasingly introduced legislation and contractual requirements that move beyond maximizing economic benefits to also ensuring 'in country value' and social development goals such as technology transfer, employment of nationals, capacity development and the development of local industries. The next chapter examines legislative and contractual trends relating to local content in Africa's extractive industries. It provides a comparative analysis and profile of local content policies n African countries, in order to determine prevailing preferences and legal risk points.

References

Alberta Energy Regulator. Compulsory pooling. https://www.aer.ca/regulating-development/project-application/application-process/compulsory-pooling. Accessed 1 June 2018

Alberta's Oil and Gas Conservation Act (OGCA) RSA 2000, c O-6

Alberta's Oil and Gas Conservation Rules (OGCR). Alberta Regulation 151/1971

[112] In Ghana, a complaints resolution committee is set up for the Ahafo Project. The resolution process involved community representatives. See World Bank Sustainable Energy – Oil, Gas and Mining Unit (2012), p. 13.

Al-Emadi T (2010) Joint Venture Contracts (JVCs) among current negotiated petroleum contracts: a literature review of JVCs development, concept and elements. Georgetown J Int Law 1:645

Allen and Overy LLP (2013) Guide to extractive industries documents. EI Sourcebook. http://www.eisourcebook.org/cms/Jan%202014/Guide%20to%20Mining%20Documents.pdf. Accessed 1 June 2018

Anaya S (2005) Indigenous peoples' participatory rights in relation to decisions about natural resource extraction: the more fundamental issue of what rights indigenous peoples have in lands and resources. Ariz J Int Comp Law 22:7

Anderson OL (1985) Oil and gas conservation on Canada lands. The Canadian Institute of Resources Law, Calgary

Andrews Kurth Kenyon LLP (2014) Unitisation – the oil and gas industry's solution to one of geology's many conundrums. Lexology 2014. http://www.lexology.com/library/detail.aspx?g=69912ce0-127d-447e-8a47-50f5d235dcfa. Accessed 28 May 2018

Ballem JB (2008) The oil and gas lease in Canada, 4th edn. University of Toronto Press, Toronto

Bankes N (1997) Compulsory pooling under the Oil and Gas Conservation Act of Alberta. Alta Law Rev 35:945

Barnes P (1995) Indonesia. The political economy of oil. Oxford University Press, Oxford

Bindemann K (1999) Production sharing agreements: an economic analysis. Oxford Institute for Energy Studies, Oxford

Black's Law Dictionary (2004) In: Garner B (ed) 8th edn. Thomson

Blinn K, Duval C, Le Leuch H, Pertuzio A, Weaver J, Anderson O (2009) International petroleum exploration and exploitation agreements: legal, economic and policy aspects, 2nd edn. Barrows Company Inc., New York

Brickner KD (2016) Community development agreements in mining projects. Denver J Int Law Policy 44(3):413

Colchester M, MacKay F (2004) In search of middle ground: indigenous peoples, collective representation and the right to free, prior and informed consent. Forest Peoples Programme, Moreton-in-Marsh

Deep Offshore And Inland Basins Production Sharing Contracts Decree 1999

Etteh N (1973) From concession to participation: restructuring the Middle East oil industry. N Y Univ Law Rev 48:774–816

Euroil Ltd v Cameroon Offshore Petroleum Sarl [2014] EWHC 12 (Comm) (06 January 2014)

Extractive Industries Transparency Initiative (2014) Mining contracts: how to read and understand them. https://eiti.org/sites/default/files/documents/mining-contracts-how-to-read-and-understand-them.pdf. Accessed 12 June 2018

Fabrikant R (1975) Production-sharing contracts in the Indonesian petroleum industry. Harv Int Law J 16:303

Freehold Owners Association (FHOA). Unit agreements. http://www.fhoa.ca/about-other-agreements/unit-agreements.html. Accessed 12 June 2018

General Conditions for Sale and Purchase of Nigerian Crude Oil. https://resourcegovernance.org/sites/default/files/GeneralConditionsforNNPCOilSalesTermContract2011.pdf. Accessed 2 May 2018

Handlan A, Sykes K (1984) Pooling and unitization: legal and ethical considerations. Tulsa Law J 19(3):309

Holland and Hart LLP (2014) The nuts and bolts of farmout agreements. Lexology

International Labour Organization's Convention on Indigenous and Tribal Peoples in Independent Countries - 169/1989

International Monetary Fund (2012) Fiscal regimes for extractive industries: design and implementation. IMF Fiscal Affairs Department. https://www.imf.org/external/np/pp/eng/2012/081512.pdf. Accessed 15 May 2017

Ithaca Energy (UK) Ltd v North Sea Energy (UK) Ltd (Rev 1) [2012] EWHC 1823

Kaiser M (2004) Fiscal system analysis: concessionary and contractual systems used in offshore petroleum arrangements. Coastal Marine Institute, Baton Rouge

Kramer B, Martin P (2006) The law of pooling and unitization, 3rd edn. Lexis Nexis, pp 1–30
Likosky M (2006) Law, infrastructure, and human rights. Cambridge University Press, New York
Likosky M (2009) Contracting and regulatory issues in the oil and gas and metallic minerals industries. Transl Corp 18(1):1
Lowe JS (1987) Analyzing oil and gas farmout agreements. Sw Law J 41:759 http://scholar.smu.edu/smulr/vol41/iss3/2. Accessed 27 May 2017
Machmud T (1993) Production-sharing contracts in Indonesia: 25 years' history. J Energy Nat Resour Law 11:179
Mann H, Danielson L, Disney K (2013) Model mining development agreement-transparency agreement, prepared by the sustainable development strategies group. International Institute for Sustainable Development, Winnipeg
Martin P, Kramer B (2009) Williams and Meyers manual of oil and gas terms, vol 203. Lexis Nexis, pp 682–704
Martin P, Kramer B (2012) Williams & Myers, oil and gas law. Lexis Nexis, New York
Mineral and Petroleum Resources Development Act 28 of 2002 (MPRDA) of South Africa
Mogale Alloys (Pty) Ltd v Nuco Chrome Bophuthatswana (Pty) Ltd and Others 2011 (6) SA 96 (GSJ) para 36
Muscolino R, Rizzo C, Mirabelli G (1993) The cost recovery oil in a production sharing agreement. Society of Petroleum Engineers. https://doi.org/10.2118/25844-MS. Accessed 21 Apr 2018
Natural Resources Governance Institute (2015) Fiscal regime design: what revenues the Government will be entitled to collect. https://resourcegovernance.org/sites/default/files/nrgi_Fiscal-Regime-Design.pdf. Accessed 12 June 2018
Neave M (1970) The conservation of oil and gas: a comparative study of on-shore legislation in Australia and America. Melb Univ Law Rev 7:201
Negotiation Support Portal for Host Governments. Mining Development Agreement. http://negotiationsupport.org/glossary/mining-development-agreement. Accessed 12 June 2018
Neto J (1985) Risk-bearing service contracts in Brazil. J Energy Nat Resour Law 3:114–120
Odumosu-Ayanu I (2014) Governments, investors and local communities: analysis of a multi-actor investment agreement framework. Melb J Int Law 15(2):473
Office of the Presidential Adviser on Petroleum and Energy (2001) Guidelines for the farmout and operation of marginal fields
Omorogbe Y (1997) The oil and gas industry: exploration and production contracts. Malthouse Press, Lagos
Onuma O (2015) Mainstreaming sustainability: an evaluation of Alberta's legal and regulatory regime for gas development. Canadian Institute of Resources Law, Calgary
Pierce DE (2007) Transactional evolution of operating agreements in the oil and gas industry. Rocky Mountain Mineral Law Foundation, Westminster
Radon J (2005) The ABCs of petroleum contracts: license-concession agreements, joint ventures, and production-sharing agreements. In: Salik S, Schiffrin A (eds) Covering oil: a reporter's guide to energy and development. Open Society Institute, New York
Rogers DR, White M (2013) Key considerations in energy take-or-pay contracts. King and Spalding, Atlanta
Ruggie J (2011) United Nations guiding principles on business and human rights: implementing the United Nations 'Protect, Respect and Remedy' framework. UN Document A/HRC/17/31
Smith EE (1992) From concessions to service contracts. Tulsa Law J 27(4):493
Special Rapporteur on the Rights of Indigenous Peoples (2009) Promotion and protection of all human rights, civil, political, economic, social and cultural rights, including the right to development: report of the special rapporteur on the situation of human rights and fundamental freedoms of indigenous people, 48, UN Document A/HRC/12/34
Sustainable Development Division of Economic Commission of Africa (2004) Improving public participation in the sustainable development of mineral resources in Africa. Africa Mining

Vision. http://www.africaminingvision.org/amv_resources/AMV/public_participation.pdf. Accessed 29 May 2017

Taylor M, Tyne S (1992) Joint operating agreements. Longman, London

Tordo S, Johnston D, Johnston D (2010) Petroleum exploration and production rights: allocation strategies and design issues. World Bank Working Paper No. 179. http://documents.worldbank.org/curated/en/785881468336848695/pdf/518400PUB0REPL101Official0use0Only1.pdf. Accessed 16 May 2017

United Nations Declaration on the Rights of Indigenous Peoples (UNDRIP) A/RES/61/295, adopted 2 October 2007

United Nations Environment Program (UNEP) (2012) UN-REDD programme guidelines on free prior informed consent. UNEP, Nairobi

United States Energy Information Administration (EIA). Definitions, sources and explanatory notes. EIA. https://www.eia.gov/dnav/pet/TblDefs/pet_pri_spt_tbldef2.asp. Accessed 12 Apr 2018

United States Energy Information Administration (EIA). EIA energy glossary. https://www.eia.gov/tools/glossary/index.php?id=T. Accessed 2 May 2018

Van Schaik J (2012) Selling the citizens' oil: how Governments sell their oil. Revenue Watch Institute, New York. https://resourcegovernance.org/sites/default/files/OilSales-HowGovtsSellOil.pdf. Accessed 30 May 2017

Wigley ML (1978) AAPL form 610-1977 model form operating agreement, Rocky MTN. MIN Law Inst, 24

Wigley M. Review of pooling and unitization. National Association of division order analysis. http://hadoa.memberlodge.org/resources/Documents/01_CDOA/Chapter%2007_Manual%20Plano%20Review.pdf. Accessed 28 May 2018

Wilkinson J (1997) Introduction to oil and gas joint ventures: United Kingdom continental shelf. Oilfield Publications Ltd., London

Williams HR, Meyers CJ (1957) The effect of pooling and unitization upon oil and gas leases. Calif Law Rev 45(4):411 http://scholarship.law.berkeley.edu/cgi/viewcontent.cgi?article=3249&context=californialawreview. Accessed 27 May 2017

Williston S (1959) A treatise on the law of contracts

World Bank Sustainable Energy – Oil, Gas and Mining Unit (2012) Mining development agreements: source book. The World Bank, Washington, DC. http://siteresources.worldbank.org/INTOGMC/Resources/mining_community.pdf. Accessed 1 June 2017

Chapter 9
Local Content and Product Mandating Requirements

Abstract Over the last decade, African countries have increasingly introduced local content legislation and contractual requirements that mandate extractive companies to give priority to nationals, indigenous communities, domestic companies and locally produced material, in the procurement of goods and services used for petroleum and mining operations. While such policies can enable national authorities to maximize in country value of extractive activities, beyond mere economic benefits, they can also serve as disincentives to foreign investors if such policies are not properly designed.

This chapter examines how local content policies have evolved across Africa. It provides a comparative analysis and profile of local content obligations in African countries, in order to determine prevailing preferences and legal risk points. It also examines how investors and national authorities can evolve collaborative and achievable local content requirements that minimise disputes and legal risks.

9.1 Introduction

African countries are increasingly introducing local content requirements (LCRs) into their legal framework, through legislation, regulations, guidelines, industry contracts and bidding practices. Despite the clear and uniform overall policy drivers of LCRs and product mandating requirements, approaches taken to enforce LCRs vary and must be carefully understood and clarified to avoid misalignments and contractual mismatch between governments and international companies. This chapter examines how LCRs have evolved across the African continent. It provides a comparative analysis and profile of LCRs in African countries, in order to determine prevailing preferences and legal risk points. It also examines how investors and national authorities can evolve collaborative and achievable LCRs that minimise disputes and legal risks.

After studying this chapter, you should be able to understand the meaning, nature and scope of LCRs; appreciate the drivers of LCRs in extractive jurisdictions in Africa; explore if and how local content is defined and codified in your country; identify the key sources of LCRs in your country; and understand how national

D. S. Olawuyi, *Extractives Industry Law in Africa*,
https://doi.org/10.1007/978-3-319-97664-8_9

governments and extractive companies (ECs) can evolve clear, mutually beneficial and realistic local content obligations.

9.2 Meaning, Nature and Scope of LCRs

LCRs are contractual provisions, laws and policies that require extractive companies to give priority to nationals, indigenous communities, domestic companies and locally produced material, in the procurement of goods and services used for petroleum and mining operations.[1] Some LCRs mandate the procurement of raw materials and goods from certain suppliers; a few mandate the use of endogenous technologies and methods that are within the capability of local contractors; while others mandate the use of pre-qualification and local registers to ensure that local contractors with the right experience are awarded work.[2] Whatever the scope, the underpinning notion of LCRs is that they require foreign companies to give priority to nationals, domestic companies and locally produced material, in the procurement of goods and services used for petroleum and mining operations.[3] LCRs measure the extent to which the extractive industry sector generates value diversification, growth and development to other sectors of the domestic economy.[4] LCRs are designed to leverage economic benefits from the extractive sector as a tool for reinforcing and developing other sectors, through the promotion of local skills development, knowledge creation, economic diversification and participation of local communities in extractive projects and activities.[5]

LCRs are generally utilized by governments, especially in oil and gas rich countries, to generate broader economic benefits for the local economy, beyond fiscal benefits, royalties and rents.[6] These economic benefits may include employment of nationals, procurement of goods and services from companies resident in the host country, partnerships with local entities, development of endogenous technology and infrastructure, and the improvement of the skills and capacity of local businesses and the domestic workforce.[7] An aspect of local content is the composite value contributed to the national economy from the purchase of bought-in goods

[1] See Olawuyi (2017a, b), pp. 1–5; also Heum et al. (2011).

[2] Olawuyi, ibid.

[3] It is estimated that over 90% of resource-rich countries have at least one form of LCR as regards their extractive industries, 50% of which are quantitative targets or requirements. See McKinsey Global Institute (2013), International Petroleum Industry Environmental Conservation Association (2011), Heum et al. (2011).

[4] Tordo et al. (2013), pp. 1–10.

[5] Organisation for Economic Co-operation and Development (2017), pp. 6–7; also Levett and Chandler (2012) and Groupo FARO (2015), pp. 1–5.

[6] Ramdoo (2015); also Gbedi and Adebisi (2013), p. 90.

[7] International Petroleum Industry Environmental Conservation Association (2011).

and services, and includes wages and benefits, materials, equipment and plant, subcontracts and taxes.

LCRs can also provide an opportunity to bring jobs and income to areas where there is considerable dissatisfaction about the overall benefits of extractive operations.[8] Despite government programs, such as price reduction, monthly stipends and subsidies, that could have been put in place to evenly distribute extractive wealth in a country, especially in resource rich communities, different groups and individuals have varying perceptions of what they may think of as commensurate benefits. This can lead to pressure from the population to increase the more tangible benefits.[9] In a more narrowly focused context, communities nearer to the oil and gas facilities, especially when it is located onshore, may suffer various adverse effects from its operation. Although these problems could be addressed by specific policies designed to consider the exact grievance, LCRs provides a tool for increasing the opportunities for local communities in resource producing communities to play more active roles in the management of resources found within their territories. By implementing policies that increase local participation in oil, gas, and mining sectors, governments aim to increase the opportunities for local businesses and individuals to play prominent roles in the extractive industry value chain, thereby increasing growth, incomes, and ultimately human development. Such policies usually aim to create an environment that will localize extractive sector operations and foster industrial transformation in the resource producing communities. Not only does greater participation in extractive industry supply chains create employment and industrial development, it could also increase the level of local support for extractive projects.[10]

The use of LCRs to support broad-based economic growth in the extractive sector is not exactly a new trend. LCRs were first introduced in the North Sea in the early 1970s and ranged from restrictions on imports to direct state intervention in the oil sector.[11] For example, in Canada, especially in the provinces of Saskatchewan, Newfoundland and Nova Scotia, LCRs have included contractual obligations that thirty five (35) per cent of all goods and services are to be sourced from aboriginal-owned businesses.[12] More recently, in 2015, Saudi Arabia's national oil company, Saudi Aramco, launched its In Kingdom Total Value Add (IKTVA) program, aimed at driving, measuring and monitoring the "added value" brought to the Kingdom by

[8] Ibid.

[9] Ibid.

[10] Ramdoo (2015) and Olawuyi (2017a, b).

[11] Tordo et al. (2013).

[12] See for example Article 4 of the Hibernia Southern Extension Benefits Agreement, requiring operators to provide a full and fair opportunity for all Canadian businesses and individuals, and first consideration for those in Newfoundland, to participate in the supply of goods and services and in employment opportunities arising from petroleum operations. See also Government of Saskatchewan (2015) Northern Benefits Summary—Saskatchewan Mine Surface Lease Agreements. See also s. 45 of the Canada–Newfoundland and Labrador Atlantic Accord Implementation Act (S.C. 1987, c. 3).

a contractor.[13] Likewise, in 2012, the United Kingdom (UK) Offshore Wind Industry Council (OWIC) approved an aspirational target of achieving 50% UK content in offshore renewable energy projects by 2020.[14] In Nigeria, the local content directive (LCD) and the Nigerian Content Policy put in place by the Nigerian Content Division of the Nigerian National Petroleum Corporation (NNPC) as far back as the early 2000s aimed to increase local content value in the provision of materials, services and equipment to the local oil and gas industry to 75% by 2010.[15] The aim of the NCD was to provide guidelines for the implementation and periodic reporting of the implementation of the Nigerian Content Policy.[16] Similarly, Regulation 26 of the *Petroleum (Drilling & Production) Regulations*, made pursuant to Section 9 of the *Petroleum Act*, is a precursor to the Nigerian Content Policy.[17] Regulation 26 provides that a licensee of an oil prospecting license and a lessee of an oil-mining lease must submit an employment plan for Nigerian employees within 12 months of the grant of a license or the grant of a lease, respectively.

LCRs have taken an even more significant dimension in African countries over the last decade, and more recently following the global fall in oil prices. Since 2014, the price of a barrel of oil has fallen more than 70%, wiping out more than $360 billion of revenue from oil producing countries in 2015 alone.[18] With official forecasts by the Organization of Petroleum Exporting Countries (OPEC), showing that a return to $100 per barrel price of oil may not be until after 2040,[19] African countries have to, more than ever, re-think how to diversify their domestic economies, and create new jobs in manufacturing and the emerging technology and innovation sectors in order to sustain current economic growth.[20] The increased introduction of more robust LCR requirements is a positive attempt by African countries to utilise oil and gas production as a key to add value to, and unlock the robust development of, other important sectors of their domestic

---

[13] See In Kingdom Total Value Add (IKTVA) program.

[14] See RenewableUK (2017) pp. 1–2.

[15] See generally, Nigerian Content Development & Monitoring Board (NCDMB), *Handbook of Operational Guideline* (2016), pp. 1–10.

[16] The Nigerian Content Division of NNPC sought to transform the oil and gas industry into the economic engine for job creation and national growth by developing in-country capacity and indigenous capabilities. It was designed to ensure that a greater proportion of the work was done in Nigeria, with active participation of all sectors of the economy. In addition, its ultimate aim is to position Nigeria as the hub for service delivery within the West African sub-region and beyond. Although the Nigerian Content Directive did not have the force of law, it had far-reaching effects with NNPC as senior partner in most oil and gas projects, and aimed to increase local content value in the provision of materials, services and equipment to the local oil and gas industry to 45% at the end of 2006, and 75% by the end of 2010. See generally, NCDMB; also OI Obasi.

[17] See the *Petroleum Act 1969* (as amended), Laws of the Federation of Nigeria, 2004, 10. [*The Petroleum Act*].

[18] International Monetary Fund (2015).

[19] Organization of Petroleum Exporting Countries (2015). Also see Mohinsky (2015).

[20] Ulrichsen (2011) and Ghafar (2016).

economies.[21] Similarly, the desire to increase domestic valued-added by substituting domestically produced goods for imported goods, and to create more local employment by substituting domestic labor for imported or foreign-based labor are among the most frequently cited objectives of LCRs.[22] In Nigeria, the passage of a comprehensive legislation, the *Nigerian Oil and Gas Industry Content Development Act* in 2010 has given the force of law to the objectives of the Nigerian Content Policy.[23] The objective of the Act, as set out in the preamble, is to provide for the development of Nigeria Content in the Nigerian Oil and Gas industry and the supervision, coordination, monitoring and implementation of Nigerian content and all incidental matters.[24] The law also establishes the Nigerian Content Monitoring Board with the mandate to provide guidelines for the implementation of the provisions of the Act.[25]

While governments in resource rich countries favour LCRs as policy tools for maximizing the degree of local benefits from the extractive sector, LCRs could also provide a basis for international oil companies (IOCs), service companies and investors to negotiate some level of freedom to procure goods and services on the basis of cost, availability and quality.[26] If carefully negotiated, LCRs can provide a basis for IOCs and indigenous communities to negotiate mutually beneficial terms

---

[21] While African countries such as Nigeria, Ghana and Uganda, have enacted comprehensive local content legislation and regulations, local content obligations in other African countries are spread across various laws and decrees passed between 2003 and 2015. See GHANA's *Petroleum (Local content and Local Participation in Petroleum Activities) Regulations 2013; and* Minerals and Mining (General) Regulations 2012 (LI 2173) which provide comprehensive legal framework for local content respectively in petroleum and mining sectors; also *Local Content and Local Participation in Petroleum Activities Policy Framework 2011.* However, in countries such as Angola, South Africa, Kenya and Tanzania for example, LCRs are spread across instruments. In Angola for example, see Law for the Promotion of Business for Local Private Companies 2003 ("Law 14/03"); General Regulatory Framework for Hiring of Services and Goods from National Companies in the Oil Industry ("Decree 127/03"); Decree 48/06 on Open Tender Procedures in the Oil Industry ("Decree 48/06"), Decree Law on the rules and procedures to observe in recruitment, integration, training and development of workers from the oil sector ("Decree-Law 17/09"). There is also the Decree 39/08 establishing the Fund for the Promotion of Entrepreneurship which is administered by the Ministry of Finance to support the creation of local companies.

[22] See Tordo et al. (2013).

[23] Nigerian Oil and Gas Industry Content Development Act 2010.

[24] Ibid.

[25] Nigerian Oil and Gas Industry Content Development Act 2010, s. 4–5. Section 70 prescribes the functions/mandate of the Board which include the following: implementation of the Act; implementation of the regulations made by the Minister in relation to any aspect of the Act; supervision, coordination, administration, monitoring and management of the development of Nigerian Content in the Industry; as well as supervision, coordination, administration and monitoring of the implementation and development of Nigerian Content as specified in the Schedule to the Act. Section 70 (p) mandates the Board to do legally, "anything necessary to be done to facilitate the carrying out of its functions."

[26] See Olawuyi (2017a, b).

that could address social, economic and environmental concerns of extractive production in indigenous communities.[27]

However, although proponents of LCRs point to its effectiveness in providing a level playing field for local industries, creating job opportunities for citizens, and fast tracking the transfer of technical expertise and skills, LCRs can lead to misalignments between governments and investors, especially in emerging countries with unclear and unspecific legal framework on LCRs.[28] For example, while LCRs may specify the portion of total expenditures that must be comprised of locally sourced goods and services, procurement procedures are frequently not well established.[29] Potentially unresolved questions include: what amounts to sufficient local content? To what extent can the host Government participate in, or exercise control and influence over, the IOCs procurement process? And is there a requirement for the IOC to report or submit its local content and procurement plans to the joint management committee for approval prior to petroleum operations? If so, what should such a report show?

Another frequently cited argument against the use of LCRs is their potential incompatibility with international trade measures applicable to members of the World Trade Organization (WTO).[30] WTO members are under obligations not to deviate from non-discrimination obligations ("national treatment" and "most-favored nation" obligations), under which foreign companies cannot be forced to buy from local suppliers or hire local service suppliers if a better alternative in terms of price or quality exists abroad.[31] While some WTO members, such as Australia, have introduced carefully worded LCRs that are in line with the NTO clause, others have in line with this obligation followed the cautious approach of not specifying any legislative obligation on LCRs. Interestingly however, the WTO rules impeding the application LCRs are rarely enforced against developing countries.[32] Regarding Agreement on Subsidies and Countervailing Measures (ASCM), for example, domestic subsidies violate WTO rules only when they adversely affect the domestic commerce of another WTO member state.[33] Few domestic subsidies meet this threshold, based in part, on the difficulty, for the complaining party, to show evidence of adverse effects as well as on the fact that the definition of subsidy under the rules is narrow.[34] This has led to a considerable increase in the use of subsidies,

---

[27] Esteves et al. (2013).

[28] See Trade Arabia (2015); S Tordo, M Warner, O Manzano and Y Anouti; see also Darling (2011), pp. 1–10; also Shared Value Initiative (2014), pp. 27–32.

[29] Olawuyi (2015), pp. 247–265; see also Olawuyi and Mercier (2015).

[30] Aladejare (2015), pp. 129–148.

[31] Under the General Agreement on Tariffs and Trade 1994 (GATT 1994) and the Agreement on Trade-Related Investment Measures (TRIMS), WTO members may not apply any domestic measure that discriminates against foreign products or that leads to quantitative restrictions, such as local content requirements. See GATT 1994.

[32] Aladejare (2015), pp. 129–148, Heum et al. (2011).

[33] Cimino-Isaacs et al. (2014).

[34] Aladejare (2015), pp. 129–148, Nwapi (2015).

rather than LCRs, as tools for boosting local value.[35] Rodrik has argued that even though LCRs breach international trade and investment principles, they are an essential part of the policy space that developing countries ought to be allowed to pursue their economic development goals.[36] This view is echoed by another commentator who notes that fair trade in the sense of a "level playing field" does not necessarily mean applying the same set of trade rules and conditions to every nation, but also means recognizing that some countries are so disadvantaged that they need reasonable accommodation under the international trade system.[37] An analogy to this is the adoption of differential income tax rates according to levels of individual income in a number of developed countries, which is not necessarily regarded as unfair because it is justified by the principle of reasonable accommodation for the poor rather than condemned as "unjustifiable discrimination" against the rich.[38] Irrespective of the theoretical viewpoint in which LCRs are analyzed and viewed, an important step is for lawyers and procurement practitioners to ensure a clear and detailed understanding of local content and its possible outcomes in order to keep a country's expectations realistic regarding the implementation of these type of measures. Complementarily, it is necessary to understand the dynamic of every stage of an oil and gas project along its value chain to assess the viability of, and the intervention through, local content requirements or policies.[39]

Lack of clarity, and resulting misalignment in understanding and expectations, between host government and an investor or contractor, on the meaning, nature, scope and methodology for implementing, measuring and reporting local content performance, can result in significant legal risks to the investor.[40] These risks include disallowance of costs for recovery, recalculation of tax obligations, fines, penalties, or even allegations of corruption and/or material breach of the host government contract; all of which may trigger the loss of the petroleum contract.[41] A misalignment of LCRs can also lead to distraction in investor-government communications, disruption in petroleum operations, and potentially costly formal dispute resolution with corrosive effects on the attractiveness of the investment and the host government to the international investment communities.[42] Procurement practitioners must

---

[35] A good example is the petroleum subsidy that has been introduced in Nigeria and other African countries. See Cimino-Isaacs et al. (2014).

[36] Rodrik (2007).

[37] Lee (2016).

[38] Lee (2016).

[39] Aladejare (2015), pp. 146–147, Hanlin (2011).

[40] Legal risks refer to the risk of financial, reputational or investment loss; legal liability; or dispute settlement costs to a company or institution that may arise from defective contractual arrangements or transaction. See McCormick (2010), pp. 1–3; see also Olawuyi and Mercier (2015).

[41] Olawuyi and Mercier, ibid. For example in Iran's legal framework expressly provides that failure to achieve a 51% local content could lead to contract termination or reduction in Remuneration Fee. See Iran's *Maximum Utilization of Production and Services Potency in Providing Country's Needs and Promotion of Exports* 2012.

[42] Esteves and Barclay (2011).

therefore understand the drivers, scope and source of local content in order to effectively anticipate and mitigate legal risks relating to non-compliance.[43]

9.3 Drivers of LCRs in African Countries

There are five key drivers of the increased adoption of LCRs in Africa. First is the desire by host governments to increase the level of domestic capabilities and competencies over time. In Africa, IOCs are often brought in mainly because local industries do not have the experience and technological capacity to effectively undertake petroleum operations.[44] LCRs proceed from the premise that domestic workforce and industries should over time develop the capacities to supply the goods, services and human resources needed to drive the oil and gas value chain, by substituting domestically produced goods for imported goods, and to create more local employment by substituting domestic labour for imported or foreign-based labour. LCRs across the African continent therefore emphasise the desire and need for IOCs to adopt practices that foster the development of a better-trained, qualified domestic workforce over the term of the petroleum contract.[45] Virtually all of the surveyed regimes mandate preference for local goods, services, consumables, works or enterprises. In all of the jurisdictions examined, the IOC has an obligation to give due and proper consideration to preferring locally-sourced services and goods when their price, quality, time of delivery and other terms are comparable to internationally available ones. They also mandate the IOC to prepare plans and programmes for training and educating nationals during the term of the contract. The aim is to ensure that the capabilities of the local workforce are enhanced over time. For example, a key objective of the comprehensive local content legislation in Nigeria is to increase the "Nigerian content" meaning "the quantum of composite value added to or created in the Nigerian economy by a systematic development of capacity and capabilities through the deliberate utilization of Nigerian human, material resources and services in the Nigerian oil and gas industry".[46] A schedule to the Act

---

[43] Ibid.

[44] Muller and Schitzer (2003); also Levett and Chandler (2012).

[45] See for example, the *Nigerian Oil and Gas Industry Content Development Act, 2010* signed into law in April 2010. With effect from this date, all regulatory authorities, operators, contractors, subcontractors, alliance partners and other entities involved in any project, operation, activity or transaction in the Nigerian Oil and Gas Industry are to consider Nigerian content as an important element of their overall project development and management. The Act is designed to give the force of law to the Nigerian Content Policy, which aims to achieve four main objectives: development of indigenous skills across the oil and gas value chain; promotion of indigenous ownership of assets and use of indigenous assets in oil and gas operations; enhancement of the multiplier effect to promote the establishment of support industries; and creation of customized training and sustainable employment opportunities. See s.2 of the Act. Also, NCDMB and Glass and Saggi (2008), pp. 3–5.

[46] Nigerian Oil and Gas Industry Content Development Act, 2010, ibid, s.106.

sets the minimum level of "Nigerian Content" to be achieved for each category of goods and services.

A second driver is the desire to create a level playing field for citizens, residents and home-based industries to participate in oil and gas exploration activities. Without creating a level playing field for new or emerging local industries and workforce to participate in oil exploration activities, and compete with international suppliers of goods and services, the cycle of excessive dependence on foreign goods and services may never be broken.[47] Governments in African countries have therefore promoted LCRs as a deliberate programme and policy aimed at ensuring that local industries are given a chance to compete with foreign suppliers. For example, the Nigerian local content law emphasises the goal of creating a level playing field for local participation by adopting uniform evaluation processes in sourcing services and materials, and by promoting uniform access to project information for local suppliers.[48] Similarly, the ultimate goal of Uganda's local content regulations is to increase the substantial combined value added or created in the Ugandan economy through the utilization of Ugandan human and material resources for the provision of goods and services to the petroleum industry in Uganda.[49] Like the Nigerian legislative framework, the Ugandan regulation introduces a joint supplier registration system as a single window system for registering suppliers in a 'common pool'.[50] The aim is to provide equal opportunities for local industries to participate in oil and gas activities.

A third driver of LCRs in Africa is the desire to maximise economic benefits to citizens through job and employment opportunities. Virtually all of the surveyed regimes in Africa mandate IOCs prioritise the employment of suitably qualified nationals.[51] By mandating the employment of nationals, the aim is to create opportunities for domestic employment, thereby contributing to growth in income, capacity development of nationals and overall increased economic growth of an oil producing country. Local employment is consistently among the topmost concerns of nationals, and often a central issue driving disputes, grievance and conflict.[52] More local jobs could result in more support for projects. Employing more citizens and community members can also help improve company-community relations, enabling IOCs to obtain the social license to operate.[53] Given the negative impacts of oil and gas development on surrounding communities, local content can help

---

[47] See Tordo et al. (2013), pp. 115–117.

[48] See s. 15 of the *Nigerian Oil and Gas Industry Content Development Act 2010,* stating that "all operators and alliance partners shall maintain a bidding process for acquiring goods and services which shall give full and fair opportunity to Nigerian indigenous contractors and companies."

[49] See The Petroleum (Exploration, Development and Production) (National Content) Regulations 2016 which regulates local content in petroleum exploration, development and production in Uganda. See also *Uganda's Petroleum (Exploration, Development and Production) Act of 2013* which contains LCRs.

[50] Ibid.

[51] See Table 9.1 below.

[52] See Ovadia (2015), pp. 37–38; also Tordo et al. (2013), pp. 7–15.

[53] Tordo et al. (2011), pp. 1–10; World Bank (2014), pp. 1–4; also Esteves and Barclay (2011), p. 205.

compensate the afflicted communities through job creation and value addition in the communities.[54]

A fourth driver is the desire of oil producing countries to improve endogenous technological capacity. All of the surveyed African regimes mandate IOCs to give preference to locally manufactured equipment, machinery and consumables when their price, quality, time of delivery and other terms are comparable to internationally available ones. They also include requirements to bring some level of technology or perform research and development (R&D) into the country, so local companies can boost their competitiveness through access to state-of-the-art technology, or benefit from technology transfer.[55] By mandating IOCs to utilise locally made technology in petroleum operations, this directly reduces the importation of technology for petroleum operations. This could in turn compel IOCs, as well as service companies, to invest in technologies and facilities for local manufacturing and service provision.[56] For example, some IOCs operating in Africa have opened technology venture arms of their operations in order to speed up the development and deployment of innovative technologies that could complement oil and gas exploration activities. By opening up technology ventures, IOCs can facilitate the domestic production and availabilities of technologies required for oil and gas exploration.

Fifth, LCRs are also used in many rentier states in Africa to mitigate and manage social and political risks that may result from rising domestic expectations for better and more equitable distribution of wealth and authority.[57] Despite the diverse programmes designed to increase the direct financial flows of oil wealth to nationals through subsidy programmes, individuals may not perceive what they may consider commensurate benefits. This can lead to pressure from the population to increase the more tangible benefits.[58] Although these problems could be addressed by specific policies designed to consider the exact grievance, governments utilise LCRs as a tool for bringing jobs and income to a specific group, area or group where there is considerable dissatisfaction with the presence of the oil and gas operations.[59] By introducing detailed LCRs, countries can ensure that access to the control of oil wealth is evenly distributed among the interest groups and tribes across the country. For example, the Nigerian legislative framework provides for increased appointment of Nigerian personnel in petroleum sector activities.[60] Such a provision allows the host country to at all times control and ensure that a greater spectrum of the society have direct access to extractive sector opportunities and wealth.

---

[54] Ado (2013), p. 142.

[55] Ulrichsen (2011) and Ghafar (2016).

[56] Muller and Schitzer (2003); also Coe et al. (2008); also Glass and Saggi (2008).

[57] Rentier states are countries that receive substantial amounts of revenues (*i.e* rents) from oil or other types of natural resources on a regular basis. See Cook and Mahdavy (1970), pp. 435–436; also Krueger (1974).

[58] See Hanlin (2011); See Ovadia (2015).

[59] Ovadia (2015).

[60] See s. 10 of the Nigerian Oil and Gas Industry Content Development Act 2010.

Despite the clear and uniform overall policy objectives of LCRs, the approaches taken by African countries in enforcing LCRs vary and must be carefully understood and clarified by an IOC to avoid contractual misalignments and disputes. While the diversity of approaches in implementing LCRs in the continent is by itself not a problem, the tendency to consider Africa as one homogenous region often creates problems for investors that fail to consider the unique LCR obligations and applicable regime in each African country before going in to do business. Furthermore, given that LCRs allocate and share a bundle of rights, obligations and risks between IOCs and the host government, IOCs need to ensure that clear, transparent and achievable LCRs are incorporated into their petroleum contracts. The next section provides a survey of the fundamental themes on LCRs found in petroleum contracts and legislation and guidelines in African countries.

9.4 Approaches to LCR Implementation in Africa

A survey of petroleum and mining legislation and contracts in African countries reveals five key variations in approaches adopted to LCR implementation. These are variations in: the definition of 'local'; local content thresholds; the role of government and/or national oil companies in procurement processes; the source of local content obligations; and monitoring/ institutional coordination for LCRs.

9.4.1 The Definition of "Local"

As presented in Table 9.1 below, all of the surveyed regimes mandate local content, whether in terms of priority or preference for local goods, services, consumables, works or enterprises.[61] However, there are widespread variations across the continent in the categories of people and entities that could come under the broad conceptualisation, and must be given priorities, as locals.[62] Generally, 'local' is several African countries means nationals, and companies owned, or majority controlled by nationals (Angola, Botswana, Ghana, Nigeria, Tanzania, Zambia).[63] In terms of

[61] See Olawuyi (2017a, b).

[62] On the importance of clarifying the meaning of 'local', see Nwapi (2015).

[63] The *Nigerian Oil and Gas Industry Content Development Act, 2010* defines Nigerian content as "the quantum of composite value added or created in the Nigerian economy by a systematic development of capacity and capabilities through the deliberate utilization of Nigerian human, material resources and services in the Nigerian oil and gas industry." Under the Act, preference must be given to Nigerian companies in the sourcing of goods and services. A Nigerian company is defined as: "a company formed and registered in Nigeria in accordance with the provisions of the Companies and Allied Matters Act with not less than 51% equity shares by Nigerians."; also in Ghana an *Indigenous Ghanaian company* means a "company incorporated under the Companies Act, 1963… (a) that has at least fifty-one percent of its equity owned by a citizen of Ghana; and (b)

Table 9.1 Survey of local content and procurement requirements in Africa

Country	Source	Provision	Comments
Angola	2008 Model PSA	Local Content	Under the Angolan regime, local content is defined in terms of maximising participation and value for local players
		Preference shall always be given to Angolan concessionaires to ensure participation of Angolan citizens in the ownership and management of the national wealth.	
	Petroleum Activity Law 2004 ("Law 10/04")	Contractor shall prepare and submit for approval of the Ministry of Petroleum, a contract program for the development of human resources within 30 days after the entry into force of the contract (for companies in the exploration phase), 60 days after the date of declaration of the first commercial discovery (for companies in the production phase), 60 days after the beginning of the activity (for companies in the downstream sector), and 30 days after the entry into force of the contract for services companies. Each contract program needs to include employment and personnel procedures showing compliance with the local content requirement	
	Law for the Promotion of Business for Local Private Companies 2003		
	General Regulatory Framework for Hiring of Services and Goods from National Companies in the Oil Industry ("Decree 127/03"); Decree 48/06 on Open Tender Procedures in the Oil Industry ("Decree 48/06")		
	Decree Law on the rules and procedures to observe in recruit-ment, integration, training and development of workers from the oil sector ("Decree-Law 17/09")		
Ghana	Model Petroleum Agreement (2000)	Local Content	Ghana's LCRs and procurement regime focus on maximizing participation and value for local players.
		Where qualified Ghanaian personnel are available for employment, Contractor shall ensure their engagement as far as reasonably possible. Only Ghanaians are to be employed in junior and middle level positions in the petroleum industry.	
		The Contractor is expected to submit an employment/local content plan prior to the performance of petroleum operations. Local content committee must approve plan prior to petroleum operations.	

		Operators must establish a project office within the district where project is located prior to carrying out any work in the petroleum industry.	Although the national oil company is not directly involved in the procurement processes, contractor is expected to submit local content and procurement plans to the Minister for approval, prior to petroleum operations. This report must demonstrate strategies aimed at complying with local content and procurement requirements, and may be audited by the Minister to verify accuracy
		Ghanaian law stipulates progressively increasing local content levels to be attained by operators. It provides for instance that at the commencement of every petroleum agreement or licence, 30% of the management staff of the operator must be Ghanaian citizens. Furthermore, local content threshold for goods and services is 10% at the start of petroleum operations, 50% after 5 years and 60–90% at 10 years.	
		Procurement	
		In tendering for goods and services, Ghanaian law requires operators to establish and implement a bidding process, which gives preference to materials, equipment, services and products produced in Ghana including shipping services.	
		Procurement entities are required to grant a margin of preference for the benefit of tenders for work by domestic contractors, the benefit of tenders for domestically produced goods or the benefit of domestic suppliers of services.	
		Contractor is expected to submit, to the Minister, a procurement proposal, with respect to goods and services obtainable in Ghana, prior to the performance of petroleum operations.	
		Procedure	
		A procurement entity shall procure goods, services or works by competitive two-stage tendering. An award of a contract shall not be solely based on the principle of the lowest bidder. An Indigenous Ghanaian Company with the capacity to execute a job shall not be disqualified exclusively on the basis that it is not the lowest financial bidder	

(continued)

Table 9.1 (continued)

Country	Source	Provision	Comments
	Public Procurement Act, 2011 (Act 663)	The Ghana National Petroleum Corporation utilizes an e-procurement system	
	Petroleum (Local content and local participation) Regulations, 2013		
	Minerals and Mining Act 2006 (Act 703) contains provisions dealing with to local content in mining operations. Also Minerals and Mining (General) Regulations 2012 (LI 2173)		
Liberia	2014 Model Production Sharing Contract	Local Content	The Liberian framework establishes a requirement to inform and provide to the Minister, procurement reports and data, for any procurement contract exceeding $100,000.
		Contractor must give preference to enterprise and goods from Liberia if price, quality, delivery time and terms of payment are similar to those from other countries.	
	Public Procurement Concession Act (PPCA).	Procurement	The Liberian regime establishes different financial thresholds as determinants for the procurement methods to be adopted.
		Contracts for the procurement of goods and services for the purpose of petroleum activities shall be subject to public tender. Contractor shall furnish the National Oil Company of Liberia information, reports and data on any procurement of goods and services in excess of $100,000.	
		Procedure	
		Public procurement shall be undertaken by means of advertised open bid proceedings, to which equal access shall be provided to all eligible and qualified bidders without discrimination. The contract shall be awarded by the Procuring Entity to the bidder that is qualified and submitted the lowest evaluated responsive bid which meets the evaluation criteria as specified in the invitation to bid, request for quotation, request for proposal or the related bidding documents.	

		The Liberian regime establishes different financial thresholds as determinants for the procurement methods to be adopted. Contract awards shall be published when the estimated value of the contract is above US$25,000 (for goods), or more than US$10,000 (for services) or US$50,000 (Works).	
	Concession Act (PPCA).	The Procuring Entity shall employ international open competitive bidding procedures when the estimated contract price of the proposed procurement is higher than $100,000 (goods) or $50,000 (for services) and $200,000 (works). The request for proposals method is to be used for the procurement of consultant services below these thresholds. Request for quotation for the procurement of small-scale goods, works and services below the threshold.	
		The Liberian Procurement Commission shall be notified of proposed contract awards over $150,000 and the contract shall not be signed until at least fourteen (14) days after the date of the notice	
		The Ministry of Finance shall take part in negotiations of contracts over US$ 250,000 for information purposes only.	
Libya	EPSA IV	Local Content	The Libyan regime imposes a 50% local content requirement on contractors. The procurement process in Libya is also not typical as it is centralised and prescribed by the Government.
		Operator must give priority to local supplies and services. Operator shall be obliged to spend at least 50% of its approved Budget on supplies, equipment and services available locally.	
		Procurement	
		Libya runs a centralised procurement system in which Government has direct control over procurement of goods and services. The Libyan Government has not yet passed a procurement law, and has only introduced a series of procurement regulations based on International Federation of Consulting Engineers (FIDIC) procurement guidelines.	

(continued)

Table 9.1 (continued)

Country	Source	Provision	Comments
Madagascar	2006 model offshore Production Sharing Contract	Local Content	The oil and gas regime in Madagascar provides the contractor with the flexibility of designing and proposing its own procurement strategies, processes and procedures, for approval by the management committee.
		Priority shall be given to the utilization of goods and services produced or delivered locally in Madagascar, as long as their price, quality, suitability, and delivery terms are comparable to those potentially imported on c.i.f terms into Madagascar.	
		Operator shall furnish OMNIS automatically and any Participant upon request with copies of any Subcontractors' Contracts.	
		Procurement	
		In arranging the supply, calls for tenders, and selecting suppliers relating to the delivery of goods and services, competitive Malagasy suppliers shall be given a genuine opportunity to secure orders	
		Procedure	
		Procurement contracts shall primarily be based on competitive bidding in accordance with transparent predefined procedures approved by the management committee (comprising three (3) members designated by OMNIS and up to three (3) members nominated by the Participants).	
		Procurement strategies, including standardized forms and terms for purchase orders and contracts, must be approved by the management committee	
Namibia	Petroleum Agreement (2012)	Local Content	The Namibian local content and procurement regime is flexible. Local content is defined in terms of preference of local goods with the qualification that this may be waived if contractor can demonstrate why preference should not be given.
	Petroleum (Exploration and Production) Act, 1991	Contractor shall use and purchase goods supplied, produced and manufactured in Namibia whenever such goods can be obtained at prices in Namibia, which are competitive in international terms.	
	Tender Board Act of 1996		

		The Company shall give preference to such supplier, producer or manufacturer, unless it is able to show good cause to the satisfaction of the Minister why such preference should not be given.	Furthermore, government is to be informed by the contractor, after the fact, through yearly local content/ procurement statements.
		Procurement	
		The Company shall prepare and submit, in respect of each Calendar Year, a local procurement statement, containing information on the amount of expenditure incurred by on goods and services supplied, produced or manufactured in Namibia.	
Nigeria	Production Sharing Contract	Local content	The Nigerian framework establishes a comprehensive list of percentage local content thresholds that must be met by the contractor for a wider range of services and good ranging from 30 to 100%.
	The Nigerian Oil and Gas Industry Content Development Act 2010		All players in the industry are required to submit a Nigerian Content Plan to the Board in bidding for any license, permit or interest and before carrying out any project in the oil and gas industry.
			Furthermore, prior approval of the national oil company must be obtained for all procurements exceeding approximately US$50,000 (locally) and $100,000 (for foreign procurements)

(continued)

Table 9.1 (continued)

Country	Source	Provision	Comments
		Operator shall give preference and first consideration to services provided from within Nigeria and to goods manufactured in Nigeria; and Nigerians shall be given first consideration for training and employment in the work programme	Similarly, the national oil company is directly involved in every stage of the procurement processes. It has the right to delete or nominate contractors to be included in the pre-qualified suppliers list.
		Nigerian framework sets the target of achieving 70% Nigerian content. It also lists various items, goods, services and activities carried out in the oil industry and stipulates the minimum specification of Nigerian content that must be demonstrated by operators when procuring or carrying out the respective activities.	
		Before carrying out any project in the Nigerian oil and gas industry, an operator shall submit a Nigerian Content Plan to the Board demonstrating compliance with the Nigerian content requirements if satisfied that the plan complies with the provisions of this Act, issue a Certificate of Authorization.	
		Procurement	
		Prior approval of the Corporation shall be obtained for all foreign contracts and foreign purchase orders awarded to third parties where the cost exceeds $50,000.	
		Prior approval of the Corporation shall be obtained for all local contracts and purchase exceeding 10 Million Naira (approximately $50,000)	
		Procedure	
		Competitive quotation processes shall be employed for all local procurement where the estimated values exceeds the equivalent USD$50,000. Such contracts, shall be transmitted to the Corporation for approval before a purchase order is issued to the selected vendor/manufacturer. Approval shall be deemed to have been given if a response has not been received from the Corporation within thirty (30) working days.	

		The Contractor shall maintain a list of approved contractors for the purpose of contracts for the Petroleum Operations. The Corporation shall have the right to delete or nominate contractors to be included in the list. Corporation and Contractor shall be responsible for pre-qualifying any contractor to be included in the Approved Contractors List.	
		For any procurement, the Contractor shall present a list of proposed bidders to the Corporation before the issuance of invitations to bid to prospective contractors. The Corporation may propose additional names to be included in the list of proposed bidders or the deletion of any one	
		Contractor shall establish a Tender Committee that would be responsible for pre-qualifying and evaluating bidder. The Corporation reserves the right to nominate its staff who shall fully participate in prequalifying such bidders	
		These procedures may be waived in emergency situations or when special circumstances warrant, upon the approval of the Corporation. In such cases, contractor must promptly inform the Corporation.	
		The principle of bid evaluation based on the lowest bidder is waived under the Act so that where a Nigerian indigenous company has the capacity to execute a contract it will not be disqualified for the sole reason that it is not the lowest financial bidder provided the value does not exceed the lowest bid price by 10%.	

(continued)

Table 9.1 (continued)

Country	Source	Provision	Comments
Tanzania	The Petroleum Act 2015 (Art. 220, 221)	Local Content	The oil and gas regime in Tanzania provides the contractor with the flexibility of designing and proposing its own procurement strategies, processes and procedures, for approval by the management committee.
		Contractor shall give preference to goods and services available in Tanzania, and when those are not available, they should be delivered by a company that is in joint venture with a local company that has a minimum share of 25% or as otherwise stipulated by the regulations	
		Contractors must annually submit a report on the execution of their employment, training and procurement and local supplier development programs, steps taken by licensee to close any identified learning gaps to the regulatory authority. Contractors are also responsible for the reporting on the procurement program by their sub-contractors	
		Procurement	
		The Draft Local Content Policy requires the inclusion of a compulsory local content requirement in each invitation to bid. Also, Contractors and subcontractors must facilitate the participation of Tanzanian companies as suppliers of goods and services; and competitiveness in every procurement process undertaken by Contractors and subcontractors.	

employment, these countries expressly aim to ensure that qualified nationals are given first consideration in matters of employment and that adequate provision is made for the training of nationals on the job. "Local" in these countries therefore refers to nationality, rather than the locality or country where the oil and gas activities take place.

However, a few African countries adopt an expansive definition by defining 'local' in terms of locally-registered firms and entities and the employment of local manpower which will include nationals and residents (Algeria, Uganda, South Africa).[64] "Local" in this case is equivalent to domestic entities or local players, meaning goods, services or labour sourced from the national locality or region where the extractive operations take place, as opposed to nationality or citizenship of the provider of the goods, services or labour.

This distinction is indeed very important in African countries and could be a legal risk point if not properly clarified and handled. Generally, in Africa, relationships between IOCs, nationals and expatriate workers could be very fragile due to consistent agitations for more direct participation of nationals in the running of petroleum operations.[65] While LCRs designed to give special recognition and priority to nationals may provide a mechanism for addressing this fragility, it could be a legal risk point for IOCs in instances whereby it is difficult to find a right fit for the required goods, services or labour from the avail-

---

that has Ghanaian citizens holding at least eighty percent of executive and senior management positions and one hundred percent of non- managerial and other positions". See *Petroleum (Local content and Local Participation in Petroleum Activities) Regulations 2013*. Also in Tanzania, *Local Companies* mean: "a company or subsidiary company incorporated under the Companies Act, which is one hundred percent owned by a Tanzanian citizen or a company that is in a joint venture partnership with a Tanzanian citizen or citizens whose participating share is not less than fifteen percent." (Petroleum Act, The Petroleum Act 2015, Art. 220. Zambia defies local in terms of citizen-owned companies. A citizen owned company means "a company where at least fifty point one per cent of its equity is owned by Zambian citizens and in which the Zambian citizens have significant control of the management of the company" See *Mines and Minerals Development Act no 7 of 2008*, Art. 2 (1).

[64] See Olawuyi (2017a, b). For example, in Uganda local content is met by sourcing goods and services from an Ugandan company. An "Ugandan company is defined as: (…) a company incorporated under the Companies Act, 2012 and which— (a) provides value addition to Uganda; (b) uses available local raw materials; (c) employs at least 70% Ugandans; and (d) is approved by the Authority under regulation." See Uganda's Petroleum (Exploration, Development and Production) (National Content) Regulations, 2016. South Africa's framework also mandates first consideration and direct beneficiation for *Historically Disadvantaged South Africans (HDSA)*, defined as "South African citizens, category of persons or community, disadvantaged by unfair discrimination before the Constitution of the Republic of South Africa, 1993 (Act No. 200 of 1993) came into operation which should be representative of the demographics of the country." By 2014, all mining companies were required to be 40% owned by HDSA and have HDSA occupy 40% of board management, senior management (EXCO), core and critical skills, middle management, and junior management positions. See Art. 2.4 of the *Broad-Based Socio-Economic Empowerment Charter for the South African Mining and Minerals Industry* 2004, also Art 9–12 of the *Mineral and Petroleum Resource Development Act 2002.*

[65] World Bank (2014).

able pool of nationals.[66] Also, conceptualising 'local' in terms of nationality could result in project delays and cost inflation for the investor where locally-produced goods or highly skilled nationals are not immediately available or only available at much higher costs. However, the benefits of maintaining social stability in a resource-bearing locality and, more broadly, the impetus to retain a social license to operate may be more important for an IOC than the loss to be occasioned by these trade-offs.[67] A more flexible model to managing such trade-offs is found in the Namibian framework which provides that Contractor shall, *when possible*, give first consideration or preference to locally manufactured or locally available goods.[68] This provides some flexibility for an IOC to consider other categories of locals if the IOC so decides for operational reasons, or in cases when suitably qualified nationals or service providers are not available.

9.4.2 Local Content Threshold

In several African countries, the minimum level of "local content" to be achieved for each category of goods and services is not expressed as a percentage or with a timeframe, rather, the obligation is simply to give priority to services, materials, equipment, consumables and other goods when their price, quality, time of delivery and other terms are comparable.[69] For example, the Liberian PSC provides that Contractor must give preference to enterprise and goods from Liberia if price, quality, delivery time and terms of payment are similar to those from other countries.[70] Similarly, the Madagascar PSC provides that Contractor shall give preference in employment to suitably qualified Madagascar nationals, provided, however, that Contractor has the right to employ its own key personnel in appropriate positions.[71] The Namibian PSC is flexible, stating that Contractor shall, when possible, select contractors who are nationals of Namibia or companies that are majority owned by Zambian nationals unless it is able to show good cause to the satisfaction of the Minister why such preference should not be given.[72] The Namibian framework also notes that in purchasing goods for petroleum operations, Contractor shall, when possible, give preference to locally manufactured or

---

[66] Ibid.

[67] Ibid.

[68] See Namibia Petroleum Agreement (2007), & Petroleum (Exploration and Production) Act, 1991.

[69] Art. 26 Yemen Model Production Sharing Contract of 2006; Olawuyi (2017a, b).

[70] Liberia Production Sharing Contract.

[71] Madagascar 2006 model offshore Production Sharing Contract.

[72] See Namibia Petroleum Agreement (2007), Petroleum (Exploration and Production) Act, 1991, also South Africa's Mining Charter (n. 66) specifying 40% local content in all mining activities in South Africa.

locally available goods so long as their technical specifications, quality and time of delivery and services are comparable to internationally available ones.

However, while local content obligations are flexible and without mandatory thresholds in countries such as Liberia, Namibia and Madagascar, in other African countries, especially in Nigeria, South Africa, Ghana and Libya, a percentage threshold of local content that must be met, and the timeframe, is expressly set out.[73] A local content indicator approach is adopted in Nigeria, Ghana and Libya, which set out a percentage rating of a company based on specific criteria defined on the basis of values ascribed to each criterion. For example, the Ghanaian framework stipulates progressively increasing local content levels to be attained by operators. It provides, for instance, that at the commencement of every petroleum agreement or licence, thirty percent of the management staff of the operator must be Ghanaian citizens. Furthermore, local content threshold for goods and services is 10% at the start of petroleum operations, 50% after 5 years and 60–90% at 10 years.[74] Similarly, the Nigerian framework provides a comprehensive list of percentage local content thresholds that must be met by the contractor for different services and good ranging from 30 to 100%,[75] while under the Libyan PSC, Operator is obliged to spend at least 50% of its approved budget on supplies, equipment and services available locally.[76] In these rare cases where a percentage is expressed, local content obligations are often indexed to overall revenue spent during the operations.

The mandatory threshold approach raises several questions and interpretational challenges. One is that given that these LCRs were recently launched and adopted, there is a need to clarify their implications for pre-existing arrangements, agreements, contracts or memoranda of understanding that existed before the mandatory thresholds.[77] Furthermore, the formulaic approach represents a shift from previous policy, where local contact was not a formal requirement, to one where local content development is mandatory and could result in contractual default, and resulting penalties, for operators or suppliers that fail to comply.

---

[73] Nigeria, South Africa, Ghana and Libya's minimum threshold approach follow the Saipem Externalities Local Content Evaluation (SELCE) Model, which provides a quantitative evaluation, in monetary terms of socio-economic effects produced by local content. The tool uses the multiplier approach to measure direct, indirect and induced socio-economic impacts generated by Saipem Operating Companies in the countries in which they are based. Three main categories are identified and quantified: (a) Economic value created, in terms of locally sourced purchases, taxes paid to local authorities and employees' salaries; (b) Employment created; and (c) Human capital development, including increased life-time earning expectancy due to the training received from the company. For each category, the total impact (given by the sum of direct, indirect and induced impacts) and the multiplier factor (i.e. numbers providing the magnitude of an impact) are calculated. See Saipem SELCE Model: Local Content for Sustainable Development.

[74] Saipem, SELCE Model: Local Content for Sustainable Development.

[75] Schedule 1 to the Nigerian Local Content Development Act, 2010.

[76] See Libyan Exploration and Production Sharing Agreement IV.

[77] See OI Obasi.

The flexible approach, adopted across Europe and in oil producing Middle East countries such as Iraq, Syria, Jordan and Qatar, is arguably a more straightforward approach that allows an operator to align the implementation of LCRs with prevailing market realities.[78] These countries do not express local content in terms of percentage or threshold formulae. Rather, they provide ECs the flexibility to purchase goods from, or employ, locals when possible. This flexibility is particularly important for operators to maintain quality when locally manufactured goods or services are unavailable or more expensive to source. The flexible approach also reduces legal risks relating to non-compliance with employment or LCR thresholds.

9.4.3 Role of the State

Government and national oil company participation in procurement processes vary. While LCRs may specify the portion of total expenditures that must be comprised of locally sourced goods and services, procurement procedures are frequently not well established. Several African regimes provide the contractor with the flexibility to design and propose its own procurement processes and procedures, so long as they comply with LCRs and are based on generally accepted standards in the international petroleum industry (for example Madagascar).[79] What constitutes international best practice, especially when there is a dispute relating to compliance with LCRs, may however be subject to debate and complex ligation. Furthermore, in some regimes, government must be informed after the fact through yearly statements, audits or mandatory performance standards that demonstrate compliance with LCRs and procurement requirements (Angola, Uganda, Namibia, South Africa)[80]; in others, procurement plans must be submitted prior to the commencement of petroleum operations (Ghana). A variation arises where the national oil company or a management committee is directly involved in an advisory capacity and, in more extreme cases, government must be informed and may participate in

---

[78] See for example Art. 23 State of Qatar Model Development and Production Sharing Agreement of 2002; Art. 26 Republic of Yemen Model Production Sharing Agreement 2006; Art. 23 Jordan Model Production Sharing Agreement of February 2007; also Art. 19.1 Oman Model Exploration and Production Sharing Agreement Dated April 24, 2004; Art. 17 INA Contract of 1998 for the Exploration, Development and Production of Petroleum Between the Government of the Syrian Arab Republic and Syrian Petroleum Company and INA-Industrija Nafte dd.- NAFTAPLIN.

[79] See Ulrichsen (2011) and Ghafar (2016).

[80] See South Africa's *Broad-Based Socio-Economic Empowerment Charter for the South African Mining and Minerals Industry* 2004 (Mining Charter) Art. 2.9; Art. 25 of the Uganda's Petroleum (Exploration, Development and Production) (National Content) Regulations, 2016; Art. 7 of Angola's *Law on the Rules and Procedures to observe in Recruitment, Integration, Training and Development of Workers from the Oil Sector Decree-Law 17/09*, stating that companies must submit annual plans and reports to the Ministry of Petroleum.

procurement activities above certain financial thresholds (Nigeria, Liberia, Libya, Egypt).[81]

Understanding the variations in the designated role of the national oil company or government in the procurement processes is critical to avoiding disputes. As governments increasingly seek to exercise some form of influence in contractors' procurement processes to ensure total value in terms of local content, it is important for IOCs to clarify, from the outset, the level of government involvement in the process. A common approach in Europe, for example, is to mandate the Contractor to submit a LCR compliance and procurement plan within sixty (60) days from the effective date of the petroleum agreement.[82] This allows the IOC to design and develop its own procurement practices, while providing the State an early opportunity to make inputs. Government participation in procurement processes, as a way of monitoring LCRs and transparency, poses significant risks and delay for petroleum operations, especially in several African countries where government approval processes could be slow and bureaucratic. Government participation in procurement processes should be avoided by IOCs concerned with the efficiency of their extractive operations in the continent. However, when necessary especially in large projects with greater transparency requirements, government participation in procurement processes could be limited to contracts with a significant financial threshold.

9.4.4 *Source of the Obligation*

Obligations relating to LCRs can change very quickly in African countries and often emanate from sources beyond domestic legislation and existing mining and petroleum agreements signed with an operator. While oil and gas producing countries such as Nigeria, Ghana and Angola have enacted comprehensive legislative frameworks dedicated to local content,[83] therefore providing clarity on the meaning, nature and scope of LCRs applicable to petroleum operations, several countries in the continent are yet to develop clear and comprehensive legislative frameworks dedicated to local content. In such countries, local content obligations are as stipu-

---

[81] The Nigerian legal framework provides for government participation in all procurement where the estimated value exceeds 10 Million Naira (approximately $50,000). See *The Nigerian Oil and Gas Industry Content Development Act* 2010, n. 16; also Production Sharing Contract between Nigerian National Petroleum Corporation and Gas Transmission and Power Limited, Energy 905 Sunera Limited and Ideal Oil and Gas Limited (2007).

[82] See for example, Croatian Hydrocarbon Agency (2015).

[83] See The Nigerian Oil and Gas Industry Content Development Act 2010; also Mexico's Hydrocarbons Act 2014 (Ley de Hidrocarburos); Ghana's Petroleum (Local content and Local Participation in Petroleum Activities) Regulations 2013; Angola's Decree Law on the rules and procedures to observe in recruitment, integration, training and development of workers from the oil sector. For a summary and overview of these laws, see Columbia Centre on Sustainable Investment, Local Content Laws & Contractual Provisions.

lated in the petroleum or mining contract and are therefore may not be clearly and comprehensively defined to clarify issues relating to meaning of local, threshold for measuring and reporting local content value; and such obligations can be expanded in routine guidelines released by the national oil company or in general public procurement laws. To avoid legal risks relating to uncertainty and misalignment with respect to the level and time frame of local content to be achieved, it is prudent for an IOC operating in the African continent to ensure that LCRs are clarified beforehand and comprehensively memorialised in the petroleum agreement.

Furthermore, while African countries such as Nigeria, South Africa, Libya and Angola, have one comprehensive local content framework applicable to both mining and petroleum operations, a few countries such as Ghana and Tanzania have separate legislative framework on local content for mining and petroleum sectors respectively. It is important for IOCs to clarify whether a country applies different LCRs and thresholds for mining and petroleum sectors respectively to minimize legal risks of non-compliance.

9.4.5 Monitoring and Institutional Coordination

One key reason for the lack of clarity on the scope and content of LCRs in the African continent is that while several African countries stipulate LCRs, only few establish institutions or mechanisms for monitoring and enforcing such requirements. For example, the Nigerian Local Content Development Board which is mandated to oversee, monitor and implement the provisions of the NCDA.[84] Ghana has followed Nigeria's example in establishing its own "Local Content Committee", mandated to monitor and coordinate all aspects of the implementation of Ghana's 2013 Regulation.[85] Tanzania's 2013 Local Content Policy similarly evinces an intention to establish a "National Local Content Committee" which will be chaired by its Ministry of Petroleum.[86]

Measuring and quantifying value-add is difficult without establishing mechanisms for effective monitoring and enforcement.[87] In addition to promoting public awareness and reducing uncertainty on the scope of LCRs, a focal local content unit can coordinate and spearhead the development of a standardised approach to local

[84] S. 4 & 70 of *The Nigerian Oil and Gas Industry Content Development Act* 2010. The Board has powers to: monitor the implementation of set Nigerian Content targets; develop interventions to ensure that these targets are met; and ensure compliance with Government directives covering the service providers and operators in the Nigerian Oil and Gas Industry.

[85] Art. 5, 8, 44 of the *Petroleum Commission Act 2011*. Also in Uganda, quarterly reports on performance on national content covering all petroleum activities are to be submitted to the Petroleum Authority. See Art. 25 of Uganda's Petroleum (Exploration, Development and Production) (National Content) Regulations, 2016.

[86] Ghana's Petroleum (Local content and Local Participation in Petroleum Activities) Regulations 2013.

[87] S. 4 of The Nigerian Oil and Gas Industry Content Development Act 2010.

content and product sourcing, in alignment with government priorities and domestic laws. Such an agency can also guide Operators and investors in formulating and developing petroleum operations or projects that are in line with the country's national vision; and more specifically that can result in real, measurable and long-term local benefits. A local content agency can also monitor and investigate companies in relation to their compliance with local content requirements. For example, the Nigerian NCDMB has clear statutory powers to prescribe penalties of at least 5% of the project total on any entity that fails to comply with LCRs.[88] Fines levied by the NCDMB are paid into a Local Content Development Fund dedicated to training and supporting local enterprises. From the perspective of an IOC, such a clear institutional approach reduces ambiguities and surprises with respect to LCR enforcement and implementation, while from a government and policy standpoint it ensures that LCRs are indeed actually monitored and enforced such that they can result in real and quantifiable benefits in the long run.

Establishing a focal LCR agency or unit can also enable it to function as a one-stop shop that can coordinate with other relevant agencies and ministries to maximise and fast track the attainment of national local content objectives. This is particularly important considering that several ministries, departments and agencies have a role to play in quantifying added value. For example, apart from agencies overseeing the oil and gas industry, the education ministry has a role in assessing and monitoring local capacity and training programmes, while labour ministries and departments have a role in measuring the quantity and quality of local jobs added by a petroleum project, same for the technology departments and ministries that have roles to play in facilitating the transfer and deployment of petroleum technologies. The wide range of stakeholders underscores the importance of coordination. A focal agency can bring together and coordinate local content initiatives across all relevant sectors of the economy. With robust intergovernmental coordination, getting approvals for local content initiatives is less cumbersome and more straightforward for an IOC. Developing a coordinated implementation approach could also simplify the process of passing information between government ministries, thereby removing inefficiencies and bureaucracy.

The diverse LCR regimes, and the unclear scope of local content obligations, across Africa raises several legal risks and uncertainties for an IOC, a very small share of which can be managed through traditional contractual documentation. Furthermore, specific problems arise due to lopsided laws or contracts that grant the national oil company or State broad discretion to participate in an Operator's procurement process as a way of monitoring local content. While it is very important to include contractual terms that specifically mitigate these risks, contractual provisions alone are not sufficient to address all of these legal risks. In order to retain IOC interest and confidence in oil activities in the region, African countries can do more to ensure that LCRs are clear, specific and aligned with global best practices to minimise contentions. Without a strong regulatory and institutional foundation that provides clarity and certainty for investors, it is difficult to compete with jurisdic-

[88] S. 4 of The Nigerian Oil and Gas Industry Content Development Act 2010.

tions that provide clearer and more realistic terms and requirements for IOCs. The next section discusses the need for a collaborative approach, between national authorities and IOCs, in minimising legal risks and disputes relating to LCRs in African extractive industries .

9.5 Mitigating Legal Risks Relating to LCRs: The Need for a Collaborative Approach

As can be seen in jurisdictions such as Nigeria, Ghana, and Angola, where LCRs have been implemented with varying levels of success, LCRs must be backed by a clear, specific and transparent legislative framework, including a robust performance monitoring mechanism.[89] While setting national requirements and targets for local content reflects a political commitment towards ensuring domestic value creation and long-term economic growth, the next step for national authorities across Africa is to develop comprehensive and holistic legal frameworks that clarify and simplify LCRs. Rather than approaching LCRs from a compliance or mandatory project requirement mind set—that demands more local content or introduces more punitive enforcement measures—governments across the African continent must adopt a more collaborative approach built on clear, transparent and attainable LCRs, with adequate institutional support for IOCs to achieve those goals.

A collaborative approach to LCR is built on creating a supportive regulatory and business-friendly economic environment for IOCs to deliver greater value in the host country.[90] Under this approach, governments have a prominent role to play in reducing regulatory and administrative barriers to domestic investments; providing fiscal incentives for IOCs to establish or support small and medium enterprises in the host country; updating intellectual property laws to provide greater protection for domestically produced technology; simplifying approval processes and fees for licenses and permits and providing and ensuring greater inter-ministerial coordination amongst key ministries and agencies that have roles to play in the employment, training and education components of LCRs.[91] Under a collaborative approach, governments will work closely with IOCs and local communities to set out realistic and achievable LCRs, and then develop supportive regulatory and institutional frameworks for the delivery of the agreed targets.

There is a strong economic case, in terms of cost, reputation and effectiveness for adopting a collaborative approach.[92] Apart from the fact that governments and the

[89] See Acheampong et al. (2016), p. 9, also Nwapi (2015), pp. 187–216.

[90] Organisation for Economic Co-operation and Development, Working Party of the Trade Committee, *Local Content Policies in Minerals-Exporting Countries*.

[91] On the importance of reducing barriers to technology absorption at national levels, see Olawuyi (2017a, b).

[92] For theoretical viewpoints on the utility of a collaborative approach to regulation, see Shapiro and Rabinowitz (1997).

public will ultimately benefit more when LCRs are achieved by an IOC, failure to effectively minimise legal risks associated with LCRs could carry significant financial, legal and reputational risks for national authorities, especially when LCRs become subjects of extensive litigation.[93] Such risks may also manifest themselves in disruptions to petroleum operations due to disputes, such as community protests over a perceived lack of benefits from a project, including potential harm to employees due to such protests. Furthermore, in a competitive oil and gas sector, a country's ability to attract investors and technologies (including financial institutions and lenders) needed to develop oil resources will depend on the processes, procedures, practices and approaches put in place to reduce contractual risks, such as those that could result from misaligned LCRs. IOCs on the other hand could avoid fines, penalties, or even allegations of corruption and/or material breach of the host government contract, which may result from misaligned LCRs.

To minimise legal contentions and risks relating to LCRs, the following action points should be considered by governments and IOCs when adopting LCRs:

9.5.1 Establish Clear and Comprehensive Local Content Laws

The starting point is for national authorities to establish clear, transparent and comprehensive local content laws that can provide guidelines on LCRs, particularly in the oil and gas sector. Such laws should, amongst other things, provide clear definitions of key concepts such as local, local content, local company, project sum and in-country value. There is also a need to clearly identify the skills, competencies, technologies and economic activities that a country wants to improve or build upon as part of local content implementation. Such clear definitions will reduce ambiguities with respect to the scope and content of LCRs.

Local content laws can also be very helpful in addressing overlaps and limitations in other domestic laws that could hinder the successful implementation of LCRs.[94] For example, procurement laws that have elaborate provisions on procurement procedures may hinder the coherent implementation of LCRs and may result in misalignments and risks if they do not clarify if such procedures apply in the oil and gas sector. The adoption of clear and specific LCRs could provide opportunities for a country to harmonise LCRs with existing laws to avoid overlap and mismatch.

Most importantly, the local content legislation should establish a designated institution or focal point as a one-stop shop for LCRs. This would help simplify the processes and procedures for seeking and obtaining regulatory permits and investment approvals; it would also serve as a rallying point that could foster intergovernmental coordination and linkages amongst the many institutions that currently play important roles in the delivery of local content initiatives.

---

[93] Franks et al. (2014).

[94] Nwapi (2015); also Ovadia (2015).

9.5.2 Establish Collaborative Focal Institution on Local Content

To provide adequate institutional support for IOCs to achieve LCRs and goals, it is important to establish a focal institution, committee or administrative unit that will coordinate the design, approval and implementation of local content plans across the life cycle of a project. While such a focal institution can be established as a supervisory committee of a petroleum contract, a more long-term approach is to establish a national local content agency or unit that will oversee LCRs in multiple sectors of the economy.

Apart from serving as a one stop shop that will streamline the approval processes for local content implementation, such an institution would also provide methodologies and tools for operators to report and monitor their compliance with LCRs so as to minimise disputes. By empowering and establishing a focal institution on projects, investors across multiple sectors can obtain relevant information and develop a standardised approach to tracking, monitoring and complying with LCRs. A coordinated approach can also reduce duplication and overlap, conflicting regulations, increased administrative costs and delays.

9.5.3 Adopt Collaborative Contract Terms on LCRs

As part of contract negotiations, national authorities and IOCs need to jointly define at the outset what constitutes success in terms of local content and value addition. The scope of objectives must be specific, measurable and achievable to avoid ambiguities and misalignments. Rather than adopting a rigid and inflexible approach, a collaborative approach that clarifies the expectations of the government, while providing the IOC with the flexibility to develop its local content plans and procurement procedures could achieve greater results in African countries. Also, rather than stipulating blanket and unrealistic timeframes, countries, in defining LCRs and ambitions, must recognise that the timeframe for delivering in-country value may vary from project to project and could be affected by political and other factors not envisaged during contract negotiation.

The negotiation stage is also a great opportunity for an IOC and the government to agree up front on the costs and trade-offs of complying with local content initiatives over time. As earlier noted for example, complying with product mandating requirements could mean project delays or higher costs on the part of the IOC, especially when suitable and reasonably priced alternatives are not immediately available locally. This can change the profit margin of a project or affect the timeline for petroleum operation activities. These trade-offs must be very well considered during contract negotiation stages to avoid long-term misalignments and contentions and to achieve a mutually beneficial and realistic contractual framework. For example, if the government insists on a local content timeline, fiscal terms such as profit oil and recovery costs, amongst others, could be amended to protect the margins of the IOC, while achieving the aggressive LCR timelines stipulated by the government. Table 9.1 below provides a survey of LCRs in frontier African extractive markets.

9.5.4 *Measure and Communicate Performance*

One of the outstanding elements of the Nigerian and Ghanaian local content frameworks is that, unlike other regimes in the region, they both establish reporting obligations and fixed key performance indicators (KPIs) that mandate suppliers to provide quarterly status updates and annual reporting on the scope and level of compliance with LCRs. Periodic reporting of targets, accomplishments and strategies provides a collaborative opportunity for both parties to mutually evaluate the feasibility and efficacy of LCRs and to explore approaches that can improve performance and deliver shared wins for both sides. Timely and effective communication can reduce the likelihood of disputes as it allows both parties to explore if, and how, a company is achieving LCRs. KPIs could also provide a basis for renegotiating or amending LCRs when established targets may be difficult to achieve. It is therefore imperative for both parties to agree upon, and incorporate, KPIs and reporting requirements with respect to LCRs during contract negotiations, rather than just stipulating that LCRs must be achieved.

9.6 Conclusion

The increased adoption of LCRs by African countries provides practical opportunities for national authorities to maximise in-country benefits and value from resource extraction activities. LCRs can also serve an investor's desire to ensure freedom to procure on the basis of cost, availability and quality, and the mutual desire for a transparent and efficient process. LCRs can, however, result in significant legal risks and misalignment between governments and investors if the procedures for implementing, reporting and measuring local content are not well clarified from the outset. Legal contentions and risks relating to LCRs in African countries can be exacerbated by the absence of robust legal and institutional frameworks on LCRs across the region. There is therefore a need for national authorities across the continent to adopt comprehensive legal frameworks that clarify the meaning, nature, scope and methodology for implementing, measuring and reporting local content performance. A clear, comprehensive and realistic legal framework on LCRs is an absolute requirement for African countries to move their local content policy goals from aspirations to realisations.

Furthermore, the compliance approach to local content, that is increasingly adopted in the continent, may be counterproductive to the policy aims of LCRs. National authorities in Africa need to adopt a collaborative approach to LCRs, built on common understanding, shared values and benefits, and propelled by collaborative and supportive institutions. Care must also be taken by both parties to develop and agree upon LCRs and procedures that are clear, practical and achievable. There is indeed a strong business sense, in terms of risks management and mitigation, for IOCs to carefully understand and assess LCRs of a host state at the earliest stage possible to determine if such requirements will reduce the profitability or commerciality of an extractive project.

The subject of diligent market risk identification and mitigation is as a whole very crucial to successful extractive investment in Africa. The next section of this book unpacks key environmental, legal and market risks that investors must carefully anticipate and manage when pursuing extractive licenses in Africa. The next chapter kicks off with an examination of the environmental challenges and risks associated with extractive operations and activities in resource rich African countries.

References

Acheampong T, Ashong M, Svanikier V (2016) An assessment of local-content policies in oil and gas producing countries. J World Energy Law Bus 9(4):282–302

Ado R (2013) Local content policy and the WTO rules on trade-related investment measures (TRIMS): the pros and cons. Int J Bus Manage Stud 2(1):137–146

Aladejare A (2015) Legal status of sections 10 and 12 of the Nigerian oil and gas industry content development act (2010) under the GATT Regime. Afe Babalola Univ J Sustain Dev Law Policy 5(1)

Broad-Based Socio-Economic Empowerment Charter for the South African Mining and Minerals Industry (2004) http://pmg-assets.s3-website-eu-west-1.amazonaws.com/1/FinalCharter.pdf. Accessed 12 May 2018

Canada–Newfoundland and Labrador Atlantic Accord Implementation Act (S.C. 1987, c. 3)

Cimino-Isaacs C, Hufbauer GC, Schott JJ (2014) A proposed code to discipline local content requirements. Peterson Institute for International Economics. Pierson Institute for International Economics. Policy Brief 14(6)

Coe D, Helpman E, Hoffmaister AW (2008) International R&D Spillovers and Institutions. Working Paper No. WP/08/104. International Monetary Fund, Washington, DC

Columbia Centre on Sustainable Investment. Local Content Laws & Contractual Provisions. http://ccsi.columbia.edu/work/projects/local-content-laws-contractual-provisions/. Accessed 10 Oct 2017

Contract of 1998 for the Exploration, Development and Production of Petroleum Between the Government of the Syrian Arab Republic and Syrian Petroleum Company and INA-Industrija Nafte dd

Cook M, Mahdavy H (1970) The pattern and problems of economic development in rentier states: the case of Iran. In: Cook M (ed) Studies in the economic history of the Middle East: from the Rise of Islam to the present day. Oxford University Press

Croatian Hydrocarbon Agency (2015) Croatian Model Production Sharing Contract, 1st Onshore licensing round

Darling R (2011) Beyond taxation: how countries can benefit from the extractive industries through local content. Revenue Watch Institute, New York

Esteves AM, Barclay MA (2011) Enhancing the benefits of local content: integrating social and economic impact assessment into procurement strategies. Impact Assess Proj Appraisal 29:205

Esteves AM, Coyne B, Moreno A (2013) Local content initiatives: enhancing the subnational benefits of the oil, gas and mining sectors. Natural Resource Governance Institute

Franks D, Davis R, Bebbington A, Saleem H, Kemp D, Scurrah M (2014) Conflict translates environmental and social risk into business costs. Proc Natl Acad Sci 111(21):7576–7581

Gbedi D, Adebisi J (2013) Managing local content policies in the extractive industries. Res J Financ Account 4(7):90

Ghafar A (2016) Will the GCC be able to adjust to Lower Oil Prices. Brookings Institution. https://www.brookings.edu/blog/markaz/2016/02/18/will-the-gcc-be-able-to-adjust-to-lower-oil-prices/. Accessed 10 Oct 2017

Glass A, Saggi K (2008) The role of foreign direct investment in international technology transfer. In: Dutt A, Ros J (eds) International handbook of development economics. Edward Elgar, Cheltenham, UK/Northampton, MA

Government of Liberia. Production Sharing Contract. http://www.eisourcebook.org/cms/Liberian%20Production%20Sharing%20Contract.pdf

Government of Libya. Libyan Exploration and Production Sharing Agreement IV

Government of Saskatchewan (2015) Northern benefits summary - Saskatchewan mine surface lease agreements. http://cvmpsk.ca/wp-content/uploads/2015-Northern-Benefits-Summary-Mine-Surface-Lease-Agreement-Final-June-2016.pdf. Accessed 12 June 2018

Groupo FARO, ACODE (2015) Oil and gas local content development strategy: the keys to success. http://www.acode-u.org/Files/Publications/PBP_38.pdf. Accessed 12 May 2018

Hanlin C (2011) The drive to increase local procurement in the mining sector in Africa: myth or reality? MMCP Discussion Paper No. 4

Heum P, Kasande R, Ekern OF, Nyombi A. (2011) Policy and regulatory framework to enhance local content: yardsticks and best practice. SNF Working Paper No. 02/11

International Monetary Fund (IMF) (2015) Global Implications of Lower Oil Prices, International Monetary Fund. https://www.imf.org/en/Publications/Staff-Discussion-Notes/Issues/2016/12/31/Global-Implications-of-Lower-Oil-Prices-43052. Accessed 12 June 2018

International Petroleum Industry Environmental Conservation Association (IPIECA) (2011) Local content strategy: a guidance document for the oil and gas industry. http://www.ipieca.org/media/1384/local_content_2016.pdf. Accessed 12 June 2018

Iran's Maximum Utilization of Production and Services Potency in Providing Country's Needs and Promotion of Exports 2012

Krueger A (1974) The political economy of the rent-seeking society. Am Econ Rev 64(3):291–303

Lee YS (2016) Reclaiming development in the world trade system, 2nd edn. Cambridge University Press

Levett M, Chandler AE (2012) Maximizing development of local content across industry sectors in emerging markets. Center for Strategic and International Studies, Washington, DC

McCormick R (2010) Legal risk in the financial markets. Oxford University Press

McKinsey Global Institute (2013) Reversing the Curse: Maximizing the Potential of Resource-Driven Economies

Mineral and Petroleum Resource Development Act 2002 of South Africa

Minerals and Mining (General) Regulations 2012 (LI 2173) of Ghana

Mines and Minerals Development Act no 7 of 2008 of Zambia

Mohinsky B (2015) OPEC: oil won't be worth $100 a barrel until after 2040. Business Insider. http://uk.businessinsider.com/opec-oil-wont-be-worth-100-a-barrel-until-after-2040-2015-12. Accessed 12 June 2018

Muller T, Schitzer M (2003) Technology Transfer and Spillovers in International Joint Ventures. Discussion Paper No. 2003–22

Namibia Petroleum Agreement (2007) http://www.eisourcebook.org/cms/Namibia%20Model%20Petroleum%20Agreement%202007.pdf. Accessed 12 Aug 2018

Nigerian Oil and Gas Industry Content Development Act 2010

Nwapi C (2015) Defining the 'Local' in local content requirements in the oil and gas and mining sectors in developing countries. Law Dev Rev 8(1):187–216

Obasi OI. Analysis of the Nigerian Oil and Gas Industry Content Act 2010

Offshore Wind Industry Investment in the UK 2017: Report on Offshore Wind UK Content (2017) Renewable UK

Olawuyi D (2015) Legal strategies and tools for mitigating legal risks associated with oil and gas investments in Africa. OPEC Energy Rev 39(3):247–265

Olawuyi D (2017a) From technology transfer to technology absorption: addressing climate technology gaps in Africa. J Energy Nat Res Law. https://doi.org/10.1080/02646811.2017.1379667

Olawuyi D (2017b) Local content and procurement requirements in oil and gas contracts: regional trends in the Middle East and North Africa. Oxford Institute for Energy Studies, Research Paper Series 8

Olawuyi D, Mercier T (2015) Local content and procurement requirements in frontier African oil and gas jurisdictions – One size does not fit all. Norton Rose Fulbright. http://www.insideafricalaw.com/blog/local-content-and-procurement-requirements-in-frontier-african-oil-and-gas-jurisdictions-one-size-does-not-fit-all. Accessed 10 Oct 2017

Oman Model Exploration and Production Sharing Agreement Dated April 24, 2004

Organisation for Economic Co-operation and Development (OECD) (2017) Local Content Policies in Minerals-Exporting Countries. http://www.oecd.org/officialdocuments/publicdisplaydocumentpdf/?cote=TAD/TC/WP(2016)3/PART2/FINAL&docLanguage=En. Accessed 12 May 2018

Organization of Petroleum Exporting Countries (OPEC) (2015) 2015 World Oil Outlook. http://www.opec.org/opec_web/static_files_project/media/downloads/publications/WOO%202015.pdf. Accessed 10 Oct 2017

Ovadia J (2015) The role of local content policies in natural resource-based development. Österreichische Entwicklungspolitik. Rohstoffe und Entwicklung

Petroleum (Local content and Local Participation in Petroleum Activities) Regulations 2013 of Ghana

Petroleum Commission Act 2011 of Ghana

Ramdoo I (2015) Unpacking local content requirements in the extractive sector: what implications for the global trade and investment frameworks? E15 initiative. International Centre for Trade and Sustainable Development (ICTSD) and World Economic Forum, Geneva

Rodrik D (2007) One economics, many recipes: globalization, institutions and economic growth. Princeton University Press, Princeton

Saipem. SELCE Model: Local Content for Sustainable Development. http://www.saipem.com/en_IT/static/documents/Saipem_Local_Content.pdf. Accessed 12 June 2018

Shapiro S, Rabinowitz RS (1997) Punishment versus cooperation in regulatory enforcement: a case study of OSHA. Adm Law Rev 49:713–720

Shared Value Initiative (2014) Extracting with purpose: creating shared value in the oil and gas and mining sectors' companies and communities. https://sharedvalue.org/sites/default/files/resource-files/Extracting%20with%20Purpose_FINAL_Full%20Report_Single%20Pages.pdf. Accessed 10 Oct 2017

State of Qatar Model Development and Production Sharing Agreement of 2002

The Nigerian Oil and Gas Industry Content Development Act 2010, n. 65. The Board has powers to: monitor the implementation of set Nigerian Content targets; develop interventions to ensure that these targets are met; and ensure compliance with Government directives covering the service providers and operators in the Nigerian Oil and Gas Industry

The Petroleum (Exploration, Development and Production) (National Content) Regulations 2016 of Uganda

Tordo S, Tracy B, Arfaa N (2011) National oil companies and value creation. Working Paper No. 218. The World Bank, Washington, DC

Tordo S, Warner M, Manzano O, Anouti Y (2013) Local content policies in the oil and gas sector. World Bank

Trade Arabia (2015) Big boost to Saudi industries as Aramco to double local market sourcing. http://tradearabia.com/news/IND_295936.html. Accessed 12 May 2018

Uganda's Petroleum (Exploration, Development and Production) (National Content) Regulations, 2016

Ulrichsen KC (2011) Insecure Gulf: the end of certainty and the transition to the post-oil era. Columbia University Press, New York

World Bank (2014) Human capital for the oil, gas and minerals industries. Science, Technology, and Skills for Africa's Development (201490) 1–4

World Trade Organization. Agreement on Trade-Related Investment Measures (TRIMS)

World Trade Organization. General Agreement on Tariffs and Trade 1994

Part III
Legal and Business Risks in African Extractive Investments

Chapter 10
Environmental Impacts and Risks of Extractive Operations

Abstract This chapter examines environmental impacts and risks associated with extractive operations and activities in resource rich African countries. After providing contextual discussions on the nature and meaning of the environment, it unpacks the key sources of environmental degradation from extractive operations in the petroleum and solid minerals sectors. It then discusses key laws designed to regulate and prevent environmental pollution in extractive industries in Africa. The chapter also provides insights on how extractive companies can avoid environmental sanctions by implementing internal policies that meet international environmental law standards on pollution prevention, remediation and restoration.

10.1 Introduction

This chapter examines environmental challenges associated with extractive operations and activities in resource rich African countries. The growth of the extractive industry in Africa, especially in regions where significant proportions of oil and solid minerals deposits are located, has resulted in significant environmental challenges to local communities.[1] In many African countries, resource extraction has been closely associated with severe environmental degradation and human rights concerns, arguably more so than in any other region. The legacy of petroleum operations and mining has left thousands of sites in Africa contaminated by oil spillage, gas flaring, discharge of heavy metals from mining, and associated mine dumps such as tailings and slag material.[2] These environmental challenges have resulted in complex human rights concerns relating to loss of arable land and property, ill health, loss of subsistence, stakeholder exclusions, and in extreme cases, loss of lives.

[1] Mapani and Kribek (2012).

[2] Ibid.

D. S. Olawuyi, *Extractives Industry Law in Africa*,
https://doi.org/10.1007/978-3-319-97664-8_10

For example, wastes generated by the mining industry typically contain high concentrations of metals and metalloids which have resulted in contamination of ground and surface water.[3] The mining of gold in the Migori Gold Belt, in Kenya, and in the Witwatersrand Basin of South Africa, for instance, have resulted in significant negative impacts on environmental and human health in these countries. Studies show that the concentrations of heavy toxic metals, mainly arsenic (As), cadmium (Cd), cobalt (Co), copper (Cu), lead (Pb), uranium (U), and mercury (Hg), from gold sites are above acceptable levels in South Africa and Kenya.[4] Most of these heavy metals are highly toxic and are not biodegradable as such contaminate waters and land for several years if not removed.[5] Similarly, salt mining in Magarini District, Kilifi County of Kenya has resulted in cases of water pollution and environmental degradation.[6] Studies also show that mining activities in Southern Katanga in the Democratic Republic of Congo has substantially exposed residents to several metals or metalloids, with residents living within 3 km of mines or metal smelters having significantly higher urinary concentrations of toxic metals than those living farther away.[7]

These concerns are not peculiar to Kenya, South Africa and D.R. Congo. Nigeria offers more worrisome examples of how resource extraction projects, if not properly regulated, could engender complex environmental, social and human rights concerns. Decades of oil spillage, gas flaring, and effluent discharge in Nigeria's Niger Delta have severely damaged the ecosystem and environment of that region.[8] According to a 2011 report of the United Nations (UN), reversing many of the environmental and social consequences of oil spillage in the Niger Delta is possible but may take 25–30 years.[9] These concerns have highlighted the need for robust environmental safeguards and policies to address environmental challenges of resource extraction.[10]

Over the last decade, international law has increasingly recognized the need to integrate environmental protection into policy frameworks on resource utilization to avoid these unintended negative consequences of resource exploitation.[11] This is

---

[3] Ibid.

[4] See Ogola et al. (2002), p. 141; also Cukrowska et al. (2012), pp. 15–18.

[5] Nkulu et al. (2012), pp. 13–14.

[6] Ocholla (2013), p. 213.

[7] Nkulu et al. (2012), pp. 13–14.

[8] Olawuyi (2015a, b).

[9] United Nations Environment Programme (2011), p. 12.

[10] See generally Oluduro (2014), p. 165 (providing an account of the nature and scope of the problems discussed); see also Human Rights Watch (1999a, b), p. 47.

[11] See The 2002 International Law Association (2002), which strongly reaffirms and concretizes the duties of states to incorporate sustainability into the management plans for natural resources and associated activities. It notes that "sustainable development is now widely accepted as a global objective and that the concept has been amply recognized in various international and national legal instruments, including treaty law and jurisprudence at international and national level"; also Schrijver (2017) (providing an overview of the ways that sustainable development principles have

part of broader theoretical debates and efforts aimed at coordinating the systemic integration of economic, social, human rights and environmental development goals under international law to ensure coherence, effectiveness, and harmonization.[12] Over the last decade, international law has promoted a rights-based understanding of the concept of corporate responsibility and risk management.[13] This requires private actors and business enterprises to have a degree of responsibility not only for the commercial or economic consequences of their activities, but also for the social and environmental implications.[14] International law has sought to underscore the importance of mainstreaming human rights norms into the design, approval, finance and implementation of resource extraction projects to avoid human rights impacts. Intricately intertwined with the concept of corporate responsibility and risk management is the idea of sustainable development which emphasizes the need for investments or resource utilization that meet human needs while preserving the abilities of future generations to meet their own needs.[15] African countries are increasingly reflecting environmental compliance requirements, penalties and sanctions in petroleum and mining contracts as a way of regulating the environmental side effects of extractive activities.[16]

Extractive operators can avoid costly environmental liability, sanctions and penalties for environmental violations by keeping up-to-date with the meaning, nature and scope of environmental protection regulation and legislation in the extractive industry. After studying this chapter, you should be able to: understand the nature and meaning of the environment; appreciate the nature and sources of environmental destruction from extractive operations in the petroleum and solid minerals sectors; outline the social and human rights impacts of environmental pollution; understand the key laws designed to manage and prevent pollution in your country; identify the key regulatory institutions with oversight functions and power to manage and prevent pollution in your country; and appreciate how extractive companies can avoid environmental sanctions by implementing internal policies that meet international standards.

---

been incorporated into and operationalized by international treaty regimes and tribunals); French (2017) (highlighting how sustainable development has been incorporated into the juridical activities of international bodies convened under the auspices of UN associated and influenced treaty regimes); [international treaty law on sustainable development]; also UN, The Human Rights Based Approach to Development Cooperation: Towards a Common Understanding Among UN Agencies.

12 Mining Working Group at the United Nations (2014); Macias (2010), pp. 10–12.

13 See Olawuyi (2016a, b).

14 See Saether and Aguilera (2008).

15 The Brundtland Report (1987).

16 For a detailed review of these efforts, see Olawuyi (2015a, b).

10.2 What Is the Environment?

In order to fully appreciate the implications of extractive operations for human and natural environments, a full understanding of the meaning, nature and scope of the term 'environment' is essential. As basic as the term sounds, there is currently no general, objective and all-embracing definition for the environment. Scholars and environmentalists have defined it in different terms. For example, Einstein is said to have argued that 'the environment is everything that isn't me'.[17] Though not exhaustive, Einstein's characterization is a good starting point for conceiving the environment as our physical and biological surroundings. The environment of an individual includes all of the other entities, including plants, animals, land and objects, with which it is surrounded. The environment has also been defined as the sum of all external conditions affecting the life, development and survival of an organism.[18] Similarly, the environment has been characterized as "all the physical, chemical and biological factors external to a person, and all the related behaviors."[19]

A more expansive view is that the environment includes all the elements of the physical and biological world (including humans), and the interactions between them. These elements may be categorized as ecosystem goods and services. Goods, meaning the actual natural resources themselves (flora, fauna, soil mineral, air, water); while services, include the harvestable products (crops, timber), processes essential to sustain the provision of these resources (nutrient cycles, climate patterns, flooding control) and aesthetic and cultural benefits of ecosystems (recreation).[20] Section 37 of Nigeria's National Environmental Standards and Regulations Enforcement Agency (NESREA) Act of 2007 defines the environment as "water, air, land and all plants and human beings or animals living therein and the inter-relationships which exist among these or any of them."[21] Notably, Section 20 of Nigeria's Water Resources Act (WRA) 1993 conceptualizes the environment to include 'all aspects of the surroundings of man whether affecting him as an individual or in his social and aesthetic factors of those surroundings.'[22]

While some of the above definitions encompass human and non-human elements (for example the above definition in NESREA), others focus only the abilities of humans to enjoy the environment and it resources (for example the WRA definition). This reflects age-long contentions amongst scholars and practitioners on the appropriate aim, boundaries and operating parameters of environmental protection.

[17] Quoted in Burges Salmon (2014), pp. 1–2.

[18] United States Environmental Protection Agency (2016).

[19] See World Health Organization (2016), pp. 23–25.

[20] Olawuyi (2015a, b).

[21] National Environmental Standards Regulatory and Enforcement Agency (Establishment) Act of 2007, Laws of the Federation of Nigeria 2004, c N164.

[22] *Water Resources Act* of 1993, Laws of the Federation 2004, c W2.

Generally, there are two different schools of thought on environment and environmental protection.[23] While some scholars view the environment from an ecocentric and nature standpoint, others assign an anthropocentric/human related connotation to the term environment.[24] From an anthropocentric or human point of view, the environment has been defined as the relationship of human beings with water, air, land and all biological forms; the combination of elements whose interrelationships make up the settings, surroundings, and conditions of life of the individuals and the society.[25] From the eco-centric or nature standpoint, the environment is defined as the sum of all external conditions affecting the life, development and survival of an organism.[26] It has also been described as the metaphor of life that provides ecological support for all organisms on earth. This means the source, support and the ingenious life-giving matrix for all living organisms.[27]

This chapter will not delve extensively into these debates.[28] Rather it takes an all-encompassing view of the environment, which includes both human and non-human elements. Both aspects have long lasting effects and desire equal attention, just as humans may find the earth unliveable if we have to live without the ecosystem; plants, human interests cannot also be sidelined in providing for nature since humans are inevitably in a better position to care for the environment. In this chapter, the environment is defined as the sum total of water, air, land, plants, animals, humans, and other living organisms that make up the ecosystem; and the cross cutting interactions between them. Generally, the environment comprises both human and natural elements. It includes all factors that surround and affect the quality of human lives on earth.

Environmental problems such as oil pollution, climate change and ozone depletion would devastate both humans and the natural habitats. The aim therefore should be to balance protection for nature with protection for human beings in extractive processes and operations. Of paramount importance to a lawyer is the role of law in managing and balancing the interactions between natural ecosystems and human beings.

Environmental law has two equally important goals: the elimination of environmental harm or pollution that can affect the natural habitats; and the elimination of environmental harm, discomfort or pollution that can shorten or imperil human life.

[23] See Olawuyi (2015a, b), pp. 8–10.

[24] Donnelly and Bishop (2007); see also Eckersley (1992), p. 58; Redgwell (1996), pp. 86–87.

[25] See Elder (1984), p. 285; Anderson (1996), pp. 10–11; Feinberg (1974). Feinberg adds that plants and animals cannot hold rights because they are incapable of claiming such rights on their own, He noted that: "…Animals are incapable of claiming rights on their own, they cannot make motion to Courts on their own to have their claims recognized or enforced; they cannot initiate on their own any kind of legal proceedings; nor are they capable of even understanding when their rights are being violated, of distinguishing harm from injury and responding with indignation and an outraged sense of justice instead of mere anger or fear".

[26] See Bodansky (2010), pp. 9–11.

[27] United States Environmental Protection Agency (2016).

[28] For detailed examination of these debates, see Olawuyi (2015a, b), pp. 8–20.

The *Stockholm Declaration of the United National Conference on the Environment* recognizes both aspects of the environment, natural and man-made as essential to well-being and to the enjoyment of basic human rights, including the right to life itself.[29] The Stockholm Declaration describes attributes that make up the environment to include natural resources of the earth, such as air, water, land, flora, fauna and wildlife. In addition, it advocates the need for conservation and sustainable use of natural resources, protection and conservation of biodiversity and endangered species, deforestation, air quality, interference with the global climate, international watercourses, quality of human life amongst others. The expansive definition in the Stockholm Declaration provides normative foundation for the linkages between environment and human rights; most importantly that environmental protection is a *sine qua non* for the enjoyment of fundamental human rights.

Environmental law therefore is that branch of law that provides the general legal framework for the regulation and protection of elements of the environment such as air, land, water, sea, amongst others. It includes the body of laws, rules, regulations and statutes that regulate the interactions between humans and the natural environment with the aim of reducing the impacts of human activity on the environment. It provides appropriate standards for measuring and apportioning liability in case of pollution or default. It may take the form of international treaties/conventions, statutes, regulations, common law, national or domestic legislation, or any other instruments that operate to reduce the impacts of human activity on the natural environment. Environmental law borrows and transplants well tested notions and principles from other traditional areas of law such as torts law, criminal law and international law. In addition to addressing issues relating to land, air, and water, the scope of environmental law also encompasses natural resource issues, pollution, land contamination, forest protection, and in recent times, issues such as renewable and non-renewable energy, biodiversity, conservation, climate change, sustainable resource management and responsible sourcing of extractive resources across global supply chains.

10.3 Sources of Environmental Pollution in the Extractives Industry

Pollution in its broadest sense is any alteration or contamination of the physical and natural environment, in such proportions that could be harmful to human life and nature. It has also been defined as any "man-made or man aided alteration of chemical, physical or biological quality of the environment beyond acceptable limits."[30] Extractive activities can pose a wide range of challenges to the environment, most

[29] Para 1, The Stockholm Declaration on the Human Environment, UN Doc A/CONF48/14/ Rev.1, (1972). [The Stockholm Declaration].

[30] Section 37 of the National Environmental Standards and Regulations Enforcement Agency (Establishment) Act 2007 Laws of the Federation of Nigeria 2004, c N164.

especially in petroleum and minerals producing communities. Residents of resource producing communities face elevated levels of pollution of water, land, air and all other important environmental media necessary for a healthy life. To effectively address these challenges, it is important to understand the specific and unique nature of pollution posed to the environment respectively by mining and petroleum operations. This section examines the nature of environmental pollution in each of the sectors that make up the extractive industry.

10.3.1 Mining Pollution

The key sources of environmental pollution from mining are: surface destruction and contamination; discharge of toxic wastes; acid mine drainage; and noise pollution and vibration.

10.3.1.1 Surface Destruction and Soil Contamination

Mining can result in a permanent contamination and destruction of land, soils, forests and landscape. Strip mining (also known as open cast, mountaintop or surface mining) involves scraping away earth and rocks to get to coal or other minerals buried under the surface.[31] It involves operations that remove a long strip of overlying soil and rock in order to reach mineral seams that are relatively near the surface.[32] Strip mining can therefore destroy landscapes, forests and wildlife habitats at the site of the mine. This could occur if the topsoil, which hitherto served as habitation for plants and animals, is scraped off to pave way for extractive operations. This could in turn lead to soil erosion and destruction of agricultural land.[33]

Another significant side effect of mining operations is deforestation and desertification. Deforestation is the clearance or clearing of topsoil, which includes the removal of a forests or trees, in order to access minerals in the subsurface. Desertification on the other hand is the process of land degradation in which fertile land becomes desert, typically as a result of deforestation, drought, or inappropriate use of land. It is the negative transformation of hitherto fertile and arable land to desert, due to destructive land use or climatic conditions. The *United Nations Convention to Combat Desertification in countries Experiencing Serious Drought/ and or Desertification, particularly in Africa* in article 1 (a) defines desertification as land degradation in and around semi-arid and dry sub-humid areas resulting from several factors, which include climatic variations and human activities.

[31] Department of Environment, South Africa, Effects of Mining on the Environment and Human Health.

[32] Gardner and Sainato (2007).

[33] Department of Environment, South Africa, Effects of Mining on the Environment and Human Health.

Desertification is one of the main causes of poverty and underdevelopment in Sahelian countries in Africa.[34] Due to the scraping of top soil during mining activities, soil loses its nutrient making it impossible for lands to grow crops or support food production for local populations. This has resulted in long-term declines in agricultural yields, livestock yields, and plant biodiversity in several mineral rich African countries. The end result is starvation and poverty in rural communities where several citizens depend on subsistence agriculture for daily survival. Desertification also has serious consequences for the environment, most especially: loss of soil surface due to wind and rain erosion, soil pollution (due to the excessive use of chemical fertilizer on damaged lands), poor water quality, landslides and dust storms. All of these effects can hurt people living in an affected country or region. Desertification-induced changes may also result in increased migration from rural areas into the urban centers. This could lead to greater pressure on the existing urban infrastructure, and subsequent social conflicts in the affected areas. In Nigeria for example, desertification has become a major development and environment problem, particularly in the Northern region. It has been estimated that about 50–75% of Bauchi, Borno, Gombe, Adamawa, Jigawa, Kano, Katsina, Kebbi, Sokoto, Yobe, and Zamfara States in Nigeria are affected by desertification. It is estimated that Nigeria is currently losing 351,000 ha of its landmass to desert-like conditions annually, and such conditions are estimated to be advancing southwards, in a haphazard manner, at the rate of about 0.6 km per year.[35]

Strip mining can also result in water and air pollution. Wastes generated during strip mining contain high concentrations of metals and metalloids can seep into streams, water table, and watersheds. Most of these heavy metals are highly toxic and are not biodegradable. Similarly, strip mining operations can cause dust and noise pollution when top soil is disrupted with heavy machinery and coal dust is created in mines.[36]

Land contamination and disturbance as a result of strip mining operations is on a vast scale in Africa. Yet, strip mining remains widely common because it is less expensive, less labour intensive and could yield more minerals than underground mining. Although many African countries have put in place laws that require reclamation and remediation plans for mining sites, reversing all of the environmental damages to land, forests, air, water, and destroyed habitats may take several years, and in some cases, may be impossible.[37]

---

[34] The name "Sahel" is derived from the Arabic word *sahil*, which means "border of the desert". The Sahel region covers nine countries Mauritania, Senegal, The Gambia, Guinea-Bissau, Mali, Burkina Faso, Niger, Chad and Cape Verde. Water is a scarce resource in all Sahel countries. Besides erratic rainfall patterns, poor soils and unfavourable socio- economic conditions are already key constraints to agricultural development in Sahelian states. See Sissoko et al. (2010) and Bailey (1979).

[35] Olawuyi (2015a, b), pp. 228–285.

[36] See Ocholla (2013), p. 213.

[37] Studies have shown that trace metals remain in the soil for a long time ranging from hundreds to thousands of years. (Pb and Cr for example may remain for several thousand years). See Johnston et al. (2008), pp. 5–20.

10.3.1.2 Discharge or Dumping of Mine Wastes

Discharge or dumping of mine processing wastes, also known as tailings, into the environment during mining operations can result in elevated levels of acidic water and air pollution in resource based communities. Mining operations such as drilling, blasting, hauling, collection, and transportation during mining can result in the discharge of particles and dusts into the air, resulting in severe air pollution. When extractive operators break up rocks during mining, the resulting particles and dust can release a variety of heavy metals commonly associated with health problems. Particulate Matters (PMs), mainly stirred up during mining processes, as well as soot released during transportation, have been linked with aggravated cardiac and respiratory diseases such as asthma, bronchitis, respiratory tuberculosis and emphysema and to various forms of heart disease in mining communities.[38] PMs also have adverse effects on vegetation and structures, and contribute to visibility deterioration and regional haze pollution. Similarly, harmful dusts such as flue dust and the asbestos-like mineral riebeckite can be absorbed into lung tissue, causing problems like pneumoconiosis and silicosis (commonly known as "black lung").[39]

The discharge of tailings into waterbodies, whether in the form of ocean dumping or submarine tailings disposal (STD); river dumping (riverine tailings disposal or RTD); and lake dumping (lacustrine tailings disposal or LTD), can significantly contaminate water bodies and marine life with toxic heavy metals and milling chemicals. STD is the dumping of waste below the ocean's surface using pipes that carry tailings from a mine into the sea; RTD is the practice of disposing of mine tailings into river valleys; while LTD is the dumping of mine tailings into natural lakes.[40] The discharge of harmful toxic metals, such as arsenic (As), cadmium (Cd), cobalt (Co), copper (Cu), lead (Pb), uranium (U), and mercury (Hg), from gold sites into lakes, rivers, seas, oceans, aquifers, and ground water, make such water unsafe and unsuitable for use. For example, mercury is commonly used as an amalgamating agent to facilitate the recovery of some precious ores.[41] However, improper disposal of mercury tailings can lead to contamination of the atmosphere and neighboring bodies of water.[42] As water takes on harmful concentrations of minerals and heavy metals, it becomes toxic, contaminated and unhealthy to use for homes surrounding the mine and beyond. Those contaminants from the tailings can also cause harm to marine life.

For example, the Lihir gold mine in Papua New Guinea allegedly dumps over 110 million cubic metres of waste, contaminated with cyanide and other chemicals,

---

[38] Paul and Campbell (2011); also Schuler et al. (2011).

[39] According to the Chinese Society of Rare Earths, every ton of REE produced generates 8.5 kg of fluorine and 13 kg of flue dust, waste materials which contain the heavy metals. Schuler et al. (2011), ibid.

[40] See Earth and Mining Watch Canada (2012), pp. 6–10; also US Department of Interior (1993); Apte and Kwong (2003), pp. 2–20; Findikakis and Law (1998).

[41] Earth and Mining Watch Canada (2012).

[42] Ibid.

into the sea each year.[43] Similarly, the Bougainville copper mine in Papua New Guinea reportedly dumped more than 500 million tonnes of tailings into the Kawerong and Jaba Rivers, before protests and civil war forced the mining company (Rio Tinto) to abandon the mine in 1989.[44] Similarly, the Simberi gold mines in Papau New Guinea leaked tailings waste, containing toxins such as arsenic, cadmium, chromium, lead, and mercury, from its tailings mixing tank in 2011. Studies show that fish diversity and abundance in the dumping areas have been significantly lower because of the tailings dumping, while arsenic and mercury concentrations in several fish species are at higher concentrations closer to the mine.[45] Estimates also indicate that mining companies globally are dumping more than 180 million tonnes of hazardous mine waste each year into rivers, lakes, and oceans worldwide, threatening vital bodies of water with toxic heavy metals and other chemicals poisonous to humans and wildlife.[46]

Water pollution from mining exposes Africans to short term and long-term diseases and death, it also disrupts the existence and survival of marine and aquatic organisms. For example, about 69 million Nigerians lack safe, sufficient and sustainable access to clean drinking water, while about 103 million Nigerians lack access to adequate sanitation facilities. Similarly, over 97,000 children die in Nigeria every year from diarrhea caused by unsafe water and poor sanitation.[47] This situation is a serious threat to the attainment and enjoyment of fundamental human rights to health, adequate standard of living, subsistence and livelihood of a significant number of Nigerians. It is therefore important to analyze and understand the web of laws that seek to prevent pollution of water bodies, and also protect and preserve water resources.

10.3.1.3 Acid Mine Drainage

Acid mine drainage (AMD), also referred to as acid rock drainage (ARD), is one of the most pervasive environmental problems in the mining industry, especially at abandoned mines in Sub-Saharan Africa.[48] AMD is the flow of polluted and acidic water from oil mining areas to surrounding soil, groundwater, and surface water. It occurs when sulfide-bearing material from mine waste rocks, tailings, and mine structures, such as pits and underground workings are exposed to water and oxygen therefore creating sulphuric acid.[49] When the produced sulphuric acid flows into surrounding water bodies, it contaminates such water bodies and marine environments. This may result in acute and chronic toxicity to both human users of water

[43] Sydney Morning Herald (2002); also McKinnon (2002).

[44] Earth and Mining Watch Canada (2012).

[45] Earth and Mining Watch Canada (2012).

[46] Ibid, 1.

[47] See United Nations (2010), Water and Sanitation Fact Sheet for Nigeria.

[48] Akcil and Koldas (2006); also Davies (2012), pp. 17–18.

[49] See Ossai et al. (1999), pp. 83–86.

and marine aquatic life. AMD is particularly damaging to lakes, streams, forests and the plants and animals that live in these ecosystems. In addition, AMD accelerates the decay of building materials and paints, including irreplaceable buildings, historical statues, and sculptures that are part of national cultural heritage. Acidic waters flowing from abandoned mines can raise acidity levels and destroy aquatic life for several decades and generations. For example, thirteenth century mining sites in England are still causing AMD more than 2000 years later.[50] AMD is therefore a complex environmental challenge with severe health impacts that may be near irreversible.

In many African countries, especially in South Africa, with several old and abandoned mines, AMD has been identified as reaching a crisis point.[51] AMD has been a major source of environmental pollution in the South African mining industry, especially from a number of mines that have ceased mining operations in early 1980s, but continue to release a considerable amount of acidic water from abandoned underground workings into the surface environment in the West and East Rand.[52] Laws and policies have been released in South Africa that mandate mining companies to stop seepage and polluted water from leaving their mine sites. Water entry into the site of acid formation may be controlled by diverting surface water flowing towards the site of pollution and controlling or preventing water infiltration or seepage into the pollution site.[53] However, despite the laws that mandate the control and prevention of AMD, it remains a complex environmental crisis in South Africa and across the African continent.[54]

10.3.1.4 Noise Pollution and Vibration

Mining, in both open and underground pits, can be a major source of noise pollution if not properly managed and controlled. Unlike petroleum operations that mostly occur offshore, mine sites tend to be close to local and residential communities. This proximity often means that vibration and noise associated with mining may pose significant disturbance and environmental challenges to individuals that reside or work near mine sites. Such vibrations can also be fatal for plants and animals. Noise pollution from mining include vibration and noise from shoveling, ripping, drilling, blasting, transport, crushing, grinding, and stock-piling operations. It also includes noise and vibrations from equipment such as continuous miners, stage loaders, shearers, compressors, fans, steel dumpers, chutes, power generation, vibrating screens, rotating breakers and mills, vehicle engines and pneumatic drilling which are commonly in use in most of the mineral processing stages. Cumulative impacts of noise on nearby residents include sleep deprivation, psychosis and

---

[50] Johnston et al. (2008), pp. 2–5.

[51] Department of Environment, South Africa, Acid Mine Drainage (AMD) in South Africa. [AMD].

[52] Ibid.

[53] Ibid.

[54] In average, AMD in South Africa has increased by nearly 40% between 1956 and 2017, ibid.

anxiety. Noise from mining is also a major occupation hazard for miners who may in the long term suffer from reduced hearing.[55] Vibration can also affect the stability of infrastructures, buildings, and homes of people living near large-scale open-pit mining operations. According to a study commissioned by the European Union in 2000, shocks and vibrations as a result of blasting in connection with mining can 'lead to noise, dust and collapse of structures in surrounding inhabited areas. The animal life, on which the local population may depend, might also be disturbed.'[56]

10.3.2 Oil Pollution

Oil pollution has been a major, and one of the most litigated, ecological concerns in oil producing communities across Africa.[57] There are three main sources of oil pollution: oil spillage, gas flaring and the discharge of effluents and wastes. This section briefly examines these sources and their key impacts.

10.3.2.1 Oil Spillage

Oil spillage is the unsafe discharge or release of oil into the human environment, for example into water bodies such as oceans, rivers and seas.[58] This occurs in the process of oil exploration and production, mainly due to accidental or negligent rupture or blow out from wellheads, flow stations, drilling rigs, pipelines and offshore platforms and facilities. Oil facilities and pipelines mostly rupture due to poor maintenance, corrosion, age or in some cases due to vandalism, sabotage, and defective installation.[59] Oil spills could also result from transportation or loading leakages.[60] Oil spills result in the release of dangerous hydrocarbons such as benzene and polynuclear aromatic hydrocarbons into soil and water sources. Oil spills often result in both immediate and long-term environmental damage, endanger public health, imperil drinking water, devastate aquatic life, lower soil fertility and cause poor growth of crops and plants. Environmental damage caused by oil spillage can last for decades.

---

[55] Sensogut (2007).

[56] MINEO Consortium (2000).

[57] See the Nigerian oil pollution case of *Nwadiaro v Shell Petroleum Development Company, Nigeria Ltd* (1990) 5 N.W.L.R. (part 150) 322; *Shell Petroleum Development Company, Nigeria ltd v Farah* (1995) 3 N.W.L.R. (part 382) 148; *Shell Petroleum Development Company Nigeria Ltd v Ambah* (1991) 3 N.W.L.R. (part 593) 1; *Shell Petroleum Development company, Nigeria Ltd v Isaiah* (2001) 11 n.w.l.r. (part 723) 168; *Elf Nigeria Ltd v Opere Sillo* (1994) 6 N.W.L.R. (part 350) 258; *Shell Petroleum Development Company, Nigeria Ltd v Tiebo* vii & Others (2005) 3-4 s.c.

[58] Olawuyi (2015a, b), pp. 177–179.

[59] See Olawuyi (2012); Emeseh (2006), p. 595.

[60] Olawuyi (2015a, b); also see Ojukwu-Ogba (2009), p. 137.

In Nigeria for example, over the years, there have been a high number of oil spills incidents, releasing copious amounts of oil into the Niger Delta. In the period between 1993 and 2007, there were 35 reported incidences of oil spills. This is aside from the unnoticed slicks and unreported cases of oil spills. For example, on December 21, 2011, Shell Nigeria announced what it describes as its worst oil spillage in a decade in the Niger Delta area. Over 40,000 barrels of crude oil spilled in one day.[61] According to the 2011 report of the United Nations (UN), many of the environmental and social consequences of oil spillage in the Niger Delta are now irreversible.[62]

The Niger Delta situation is not an isolated one; the tragic consequences of oil spills are apparent in many nations of the world. One of the largest oil spills in history occurred in 2010, when up to 4.9 million barrels of oil was spilled from BP's oil well into the Gulf of Mexico over a three-month span.[63] The BP spill resulted in the loss of animals, plants and human lives in monumental proportions. These events emphasize the need for appropriate legal frameworks to prevent the environmental and health impacts of oil spills on the residents of oil producing areas.

10.3.2.2 Environmental Impacts of Oil Spills

Oil spills can result in negative environmental impacts in many resource-based countries. Oil spillage is a violation of a multitude of national and international laws and regulations on environmental pollution and damage. The situation in Nigeria's Niger Delta illustrates how oil spillage negatively impacts the environment. As a result of the high frequency of oil spills in the Niger Delta, many local communities have been destroyed, drinking water have been contaminated, and the ecosystem and soil composition have been altered irreversibly. Studies show that in the Niger Delta, a year's supply of food is often destroyed by only a minor leak of oil, debilitating the farmers and their families who depend on the land for their livelihood. Drinking water have also been contaminated, causing serious health problems to local communities. Offshore spills, which are usually much greater in scale, taint coastal environments in the Niger Delta, causing a decline in local fishing production. The rainforest, which previously occupied 7400 km^2, has disappeared. Similarly, oil spillage in the Niger Delta has destroyed its mangrove forests. Estimates suggest that 5–10% of Nigerian mangrove ecosystems have been wiped out by oil, which acidifies the soils, thus halting cellular respiration and starving

[61] Vidal (2011).

[62] United Nations Environment Programme (2011).

[63] Former U.S President Obama declared it as the 'worst environmental disaster America has ever faced'. See Nat'l Comm'n On The Bp Deepwater Horizon Oil Spill & Offshore Drilling, Report To The President, Deepwater: The Gulf Oil Disaster And The Future Of Offshore Drilling, vi (2011).

plant roots of oxygen.[64] The overall adverse effects of oil exploration in the Niger Delta, Nigeria, have been summarized as follows:

We witnessed the slow poisoning of the waters of this country and the destruction of vegetation and agricultural land by oil spills which occur during petroleum operations... We don't have pipe borne water here, our only source of drinking water is the surrounding stream and creeks, but we can no longer drink the water because it has been polluted. Everywhere you go, you see dead fishes and layers of crude oil.[65]

Oil spills bring about devastating and sometimes irreversible damage to wetland ecology, wildlife, and biodiversity, most of which have been accorded legal protections in national and international law. Major oil spills violate international environmental principles of sustainable development, particularly the prevention principle and the precautionary principle. International law aside, oil spills also violate domestic laws enacted to prevent environmental pollution and provide penal consequences for violation.

10.3.2.3 Loss of Business Profits and Subsistence Rights

Oil spillage leads to loss of income and means of subsistence for individuals and companies in the commercial fishing, shrimp, and oyster industries.[66] This affects fishermen and women, charter boat operators, owners of hotels, tourist management agencies, rental property owners, and other businesses in coastal resort areas. Thus, oil spillage is a violation of some of the international and national law principles, which seek to protect the rights of individuals to their means of subsistence.

A number of international treaties recognize subsistence rights as inherent human rights, which must be protected by States.[67] The right to food is expressly recognized in Article 11 of the International Covenant on Economic, Social and Cultural (ICESCR) as part of the right to an adequate standard of living. The 2009 Optional Protocol to the International Covenant on Economic, Social and Cultural Rights, makes the right to food justiciable at the international level before the Committee on Economic, Social and Cultural Rights. Furthermore, though they do not enjoy any binding legal status under international law, several UN general comments, declarations and resolutions provide significant and influential normative interpretations of subsistence rights to include access to traditional foods. For example, Article 25 of the 1948 Universal Declaration of Human Rights recognizes the right to food as part of the right to an adequate standard of living.[68] Summarily, when a community's food-growing capacity is affected by oil spillage or when their abilities to grow or

[64] Greenpeace International (1994).

[65] Quoted in Greenpeace International, ibid; see also Dappa-Biriye et al. (1992), p. 52.

[66] See Olawuyi (2015a, b), pp. 179–185.

[67] See Article 23 of the Universal Declaration of Human Rights (UDHR) and Article 6 of the International Covenant of Economic and Social Rights (ICESR).

[68] Ibid.

harvest foods that are traditionally part of their diet is undercut, elements of their culture may also be at risk. Subsistence rights were recognized in the *Yanomami Case* in which the Inter-American Commission called on States to restore, protect, and preserve the rights of indigenous peoples to their ancestral territories because they depend upon them for farming, hunting, subsistence, and continuity of life.[69] Similarly, in the *Toledo Maya* case, the Commission noted that "development activities must be accompanied by appropriate and effective measures to ensure that they do not proceed at the expense of the fundamental rights of persons who may be particularly and negatively affected, including indigenous communities and the environment upon which they depend for their physical, cultural and spiritual well-being."[70]

To protect subsistence, international environmental law places a duty on the polluter corporation to pay victims of oil pollution adequate compensation for economic and pecuniary losses incurred. For example, in line with this obligation, after the Deepwater Horizon explosion, BP announced a $20 billion escrow fund that would be used to compensate businesses and workers in Louisiana, Mississippi, Alabama, Florida, and Texas, whose financial livelihood suffered as a result of the oil spill.

10.3.2.4 Forced Displacements and Loss of Property

Closely intertwined with the loss of subsistence is the loss of property or damage to property occasioned by major oil spills. Oil spills most times result in damage to residential and commercial properties located along the Niger Delta area where major oil spills occur. Property damage also arises from the use of toxic chemicals and oil dispersants to aid the clean-up of major oil spills. This also results in forced displacements and relocation for individuals in affected areas. For example, over 200,000 Niger Deltans have been forcefully displaced from their homes due to oil spillage. Such loss of property violates some of the constitutional rights to free and exclusive ownership and enjoyment of property in national constitutions.[71] It is also a violation of the right to exclusively own and possess property as guaranteed under international law.

The right to property is one of the rights recognized in John Locke's famous human rights tripod of "life, liberty and possessions."[72] The Universal Declaration

---

[69] *Yanomami Case*, Rep. No. 12/85, Case 7615 (Brazil), (1985).

[70] Case of the Mayan Indigenous Communities from the Toledo District, Belize Report No. 40/04, Case 12.053, Merits, Mayan Indigenous Communities from the Toledo District, Belize, 12 Oct. 2004.

[71] See for example, section 43 of the1999 Constitution of the Federal Republic of Nigeria which recognizes the right of every citizen of Nigeria to acquire and own immovable property anywhere in Nigeria. Section 44 provides that no moveable property or any interest in an immovable property shall be taken possession of compulsorily in any part of Nigeria without just compensation'.

[72] J Locke, *Two Treatises of Government*.

of Rights (UDR) recognizes the right of individuals to the exclusive possession and use of property.[73] The African Charter on Human and Peoples Rights also recognizes this right.[74] The African Court on Human Rights upheld this right in the case of *SERAC v Nigeria* in which the court found the government of Nigeria in violation of right to property of the Ogoni people in Nigeria's Niger Delta due to its condoning and facilitating the operations of oil corporations in Ogoniland. The Commission also held that the implied right to housing, including protection from forced eviction, which is derived from the express rights to property, health and, family, was violated by the destruction of housing and forceful eviction of residents from their homes.[75]

10.3.2.5 Health Risks

Oil spills affect human health and well-being, an important tenet of environmental sustainability. Residents of oil-producing areas in the Niger Delta most times have had to cope with drinking polluted water that contains residual oil, even many years after clean-up. There are also the lasting health effects of chemical dispersants used during clean-up. For example, many residents of the Niger Delta have been diagnosed of asthma, breathing difficulties and pain, headaches, nausea, and throat irritation as well as chronic bronchitis.[76] Such health concerns can bring about substantive causes of action in toxic tort for exposure to dangerous substances and chemicals.[77] It has also been held to be a violation of the fundamental right to health in the Nigerian case of *Jonah Gbemre v Shell*, in which the Nigerian federal court held that gas flaring and oil spillage by Shell in the course of their oil exploration and production activities in the applicant's community were violations of the fundamental right to a healthy environment and dignity of human persons. This right was guaranteed by the Constitution of the Federal Republic of Nigeria and the African Charter on Human and Peoples Rights, which provides that every individual shall have the right to enjoy the best attainable state of physical and mental health.[78]

The right to health can be found in a number of international instruments. Though not legally binding, the UDR provides that everyone has the right to a standard of living adequate for the health and well-being of himself and of his family. Similarly,

---

[73] See Article 17 'Everyone has the right to own property alone as well as in association with others' 17 (2) No one shall be arbitrarily deprived of his property. See UDHR, adopted 10 Dec.1948, G.A Res.217A9 (III), 3 UN GAOR, UN Doc.(1948).

[74] The right to property shall be guaranteed. It may only be encroached upon in the interest of public need or in the general interest of the community and in accordance with the provisions of appropriate laws. Charter on Human and Peoples Right, adopted 27 June 1981, entered into force 21 Oct. 1986, O.A.U Doc. CAB/LEG/67/3 Rev. 5.

[75] (2001) AHRLR 60 (ACHPR 2001). Communication 155/96.

[76] See Environmental Rights Action (2005).

[77] Olawuyi (2015a, b).

[78] See *Jonah Gbemre v. Shell Petroleum Development Company Nigeria Ltd and Others*. No. FHC/B/CS/53/05, (Nov. 14, 2005) Federal High Court of Nigeria.

the International Covenant on Economic, Social, and Cultural Rights (ICESCR) recognizes the "right of everyone to the enjoyment of the highest attainable standard of physical and mental health."[79] One of the steps listed in the ICESCR for States to achieve this is the improvement of all aspects of environmental and industrial hygiene. This would include environmental sustainability and the prevention of industrial and environmental health hazards.

10.3.2.6 Loss of Life and Wrongful Deaths

Thousands of lives have been lost in the Niger Delta due to oil spillage, gas flaring and militancy. Many residents of the Niger Delta have died from the direct impacts of oil spillage, for example, drinking oil polluted water, or eating contaminated seafood; while others have died as a result of instability and uprisings arising from oil production in the region.[80]

Under international law, the right to life enjoys maximum guarantee. It is a supreme human right from which no derogation is permitted, even in times of war or public emergency.[81] As such, though not legally binding, the *United Nations' 'Protect, Respect and Remedy' Framework* places a responsibility on multinationals to protect and respect the right to life of their workers and of the residents of any country where they do business.

10.3.2.7 International Liability Issues

Major oil spills could affect water and environmental quality in nearby countries. For example, some areas in neighboring Cameroun have argued that oil spills in the Niger Delta have produced cross-border effects.[82] Such transboundary pollution will be a violation of Article 194(2) of the Law of the Sea Convention, which requires States to take all measures necessary to ensure that activities under their jurisdiction or control are so conducted as not to cause damage by pollution to other States and their environment, and that pollution arising from incidents or activities under their jurisdiction or control does not spread beyond the areas where they exercise sovereign rights in accordance with this Convention.[83]

This Convention places a duty on States to respect the territorial integrity and sovereignty of other States in the use of resources and to prevent trans- boundary pollution. Principle 21 of the *1972 Stockholm Declaration* also provides that States have the responsibility to ensure that activities within their jurisdiction or control do

---

[79] Article 12 of the ICESCR.

[80] See UNDP (2006), p. 17, see also Human Rights Watch (1999a, b) Chapter VI.

[81] Nowak (2000), p. 75.

[82] See Nwilo and Badejo (2006).

[83] UNCLOS Convention, United Nations Convention on the Law of the Sea, 1982, 21 ILM (1982), 1261 (adopted on 10 December 1982, and entered into force on November 16 1994).

not cause damage to the environment of other States or of areas beyond the limits of national jurisdiction.[84] An aggrieved neighboring state can bring an action before the International Court of Justice (ICJ) under the court's jurisdiction in Article 36(2)(c) for violation or breach of an international obligation to prevent trans-boundary pollution.[85]

10.3.2.8 Clean-Up Obligations and Costs

Major oil spills can take several years to clean up. For example, the United Nations estimates that the restoration process in Nigeria's Niger Delta could take up to 30 years, with a projected cost of $1 billion (USD) in the first 5 years.[86] This illustrates the fact that the costs of clean-up, restoration, and reclamation are often enormous and far-reaching.

International environmental law therefore places a duty on responsible multinational corporations to pay for such clean-up costs. The polluter-pays principle advocates that the polluter should bear the cost of preventing damage to the environment. The object of this principle is to channel the costs of prevention and reparation of environmental damage to the legal or juristic person who is in the best position to prevent such damage and thus "internalize" the costs of pollution damage.[87] For example, the UN recommended that Shell should set up an Environmental Restoration Fund to support clean-up and restoration in the Niger Delta.[88]

10.3.3 Gas Related Pollution: Gas Flaring

Gas flaring is the burning or release of natural gas, that cannot be processed or sold, into the atmosphere. It should be noted that in the course of oil production worldwide, gas is routinely flared.[89] There are several reasons why it may be necessary to flare gas during drilling, production or processing. For example, flaring is used to release gas during an emergency situation where equipment or piping becomes over-pressured. Flaring can also serve as an outlet for gas during maintenance and equipment repairs. In these scenarios, the flare is operated temporarily until the emergency situation is resolved, or until maintenance activities have been

---

[84] Stockholm Declaration.

[85] See for example *Trail Smelter Arbitration* (U.S.A. v. Canada), 1938-1941, 9 I.L.R. 315; *LacLanoux Arbitration* (France v. Spain), 1957, 24 I.L.R. 101; *Case Concerning the Gabcikovo-Nagymaros Project* (Hungry v Slovak Republic), 32 ILM (1993) 1293.

[86] United Nations Environment Programme (2011).

[87] European Environmental Bureau (1997), Chapter 1.

[88] Ibid.

[89] Tiratsoo (1972), p. 17.

completed. Flaring is also used to dispose of volumes of sour gas containing hydrogen sulfide (H_2S) and waste gas containing contaminants such as carbon dioxide (CO_2).

However, a great deal of gas flaring in many oil and gas production sites in developing countries, such as Nigeria, have little or nothing to do with protection against emergency situations or over-pressuring. In Nigeria, as well as in other countries lacking gas processing and transportation pipelines and infrastructure, vast amounts of such associated gas are commonly flared as waste gases.[90] Nigeria's flaring attracts significant attention because of the high volume of the gas flared since the beginning of commercial oil production in the country, and especially when compared with the flaring volumes in other countries.[91] Nigeria holds the unenviable record as one of the countries with the highest gas flaring statistics on earth, with the World Bank ranking it as the country with the second highest gas flaring rates in the world.[92] Estimates suggest that of the 226.255 billion Standard Cubic Feet (SCF) gas produced monthly, 187.85 billion SCF are utilized and while 39.070 billion SCF or about 18% is wasted via flaring. This equals about 25% of the UK's total natural gas consumption, and is the equivalent to 40% of the entire African continent's gas consumption.[93] Nigeria's loses an average of $170.166 million, around N27.227 billion to gas flaring every month, (about 326 billion Naira a year). Although the Nigerian Government has severally proposed target dates to end flaring, this economic and environmental waste continues in Nigeria till date. Gas flaring is a major contributor to environmental degradation and climate change, and has resulted in incalculable damage to Nigeria's environment.[94] The main effects of gas flaring have been in the form of acidic precipitation. The incineration of sour gas in the crude oil production process produces sulphur oxide, which is released into the atmosphere. The end result of these compounds, when they combine with atmospheric compounds, namely oxygen and water is what is commonly referred to as acid rain, which produces a lot of negative environmental effects. Gas flare sites, which often times are situated close to villages, produce "soot", which is deposited on building roofs of nearby villages. When it rains, this soot runs off the roofs of building and pollutes the soil and water aquifers of the people.

An adjunct to this is the colossal destruction of the environment through the charring of vegetation in gas flare sites. Furthermore, gas flaring is a major cause of the progressive destruction of mangrove and rainforest vegetation, with its attendant loss of biodiversity, numerous trees and plants that carry high economic and pharmaceutical

---

[90] See Olawuyi (2015a, b), pp. 185–189.

[91] It is estimated that the volume of gas flared in Nigeria accounts for over 75% of the gas flared in Africa. See generally, Aghalino (2009), p. 220.

[92] The five countries with the highest flaring amounts in billion cubic meters (bcm) are Russia (35), Nigeria (15), Iran (10), Iraq (10) and USA (5). See Elvidge et al. (2016), also World Bank (2009).

[93] See Social Development Integrated Centre (Social Action) (2009). See also Eboh (2014).

[94] For a detailed discussion of the effects of gas flaring on lives, health and the environment of local communities in Nigeria's Niger Delta, as well as its contribution to climate change, see Olawuyi (2015a, b), pp. 186–190.

values. Flaring also releases gases that deplete the stratospheric ozone layer, and filter ultraviolent radiation to the earth surface. Again, because of the heat emitted from flare sites, soil elements in those areas are highly degraded resulting in a huge loss in arable lands from the over 1000 flare sites in the Niger Delta region.

Gas flaring results in a wide range of health effects including respiratory problems, such as asthma, bronchitis, hearing loss, skin problems and serious childbirth problems. Another major problem is noise pollution, which is a challenge because exposure to intense noise level, may damage ear cells temporarily or permanently or lead to speech impairment. Environmental noise pollution has been recognized in recent years as a serious threat to the quality of life enjoyed by the people in oil producing areas. Furthermore, the Niger Delta as a coastal area is among the places most likely to become vulnerable to the effects of global warming and the resultant climate change. Of particular concern is the likely impact of rising sea levels, tidal waves and flood. If sea level rises, inundation could occur along more than 30% of the Nigerian coastline. This may place arable land and farming opportunities at risk. Flaring has also resulted in the "loss of night time" in many part of Niger Delta, due to a perpetual glow or reflection of light and burning flames from flare sites.[95] Flare sites also generate excessive heat, which is felt over large areas. This results in sleep deprivation, insomnia and physiological problems for residents of such areas.[96]

10.3.4 Effluent and Waste Discharges

Another source of oil related pollution is the disposal or discharge of wastes and effluents into the surrounding environment, sometimes into the water, by oil companies. For instance, during exploration or seismic surveys by oil companies, drill cuttings, drilling mud and fluids are used for stimulating production, chemicals are also used during seismic activities. When the major constituents of drill cuttings, such as baryotes and bentonitic clays, are dumped on the ground, they prevent local plant growth. In the water, these materials may kill aquatic plants and animals.

In addition to the pollutants introduced into the environment from exploration and exploitation operations, there is also the risk of unsafe decommissioning or abandonment of disused oil and gas platforms and facilities. Simply put, decommissioning refers to the process of removing, disposing or abandoning obsolete or disused oil exploration facilities and installations from oil exploration fields. Decommissioning is the process of shutting down operations at the end of oil field's life. It covers closing the oil wells, removing some or all of the facilities and disposing them. It also includes flushing, plugging and cementing oil wells to make

[95] According to Aniete Aniete, a fisherman in the Niger Delta. "Because of the flares, it is so hot, it is smoky, the air is thick and it is constantly daytime here…Our rivers are black and the [acid] rain eats our houses. Our bodies are covered in oil. You feel that if you live to old age here, it is a miracle." See Monica (2012). See also Ajugwo (2013), pp. 6–8.

[96] Climate Justice Programme (2005).

them safe for rig removal and the shutting down of operations at the end of a field's life.[97] Being the final stage of oil exploration activities, decommissioning happens mostly after a long period of oil production when platforms have reached the end of their productive lives. Studies show that on average, offshore fields mature for decommissioning between 20 and 50 years after they are opened for exploration activities.[98]

While decommissioning is an inevitable end process of oil exploration and production, it becomes a source of complex and far-reaching environ-mental problems when offshore oil exploration facilities are disposed in an environmentally unsound manner. Scientific studies have shown that the potential hazardous effects of the deep sea disposal of oil facilities includes the release of highly toxic polychlorinated biphenyls (PCBs), hydrocarbons, low specific activity (LSA) scale, heavy metals and other toxic pollutants into the marine environment with uncertain long-term impacts.[99] These organic compounds are resistant to environmental degradation, they therefore persist in the water, causing bioaccumulation through the food web, and posing a risk of causing adverse effects to human health and the environment in general. Human exposure to these residual wastes through fishing or water consumption could pose serious health concerns including the disruption of the reproductive, and immune systems; neurobehavioral disorders; and cancers.[100] Studies have shown an increase in reproductive problems amongst rural farmers due to their exposure to PCB contaminated waters.[101] Secondly, the physical act of disposal itself has its negative environmental side effects. The use of explosive cutting materials to free the legs of the oil rigs creates shock waves that harm marine and aquatic life. According to scientific studies, these shock waves are a major threat to fishes and organisms that live in the seabed. The shock waves destroy the equilibrium in the water, altering aquatic life in no small ways.[102]

Deep-sea disposal of offshore oil and gas facilities also causes blockage to ship movement.[103] The metals from the facilities pose navigational risks for ships and boats. Similarly, the falling of these metals into the sea can crush and kill the living organisms beneath the offshore facilities. Similarly, the presence of these offshore facilities on the seabed in perpetuity limits the spaces available to marine life and other sea users like fishermen. The presence of redundant installations and pipelines on the seabed could also result in a situation whereby the contaminant leach into the surrounding ecosystem thereby accumulating in fishes posing health risks to the human consumers. The process of decommissioning oil platforms especially the underwater cutting and handling of large quantities of structural steel is a potential

[97] Ayoade (2002), p. 2.

[98] Ibid, 6–7, also Gorman and Neilson (1997).

[99] Ritter et al. (1995).

[100] Ritter et al. (1995).

[101] Patin (1999), p. 34.

[102] Macdonald (1994), p. 94.

[103] Ibid.

risk for safety and health personnel such as mechanics, welders, electricians and divers.[104]

10.4 Laws and Institutions for Environmental Protection

Previous chapters of the book have discussed the sources of laws and regulations that govern the extractive sector in general. These include international law, domestic legislation on the environment as well as secondary legislation on environmental impacts assessments for all forms of development projects which include mining and petroleum operations. This section highlights some of the key and most applicable instruments that address water and air pollution in international and national levels.

10.4.1 International Law

International law has, over the last decades, provided robust instruments that address the need to protect air, water and marine resources from all sources of pollution. Although countries are the main actors under international law, private sector actor such as extractive operators can be held liable or fined by a resource based country if found to be in violation of core international law instruments that seek to prevent pollution from petroleum and mineral operations and activities. Similarly, many extractive sector agreements and contracts now increasingly incorporate elements of international law obligations on environment to ensure that corporations and business enterprises also comply with international best practices on environmental safety, compliance and pollution.

Governance and protection of the marine environment from pollution began in the mid-twentieth century as a result of disputes over rights to fish outside territorial boundaries and waters; extraction of minerals and hydrocarbons; as well as the demands borne out by increasing populations and technological advancements. The foregoing issues moved the United Nations (working with the International Law Commission) to launch efforts aimed at codifying customary principles of law relating to the use of the sea. This gave rise to the United Nations Conference on the Law of the Sea in 1958, which resulted in the adoption of four separate conventions. The Conventions include the *Convention on the Territorial Sea and Contiguous Zone*, the *Convention on the High Seas*, the *Convention on Fishing and Conservation of the Living Resources of the High Seas* and the *Convention on the Continental Shelf.*[105] These Conventions respectively address broad range of issues dealing with

[104] See UKOOA (1995).

[105] For more on these conventions, see generally, the United Nations Diplomatic Conferences—United Nations Conference on The Law of The Sea 1958, UN Treaty.

high seas, territorial seas, contiguous zone, continental shelf, fishing and conservation of living resources, and settlement of disputes amongst others. They however did not address the issue of marine pollution. This led to a second conference on the Law of the Sea in 1960 to consider the topics, which had not been previously agreed upon. A third conference was held from 1973 to 1982 resulting in the adoption of the *United Nations Convention on the Law of the Sea*. Its key provisions that relate to the prevention of marine pollution are examined below.[106]

10.4.1.1 The United Nations Convention on the Law of the Sea ("UNCLOS")

UNCLOS was in part, a direct outcome of discussions from the Stockholm Conference on the Human Environment of 1972, and its *Stockholm Declaration on the Environment*.[107] Its scope was intended to be wide, far-reaching and comprehensive.[108] The Convention was adopted to establish a legal order for seas and oceans that will facilitate international communication; promote peaceful uses of seas and oceans; foster equitable and efficient utilization of their resources; and more importantly encourage the study, protection and preservation of the marine environment.[109] The Convention covers several subject maters such as delimiting the territorial sea, contiguous zone, exclusive economic zone, continental shelf, the high seas, and setting out the rights and obligations of states with regard to them. Our focus will be on the provisions of Part X11, which deals with the protection and preservation of the marine environment.

Article 192 of UNCLOS places an obligation on countries to protect and preserve the marine environment. It covers a wide range of pollution sources and provides a global framework for the rational exploitation and conservation of the sea's resources and the protection of the marine environment. Furthermore, it gives special recognition to the interests of developing countries. Under *UNCLOS*, control of pollution is a matter of legal obligation. In addition, it has also moved emphasis to international regulation and cooperation for the protection of the marine environment.

There are six main sources of ocean pollution addressed in the Convention, namely: land-based sources and coastal activities; continental shelf drilling; potential seabed mining; ocean dumping; vessel-source pollution; and pollution from or through the atmosphere. With regards to these various sources of pollution, States are to take various measure, which include the adoption of national regulations to reduce and control pollution in the areas under their jurisdiction and concerted

---

[106] Ibid.

[107] Stockholm Declaration.

[108] See the *United Nations Convention on the Law of the Sea*, 10 December 1982, 1833 UNTS 3, UN, Oceans and Law of the Sea, preamble.

[109] See the preamble, ibid.

international effort to control pollution in international waters.[110] The broad measures taken by UNCLOS to protect the marine environment, by preventing, reducing and controlling its pollution, are set out in article 194. Under this article, States are to employ measures consistent with the Convention that are necessary to prevent, reduce and control pollution of the marine environment from any source, using the best practicable means at their disposal and in accordance with their capabilities. States are also to take all measures necessary to ensure that activities under their jurisdiction or control are conducted in a manner that does not cause pollution damage to other States and their environment, and that pollution arising from incidents or activities under their jurisdiction or control does not spread beyond the areas where they exercise sovereign rights. In addition, the measures to be taken shall include, those designed to minimize to the fullest possible extent: (a) the release of persistent toxic, harmful or noxious substances, from land-based sources, from or through the atmosphere or by dumping; (b) pollution from vessels, in particular measures for preventing accidents and dealing with emergencies, ensuring the safety of operations at sea, preventing intentional and unintentional discharges, and regulating the design, construction, equipment, operation and manning of vessels; (c) pollution from installations and devices used in exploration or exploitation of the natural resources of the seabed and subsoil, or operating in the marine environment in general, in particular measures for preventing accidents, dealing with emergencies, ensuring the safety of operations at sea, and regulating the design, construction, equipment, operation and manning of installations or devices.[111]

The measures to be taken include those necessary to protect and preserve rare or fragile ecosystems as well as the habitat of depleted, threatened or endangered species and other forms of marine life. In addition, States are take all necessary measures to prevent, reduce and control pollution of the marine environment resulting from the use of technologies under their jurisdiction or control, or the intentional or accidental introduction of species, alien or new, to a particular part of the marine environment, which may cause significant and harmful changes. However, in recognition of the sovereign rights of States, the Convention provides that States shall refrain from unjustifiable interference with activities of other States, while taking their own measures to prevent, reduce or control pollution of the marine environment. Other tools include global and regional cooperation, notifying other state parties of imminent or actual damage to marine environment caused by pollution, making contingency plans to eliminate the effects of pollution, or to prevent and minimize damage, as well as monitoring, observing, measuring, evaluating and analyzing, the risks or effects of pollution of the marine environment, and also keeping under surveillance the effects of permitted and ongoing activities in order to determine whether these activities are likely to pollute the marine environment.[112]

---

[110] arts 207–212, *ibid.*

[111] art 194, *ibid.* 14 art 196, *ibid.* 15 *ibid.*

[112] See the *United Nations Convention on the Law of the Sea*, 10 December 1982, 1833 UNTS 3, arts 197–199, 204.

The Convention, as earlier highlighted, makes special concessions for developing country parties. Under article 202, States are to provide scientific and technical assistance to developing state parties. In addition, they are to provide appropriate assistance, for the preparation of environmental assessments and the minimization of the effects of major incidents, which may cause serious pollution of the marine environment, as well as promote programmes of scientific, educational, technical and other assistance to developing states for the protection and preservation of the marine environment and the prevention, reduction and control of marine pollution. Such assistance shall include, *inter alia*: (1) training of their scientific and technical personnel; (2) facilitating their participation in relevant international programmes; (3) supplying them with necessary equipment and facilities; (4) enhancing their capacity to manufacture such equipment; (5) advising on and developing facilities for research, monitoring and educational programmes. Again, under article 203, developed state parties and international organizations are to extend preferential treatment to developing states in the allocation of appropriate funds, technical assistance and the utilization of their specialized services for the prevention, reduction and control of pollution of the marine environment or minimization of its effects.

Enforcement under the Convention varies according to the different sources of pollution covered by the Convention. Enforcement powers are exercised by state parties, such as the coastal state, flag state, and port states save in the case of enforcement of the Area, where dispute resolution is by the Seabed Disputes Chamber.[113] Under article 235, States are responsible for the fulfillment of their international obligations concerning the protection and preservation of the marine environment. They shall be liable in accordance with international law. States shall also ensure that recourse is available in accordance with their legal systems for prompt and adequate compensation or other relief in respect of damage caused by pollution of the marine environment by persons under their jurisdiction. To ensure prompt and adequate compensation for marine pollution damage, States shall cooperate in the implementation of existing international law, and the further development of international law relating to responsibility and liability for the assessment of and payment of adequate compensation for damage and the settlement of related disputes. Under the Convention, States have a duty to immediately notify other States, who may likely be affected and relevant international organizations, once they become aware of cases in which the marine environment is in imminent danger of being damaged or has been damaged by pollution.[114]

[113] Arts 215–222.

[114] See the *United Nations Convention on the Law of the Sea*, 10 December 1982, 1833 UNTS 3, art 198.

10.4.1.2 The International Convention for the Prevention of Pollution from Ships 1973/78 ("MARPOL")

The importance of shipping to the extractive industry cannot be underestimated. However, shipping can also be a significant source of water and marine pollution. *MARPOL* is the primary Convention that addresses marine pollution from ship/vessel based sources.[115] The *MARPOL*, entered into under the aegis of the International Maritime Organization, came after the *International Convention for the Prevention of the Pollution of the Sea by Oil ("OILPOL")*.[116] The *OILPOL*, applied to tankers engaged in the transportation of oil, and prohibited the discharge of oil into the sea save in exempted circumstances. The *OILPOL* however only dealt with the pollution of the marine environment by oil, leaving out all the other pollutants that may be discharged into the marine environment during sea transportation activities. The need to develop more comprehensive measures to address marine pollution beyond oil led to the adoption of *MARPOL*.

The MARPOL Convention is the main international convention governing prevention of pollution of the marine environment by ships from operational or accidental causes. The Convention was adopted on 2 November 1973 under the auspices of the International Maritime Organization ("IMO"). The 1978 Protocol was adopted in response to the spate of tanker accidents in the period between 1976 and 1977. As the 1973 MARPOL Convention had not yet entered into force, the *1978 MARPOL Protocol* was absorbed into the parent Convention. The combined instrument entered into force on 2 October 1983.[117] MARPOL was adopted to preserve the human environment in general and to protect the marine environment in particular. Its primary objective is to achieve the complete elimination of intentional and accidental pollution of the marine environment by oil and other harmful substances.[118] "Harmful substance" means any substance, which, if introduced into the sea, could create hazards to human health, to harm living resources and marine life, to damage amenities or to interfere with other legitimate uses of the sea, and includes any substance subject to control by the present Convention.[119] The Convention includes provisions aimed at preventing and minimizing accidental and routine pollution from ships. It includes six technical Annexes on prevention of pollution by oil, by noxious liquid substances in bulk, harmful substances carried by sea in packaged form, pollution by sewage and garbage and air pollution from ships.

---

[115] *The International Convention for the Prevention of Pollution from Ships 1973/78,* 2 November 1973, 1340 UNTS 184.

[116] International Convention for the Prevention of the Pollution of the Sea by Oil, 12 May 1954, 327 UNTS 3.

[117] See generally, International Maritime Organization, *International Convention for the Prevention of Pollution from Ships 1973/78*, 2 November 1973, 1340 UNTS.

[118] See the preamble, ibid.

[119] See the International Convention for the Prevention of Pollution from Ships 1973/78, 2 November 1973, 1340 UNTS 184.

In article 1, the parties to the Convention undertake to give effect to the provisions of the Convention necessary to prevent the pollution of the marine environment by the discharge of harmful substances or effluents containing such substances Article 3 sets out the scope of the Convention, specifying that it will apply to ships flying the flags of state parties or operating under their authority. It excludes warships, naval auxiliary or other ship owned or operated by state parties and used, for government non-commercial service. The regulatory tools employed by the Convention include the issuance of an International Oil Pollution Certificate detailed in regulation 5 to the Convention. The Convention and also sets out special rules for the inspection of ships. In the absence of a certificate, a ship will be inspected to ensure that the ship shall not sail until it can proceed to sea without presenting an unreasonable threat of harm to the marine environment.[120] Under article 6 parties to the Convention shall cooperate in the detection of violations and the enforcement of the provisions of the Convention by using all appropriate and practicable measures of detection, environmental monitoring, adequate procedures for reporting and accumulation of evidence. Under article 8, reports of incidents are to be made without delay to the fullest extent possible Settlement of disputes is in the first place by negotiation between the Parties involved has not been possible, and if these Parties do not otherwise agree, be submitted upon request of any of them to arbitration. Under article 17, state parties, in consultation with other international bodies, and the Executive Director of the United Nations Environment Programme, are to promote support for parties, which request technical assistance to train scientific and technical personnel; for the supply of necessary equipment and facilities for reception and monitoring; and the facilitation of other measures, such as research to prevent or mitigate pollution of the marine environment by ships.

10.4.1.3 Protocols to the MARPOL Convention

The first protocol, the Protocol on Provisions Concerning Reports on Incidents Involving Harmful Substances was adopted pursuant to article 8 of the convention. It places an obligation on parties to make reports of incidents involving harmful substances.[121] Article 1 imposes a duty on the master of a ship involved in an accident, which results in the pollution of the marine environment to report the particulars of the incident without delay and to the fullest extent possible. It also sets out the mode of reporting and the nature of incidents that should be reported and the contents of the report. Secondly, the Arbitration Protocol was adopted pursuant to

---

[120] See the International Convention for the Prevention of Pollution from Ships 1973/78, 2 November 1973, 1340 UNTS 184, art 5.

[121] See the Protocol on Provisions Concerning Reports on Incidents Involving Harmful Substances to the International Convention for the Prevention of Pollution from Ships 1973/78, 2 November 1973, 1340 UNTS 184, arts 2–4.

article 10 of the Convention. It sets out the arbitration procedure to be followed in the event of a dispute.[122]

With respect to air pollution from extractive operations, a number of international treaties have been adopted to govern and reduce the emissions of sulphur and other toxic substances into the air.

10.4.1.4 The 1979 Geneva Convention on Long-Range Transboundary Air Pollution (the "LRTAP")[123]

Events that led to the *LRTAP* can be traced back to the 1960s, when scientists demonstrated the interrelationship between sulphur emissions in continental Europe and the acidification of Scandinavian lakes. In the intervening period between 1972 and 1977, several studies confirmed the earlier hypothesis that air pollutants could travel thousands of kilometers before they were deposited and caused damage. This signaled that international cooperation was necessary to solve this challenge. This desire for a regional response to transboundary air pollution resulted in the adoption of the *LRTAP* under the umbrella of the United Nations Economic Commission for Europe (UNECE) in 1979. UNECE member states, mainly European and North American countries, are Parties to the LRTAP.[124] Although the membership and coverage of the LRTAP does not extend to African countries, many of the Parties to LRTAP have extractive operations and interests in Africa, especially countries such as United Kingdom, France, Canada and Netherlands amongst others that are home to some of the largest extractive sector operators in Africa. International law is increasingly recognizing the obligations of extractive companies from these countries to, within their respective spheres of activity and influence, comply with provisions of all applicable international treaties, such as LRTAP, while investing abroad.[125]

The primary purpose of the *LRTAP* is to prevent, reduce, and control transboundary air pollution both from new and existing sources. It adopts the form of a framework convention and adopts a number of tools aimed at gradually reducing and preventing long range transboundary air pollution. The tools employed by the *LRTAP* are the exchange of information, consultation, monitoring, research and development to control and reduce transboundary air pollution. Parties to the Convention meet annually at sessions of the Executive Body to review ongoing work, assess levels of implementation, and plan future activities including a work plan for the coming year.

A number of protocols have been adopted to reinforce, enforce and extend the objectives of the Convention. These include the Gothenburg Protocol to Abate

---

[122] Arbitration Protocol to the International Convention for the Prevention of Pollution from Ships 1973/78, 2 November 1973, 1340 UNTS 184, arts 1–10.

[123] *The Convention on Long-range Transboundary Air Pollution*, 13 November 1979, 1302 UNTS 217, UNECE.

[124] For a list of countries that are members of the UNECE, see UNECE (2018a).

[125] Kamminga and Zia-Zarifi (2000); also United Nations Guiding Principles (2011).

Acidification, Eutrophication and Ground-level Ozone; Aarhus Protocol on Heavy Metals; and the 1998 Protocol on Persistent Organic Pollutants. [126]

10.4.1.5 The Stockholm Convention on Persistent Organic Pollutants (the "POP Convention")[127]

Persistent Organic Pollutants ("POPs") are chemical substances that persist in the environment, accumulate and cause adverse effects to human health and the environment. They are persistent (i.e., they remain in the environment for several years). POPs have been widely used in coal mines, smelters, pulp and paper mills, pest and disease control in agriculture, power generation, heating and manufacturing. They include the highly toxic polychlorinated biphenyls (PCBs) which are used industrially; dichlorodiphenyltrichloroethane (DDT) which is commonly used as a cheap insecticide; and dioxins which are normally formed during combustion processes, such as waste incineration, forest fires, backyard trash burning, and manufacturing processes (including smelting, chlorine bleaching of paper pulp and the manufacturing of some herbicides and pesticides). POPs have been linked to adverse effects on human health, such as cancer, damage to the nervous system, reproductive disorders, and disruption of the immune system.[128] Given that POPs can be transported by wind and water, POPs generated in one country can affect people and wildlife far from where they are used and released.

Evidence of long-range transport of POPs to regions where they were not being produced or used, as well as the consequent threats they posed to the global environment led to calls for urgent global actions to reduce and eliminate releases of these chemicals in the international community.[129] The POP Convention was adopted on 22 May 2001 in Stockholm, Sweden, and entered into force on 17 May 2004. This Convention was adopted in recognition of the need for global action to address the transportation and deposition of persistent organic pollutants, which possess toxic properties and resist degradation, across international boundaries, where they accumulate and affect both terrestrial and aquatic ecosystems.[130] In keeping with the

---

[126] For more on these protocols, see generally, UNECE (2018b).

[127] The *Stockholm Convention on Persistent Organic Pollutants* (adopted 22 May 2001, effective 17 May 2004) 2256 UNTS119.

[128] The United States Environmental Protection Agency has categorized dioxin as a "likely human carcinogen" causing cancer. See United States Environmental Protection Agency.

[129] See the United Nations Environment Programme-Chemicals: Persistent Organic Pollutants.

[130] POPs persist in the environment, bioaccumulate in human and animal tissue, bio magnify in food chains, and are such capable of having potential significant impacts on human health and the environment. Because they circulate globally via the atmosphere, oceans, and other pathways, POPs released in one part of the world can travel to regions far from their source of origin. Thus, under the Convention, countries have committed to reduce and/or eliminate the production, use, and/or release of the 12 POPs of greatest concern to the global community and to establish a mechanism by which additional chemicals may be added to the Treaty in the future. The *Stockholm Convention on Persistent Organic Pollutants*, 22 May 2001, 2256 UNTS119, preamble.

precautionary approach, the objective of the Convention is to protect human health and the environment from persistent organic pollutants. As part of its measures to eliminate the release of POPs from intentional, and unintentional production and use, as well as stockpiles and wastes, the Convention requires that parties take several measures, which include: to prohibit and/or take the legal and administrative measures necessary to eliminate production, use and import and export of the chemicals listed in Annex A; to restrict production and use of the chemicals listed in Annex B; with the Convention setting out exceptions to the absolute prohibition on imports, when the imports are for environmentally sound disposal or for permitted use. In addition, parties are also to develop action plans to identity, characterize and address releases of chemical from anthropogenic sources; promote the application and development of measures to achieve reduction of releases, and eliminate POPs from their sources as well as the development of alternatives. For stockpiles and wastes, parties are to ensure that they are handled, collected, stored in an environmentally sound manner, or disposed of in a way that destroys or irreversibly transforms the POPs.[131]

Parties are also to develop and carry out plans detailing how they will implement their obligations under the Convention.[132] One of the enforcement tools utilized by this Convention is that annexes A, B and C respectively produce comprehensive lists of chemicals that are to be eliminated, restricted; and the unintentionally produced pollutants.[133] In addition, the Convention contains comprehensive provisions on the exchange of information, public information and education, research, development and monitoring, provision of technical assistance to developing countries and economies in transition, provision of financial resources and mechanisms, and reporting.[134] Article 19 establishes a Conference of the Parties, whose main functions include the continuous review and evaluation the implementation of the Convention, the establishment of necessary subsidiary bodies cooperation with competent international organizations and intergovernmental and non-governmental bodies; and regular review of information given to the Parties. The Secretariat on the other hand, arranges and services meetings of the COP and facilitates assistance to parties in the implementation of the Convention.[135]

10.4.1.6 The United Nations Framework Convention on Climate Change ("UNFCCC")[136]

Climate change is an increase in the average temperature of the atmosphere at an abnormal rate caused by the anthropogenic emission of gases, known as greenhouse

[131] See arts 3, 5 and 6, ibid.

[132] Art 7, ibid.

[133] Annex A, B and C, ibid.

[134] See generally, arts 8, 9, 19, 11, 12, 13 and 15.

[135] For the mandate and functions of the secretariat, see Art 20, *ibid.*

[136] The *United Nations Framework Conference on Climate Change* (Adopted 9 May 1992, effective 21 March 1994) 1771 UNTS 107.

gases (GHGs), that trap the sun's heat in the atmosphere.[137] GHGs that contribute to climate change include: carbon dioxide ($C0_2$), methane (CH_4), nitrous oxides (N_20), water vapour (H_20), ozone (0_3); and high global warming potential gases (high GWP gases) such as chloroflourocarbons (CFCs), hydroflourocarbons (HFCs), perfluorocarbons (PFCs), and sulfurhaxaflouride (SF6).[138] Several of these GHGs are introduced to the atmosphere during oil, gas and mineral development operations. For example, gas flaring during oil operations is a significant contributor to climate change in Africa, with Nigeria being a major source.[139] Excessive dependence on oil, gas and other mineral resources that release GHGs, for industrial, manufacturing processes and transportation, is also a chief contributor to global climate change.

The UNFCCC was adopted in 1992 as the first global legal agreement on climate change. It has the ultimate objective of stabilizing greenhouse gas concentrations in the atmosphere at a level that would prevent dangerous anthropogenic (human-induced) interference with the climate system. In its preamble, the UNFCCC reiterates the right of States to exploit their own resources pursuant to their environmental and developmental policies, and the overarching responsibility to ensure that their activities do not cause damage to the environment of other States. It also *urges* States to enact effective environmental legislation to reduce GHG emissions from all sources. The preamble places the onus of taking immediate action to towards the remediation of climate change and its effects on developing countries, based in part on the consideration of their contributions to the enhancement of the greenhouse effect from their industrialization processes. The preamble also reiterates the effect of climate change on vulnerable and developing countries, as well as the need for considerations in dealing with developing countries whose economies are particularly dependent on fossil fuel production, use and exportation.

Article 1 defines climate change as a change of the global climate, which is attributed directly or indirectly to human activity that alters the composition of the global atmosphere. This is in addition to natural climate variability observed over comparable time periods. Climate change mitigation therefore refers to efforts designed to reduce or prevent the emission of greenhouse gases, while climate change adaptation refers to efforts aimed at coping with those impacts that cannot be avoided. It includes policy adjustments, physical projects and infrastructural changes, proposed as means for ameliorating the anticipated adverse consequences associated with climate change.

---

[137] There is now sufficient scientific consensus on the fact that climate change is real and that anthropogenic GHG emissions are a cause. See Intergovernmental Panel on Climate Change (IPCC), *Climate Change 2007*, the Fourth Assessment Report (AR4), in which the IPCC stated that the likelihood was 90–99% that emissions of heat- trapping greenhouse gases such as carbon dioxide were the dominant cause of the observed warming of the last 50 years. According to the Panel, it is 'very likely' that human activity is the main driver of the rise in temperatures since 1950. See also Stevens (2007) and Rosenthal and Revkin (2007). Also see Tickell (1968). See also The European Renewable Energy Council (2007).

[138] See Intergovernmental Panel on Climate Change (2007).

[139] Elvidge et al. (2016), Knippertz et al. (2015), also Social Development Integrated Centre (Social Action) (2009).

The primary objective of the Convention as set out in article 2 is the stabilization of atmospheric greenhouse gas concentrations at levels that will prevent dangerous anthropogenic interference with the climate system. This level is to be achieved within a timeframe that allows ecosystems to adapt naturally to climate change, in order to ensure sustainable food production and economic development. In article 3, the Convention sets out the foundational principles, which guide the Convention to include the benefit of present and future generations, common but differentiated responsibilities, the precautionary principle and sustainable development. The regulatory tool employed by this Convention to address climate change is the use of commitments on the part of state parties to reduce their GHG emission levels. Article 4 sets out the commitment of all parties in to include the development, and publication of national inventories of sources anthropogenic emissions and removals all greenhouse gases; the formulation and implementation of national and regional programmes to mitigate climate change, measures to facilitate adequate adaptation to climate change; promotion and cooperation in the development and transfer, of technologies, that control, reduce or prevent anthropogenic emissions of greenhouse; the promotion of sustainable management, and cooperation in the conservation and enhancement of sinks and reservoirs of all greenhouse gases; cooperation in adapting the impacts of climate change; promotion and cooperation in scientific, technological, technical, socio-economic and other research and exchange of technology and information, systematic observation and development of climate change related data; promotion of and cooperation education, training and public awareness related to climate change; and communication to the Conference of the Parties of information related to implementation, in accordance with article 12.

Specifically, as it relates to developed country parties, which are set out in Annex 1 and II to the Convention, they commit themselves to the adoption of national policies and taking corresponding measures on the mitigation of climate change, by limiting their anthropogenic emissions of GHGs, protecting and enhancing greenhouse gas sinks and reservoirs. These parties may implement such policies and measures individually, or jointly with other Parties. In addition, the parties included in Annexes I and II shall provide financial resources to meet the adaptation and compliance costs, incurred by developing and vulnerable country parties in meeting their article 12 obligations. They are also required to provide the transfer of, and facilitate access to technology, needed by the developing country Parties to meet their obligations under the Convention.

A number of protocols and agreements have been adopted to further elaborate and expand the objectives of the UNFCCC. These include the Kyoto Protocol which was adopted in 1997, and entered into force in 2005, as the first international agreement that establishes legally binding emission reduction targets on climate change.[140] The Kyoto Protocol *commits* industrialized countries to stabilize greenhouse gas emissions by setting legally binding emission reduction targets and obligations. The

[140] See the *Kyoto Protocol to the United Nations Framework Convention on Climate Change.*

major feature of the Kyoto Protocol is that it sets legally binding targets for the reduction of greenhouse emissions for 37 industrialized countries and the European community. These targets add up to an average 5% emissions reduction compared to 1990 levels over the first 5-year period 2008–2012. At the 17th Conference of the Parties (COP 18) held in Durban, South Africa, parties agreed to establish an Ad Hoc Working Group on a Durban Platform for Enhanced Action (AWG-DP). The AWG-DP has the mandate to develop a protocol, another legal instrument or an agreed outcome with legal force under the Convention applicable to all Parties, that would be ready for signing in Paris in 2015.[141] This was to give parties enough time negotiate a new climate change agreement to come into effect by 2020. Furthermore, at the 18th Conference of the Parties (COP 18) held in Doha in November 2012, it was agreed that the second commitment period under the Kyoto Protocol would run from 1 January 2013 until the end of 2020 (2013–2020).[142] A significant achievement of the Doha Conference is that it served as the transitional platform that ensured the continuity of the Kyoto Protocol and its mechanisms without any gaps.

It was at the 21st session of the Conference of the Parties to the UNFCCC in Paris, France, that the Paris Climate Agreement was adopted.[143] The agreement sets clear targets and goals to reduce GHG emissions and keep global temperature under 2 °C (3.6 °F), and as close as possible to 1.5 °C starting in the year 2020.[144] As obligations under the Paris Agreement continue to be elaborated and defined, it has become increasingly important for companies in the petroleum and mining sectors to reduce the emission of GHGs in their operations. Efforts are underway across many African countries to develop domestic climate change legislation that establish penalties for companies that fail to reduce their GHG emissions in line with nationally determined emission reduction targets of the country.

10.4.1.7 The West and Central Africa Regional Framework Agreement on Air Pollution (the "Abidjan Agreement of 2009")[145]

The Abidjan Agreement of 2009 is a sub-regional agreement in West and Central Africa, that seeks to address and reduce the release of toxic and air polluting

---

141 UNFCCC, Outcome of the Durban Climate Change Conference (2011).

142 See Doha Amendments to the Kyoto Protocol (Adopted 8 December 2012) C.N.718.2012. TREATIES-XXVII.7.c. Pursuant to Article 20, paragraph 4, and Article 21, paragraph 7 of the Kyoto Protocol, the Amendment shall enter into force for those Parties having accepted it, on the ninetieth day after the date of receipt by the Depositary of an instrument of acceptance by at least three fourths of the Parties to the Kyoto Protocol.

143 Conference of the Parties to the United Nations Framework Convention on Climate Change, Paris Agreement (2015).

144 Article 2, ibid.

145 See the *West and Central Africa Regional Framework Agreement on Air Pollution* (2009).

substances into the atmosphere, most especially from forest burning practices, truck and motor transportation and other pollution from extractive and industrial processes in the sub-region that elevate the release of air pollutants (especially particulate matter, sulphur dioxide, oxides of nitrogen, ammonia, volatile organic compounds, persistent organic pollutants, carbon monoxide and heavy metals). Though not legally binding, the agreement develops actionable targets to address air pollution issues in transport; industry and mining; household pollution; waste disposal; bush fires, uncontrolled burning and deforestation; urban planning; and management and national and regional environmental governance.

The Abidjan Agreement brings together 21 African countries namely: Angola, Benin, Burkina Faso, Cameroon, Cape Verde, Chad, Republic of Congo, Côte d'Ivoire, Democratic Republic of Congo, Equatorial Guinea, Gambia, Ghana, Guinea, Guinea Bissau, Liberia, Mali, Niger, Nigeria, Senegal, Sierra Leone and Togo and the Economic Community of West African States (ECOWAS).[146] In setting out its targets, the agreement took cognizance of the increase in atmospheric pollutants from socio-economic development in the sub-region, the growing incidence of transboundary air pollution; the atmospheric transformations associated with emissions of volatile organic compounds and oxides of nitrogen that may have adverse impacts on human health and the environment; the linkage between air pollution and climate change amongst other things. In order to achieve the set targets, the state parties agree to cooperate in the preparation of flexible and differentiated agreements to control and reduce agreed air pollutants; consider the benefits of taking joint measures against the emission of air pollutants and greenhouse gases; harmonize national air quality management legislation, standards, where practicable; and promote the exchange of educational and research information on air quality management.[147] Other governance tools include setting standards with respect to motor vehicles, which include restricting the age of imported vehicles and ensuring that all imported gasoline vehicles are equipped with functioning converters by 2011, establishing harmonized emission standards, adopting modern technologies that promote vehicle fuel efficiency and reduce emissions.

With regards to the activities in industries and the mining sectors, the parties agree to elaborate and adopt policy and legal frameworks to promote adoption of best available technology and standards, facilitate and promote the use of cleaner fuels and efficient utilization of energy; conduct environmental and social impact assessment and environmental audit process on activities that impact on air quality;

[146] Similar agreements aimed at increasing air quality have been put in place in other African sub-regions. See Eastern Africa Regional Framework Agreement on Air Pollution (Nairobi Agreement); Southern Africa Regional Framework Agreement on Air Pollution (Lusaka Agreement) and the North African Framework Agreement on Air Pollution.

[147] Art 1, ibid.

encourage air pollution and climate change mitigation projects; enact emission standards and regulations for industrial activities; build capacity for national monitoring of emissions from industrial and mining activity; and harmonize industrial and mining emission standards and regulations to address the effects of transboundary air pollution.[148] The parties also agree to, on a national and regional level, to promote the coordination of the development and implementation of air quality policies and emission management strategies.[149] In addition, the state parties will also enhance public awareness on air pollution issues; enhance stakeholder participation in air quality management; promote access to and exchange of information on air pollution including.[150] Other measures include capacity building and also research development.[151]

10.4.2 Domestic Environmental Law

In addition to the web of international treaties and instruments that establish the need to reduce environmental pollution in the extractive sector, African countries have adopted specific national legislation that prescribe environmental standards. There is strong business sense, in terms of cost, reputation and effectiveness for extractive companies to familiarize themselves with environmental legislation and guidelines in an applicable African jurisdiction to avoid sanctions, fines and penalties that may result from pollution and noncompliance.

10.4.2.1 Primary Environmental Legislation

The right to enjoy a clean and healthy environment, that is free from all sources of pollution, has been recognized in several African countries either through their constitutions, environmental legislation, or court decisions.[152] Regionally, the African Charter on Human and Peoples' Rights also recognizes this right. Article 24 provides that "all peoples shall have the right to a generally satisfactory environment favourable to their development."[153] Similarly, African countries such as

[148] Art 4, ibid.

[149] Art 7, ibid.

[150] Art 8, ibid.

[151] Art 9, ibid.

[152] See Boyd (2012), p. 2.

[153] *African Charter of Human and Peoples' Rights*, 27 June 1981, Organization of African Unity Doc CAB/LEG/67/3 Rev 5, 21 ILM 58 (providing a "right to a general satisfactory environment favorable to…development").

Benin,[154] Ethiopia,[155] Niger,[156] Mali,[157] Cameroon,[158] South Africa,[159] Togo,[160] Burkina Faso,[161] Nigeria,[162] and most recently, Kenya,[163] have provided for environmental rights in their constitutions. However, except for the Constitution of Kenya, the other referenced constitutional rights are placed in the non-justiciable sections of the Constitution, making them unenforceable in courts.[164] Despite the non-justiciability concern in many African countries, the geometric recognition of the environmental protection as a human right concern places increased obligations on

[154] *Constitution of the Republic of Benin,* 1990, arts 27–8 ("Every person has the right to a healthy, satisfying and lasting environment").

[155] *Constitution of the Federal Democratic Republic of Ethiopia,* 1995, art 44(1) ("All persons have the right to a clean and healthy environment").

[156] *Constitution of the Republic of Niger*, 2010, art 35 ("Any person has the right to a healthy environment"). At time of writing (August 2015), currently suspended by the military junta of February 2010.

[157] *Constitution of the Republic of Mali*, 1992, art 15 ("Every person has a right to a healthy environment").

[158] *Constitution of the Republic of Cameroon*, 1996, Preamble ("Every person shall have a right to a healthy environment"). Note that art 65 provides that the "Preamble shall be part and parcel of this Constitution."

[159] *Constitution of the Republic of South Africa*, 1996, art 24 ("Everyone has the right to (a) an environment that is not harmful to their health and well being; and (b) to have the environment protected, for the benefit of present and future generations through reasonable legislative and other measures").

[160] *Constitution of the Republic of Togo*, 1992, art 41 ("Anyone has a right to a healthy environment").

[161] *Constitution of the Republic of Burkina Faso*, 1991, art 29 ("The right to a healthy environment is recognized; the protection, the defense and the promotion of the environment are a duty for all"), art 30 ("Every citizen has the right to initiate an action or to join a collective action under the form of a petition against the act…affecting the environment…").

[162] *Constitution of the Federation of Nigeria*, 1999, s 20 ("The State shall protect and improve the environment and safeguard the water, air, land forest and wildlife of Nigeria").

[163] *Constitution of Kenya*, 2010, art 42 ("Every person has the right to a clean and healthy environment, which includes the right a) to have the environment protected for the benefit of present and future generations through legislative and other measures, particularly those contemplated in art 69, and b) to have obligations relating to the environment fulfilled under art 70").

[164] In Nigeria for example, the environmental right provision is placed in Chapter II of the *Constitution*, titled *Fundamental Objectives and Directive Principles of State Policy*. Section 6(6)(c) of the Constitution however strips all courts in Nigeria of powers to adjudicate on all rights listed in the Fundamental Objectives and Directive Principles of State Policy. It provides that judicial powers vested in Nigerian courts:

> shall not…extend to any issue or question as to whether any act of omission by any authority or person or as to whether any law or any judicial decision is in conformity with the Fundamental Objectives and Directive Principles of State Policy set out in Chapter II of this Constitution.

This provision removes the rights of the public to enforce constitutional environmental rights in Nigeria. Constitutions of several of the countries referenced above contain similar language that directly declare that environmental right provisions shall not be enforceable (*Constitution of the Federation of Nigeria*, *supra* note 63).

transnational corporations and business enterprises to integrate environmental safeguards into their operations to avoid sanctions and penalties.[165] Human rights principles should not only come to the table when there is a violation. Rather, human rights should become part of the rule of the game of approving and designing resource extraction projects and actions.

In addition to the widespread constitutional recognition of environmental protection, specific environmental legislation have been developed across the continent to address questions relating to water, air, land and noise pollution.[166] For example, Nigeria's *National Environmental Standards Regulatory and Enforcement Agency (Establishment) Act of 2007* is the principal legislation on environmental protection. It establishes an Agency entrusted with the regulation and enforcement of environmental standards, regulations, laws, policies and guidelines.[167] The broad functions and powers of the Agency as outlined in sections 7 and 8 include enforcing compliance with laws, guidelines, policies environmental regulations and standards on air, land, development of natural resources and pollution abatement among others. These regulatory functions come with a notable exception; none of them extend to the oil and gas sector. The *NESREA* specifically exempts the oil and gas sector from its sphere of operation or regulation on a myriad of issues.[168] For example, the Agency's powers to: enforce compliance with regulations on the importation, exportation, production, distribution, storage, sale, use, handling and disposal of hazardous chemicals, enforce the environmental regulations and standards on noise, air, land, seas, oceans and other water bodies, enforce environmental control measures through registration, licensing and permitting systems, conduct environmental audit and establish data bank on regulatory and enforcement mechanisms of environmental standards, promote private sector compliance with environmental regulations all exclude operations in the oil and gas sector.[169]

In addition, the Agency's powers to conduct public investigations on pollution and the degradation of natural resources, and review existing guidelines, regulations

[165] For a detailed review of these efforts, see Olawuyi (2015a, b).

[166] See for example ANGOLA's Framework Environment Law—Law No. 5/98, of 19 June 2015; EGYPT's *Law* 4/1994 for the Protection of the *Environment*; GHANA's *Environmental* Protection Agency *Act*, 1994; ALGERIA's Law no. 82-10/1982 on Fishing, Law no. 83- 16-17 July 1983 on Water, and Law no. 84 -12/1984 on Forestry Order; KENYA's Environmental Management and Coordination Act no. 8 of 1999; the NIGERIAN National Environmental Standards and Regulations Enforcement Agency (NESREA) Act *of 2007*, Laws of the Federation of Nigeria 2004, c N164, s 7; the Niger Delta Development Commission (NDDC) (Establishment etc) (Amendment) Act of 2017, the National Oil Spill Detection and Response Agency (NOSDRA); SOUTH AFRICA's National Environmental Management Act No. 107 of 1998 National Environmental Management Laws Amendment Act, No. 25 of 2014 and ZIMBABWE's *Environmental* Management *Act* (EMA), *Act* No.13 of 2002 (Chapter 20:27), revisions under *Act* No.5 of 2004 (s.23) and *Act* No. 6 of 2005 (s.28).

[167] NESREA Act, ibid, see 3 (1), also ss 7 (g)-(l), *ibid.*

[168] Ibid.

[169] Ibid.

and standards on the environment exclude oil and gas sector.[170] While the Agency is specifically charged with making regulations setting specifications and standards to protect and enhance the quality of Nigeria's air resources, and prohibit the discharge of hazardous substance into the air in harmful quantities, these general rules exempt discharges permitted or authorized under any law in force in Nigeria.[171] The Agency is also required to co-operate with other Government agencies for the removal of pollutants discharged into the environment, "excluding oil and gas related ones."[172]

Another important legislation in Nigeria is the *Niger Delta Development Commission (NDDC) (Establishment etc) (Amendment) Act of 2017* ("*NDDC Act*").[173] The NDDC Act sets up a commission, which is primarily charged with using allocated and designated funds to tackle ecological problems arising from the exploration of oil minerals in the Niger Delta.[174] The broad powers of the Commission, as set out in section 7, include: formulating policies, guidelines and implementing programmes for the development of the Niger Delta, addressing the ecological and environmental problems that arise from the oil exploration in the Niger Delta area, advising government on measures to be taken to prevent and control of oil spills, gas flaring and environmental pollution, and also liaising with oil mineral and gas prospecting and producing companies on all matters of pollution prevention and control. In addition, section 14 establishes a fund, into which a portion of government revenue, as well as a percentage of the budget of operating companies is paid into for addressing environmental problems associated with oil and gas development, preventing and controlling future pollution amongst other things.[175] This fund created by the NDDC Act is very specific to the development of the Niger Delta region alone.

Similarly, the *National Oil Spill Detection and Response Agency (Establishment) Act 2006* ("NOSDRA Act") establishes the National Oil Spill Detection and Response Agency, which is responsible for detecting and responding to oil spills in Nigeria.[176] The objectives of the Agency include coordinating and implementing the National Oil Spill Contingency Plan and undertaking safe, timely, effective and appropriate responses to major or disastrous oil pollution, identifying high-risk and priority areas for protection and clean up, establishment of mechanisms to monitor, assist and where expedient mobilize the necessary resources to save lives, protect threatened environment, and clean up the impacted site to the best practical extent possible, providing a programme of activation, training and drill exercises to ensure readiness for oil pollution response, cooperating with the International Maritime Organization and other national, regional and international organizations in the pro-

---

[170] s 8 (g)-(k), ibid.

[171] Ss 20, 27(1). ibid.

[172] Ss 21 (1)-(2), 29.

[173] *Niger Delta Development Commission Act*, Laws of the Federation of Nigeria 2004, c N87.

[174] See the preamble, s 1, ibid.

[175] Ibid.

[176] The National Oil Spill Detection and Response Agency (Establishment) Act 2006.

motion and exchange of results of research and development programme relating to the enhancement of the state of the art of the oil pollution preparedness and response, including technologies, techniques for surveillance, containment, recovery, disposal and clean up to the best practical extent as well as determining and prepositioning vital combat equipment at most strategic areas for rapid response ahead of spills. The functions of the Agency include surveillance and ensuring compliance with existing environmental legislation, the detection of oil spills in the petroleum sector and coordinating oil spill response activities in Nigeria. Failure to report spills and to clean up affected areas are offences, which are punishable with fines.[177] The Act in section 18 establishes a National Control and Response Centre, whose functions include acting as a report processing and response coordinating center for all oil spill incidents in Nigeria, receiving reports of oil spills from its zonal offices and control units, serving as the command and control center for compliance monitoring of all existing legislation on environmental control, surveillance for oil spill detection, monitoring and coordinating responses required in plan activations.

Section 19 of the NOSDRA Act sets out the power of the Federal Government to intervene in the event of a major or disastrous oil spill. At such times, the Agency will collaborate with other Agencies to assess the extent of damage to the ecology by matching conditions following the spill against what existed before, undertake a post-spill impact assessment to determine the extent and intensity of damage and long-term effects, advise the federal and state governments on possible effects on the people's health and ensure that appropriate remedial action is taken for the restoration and compensation of the environment, assist in mediating between affected communities and the oil spill, monitor the response effort during an emergencies, with a view to ensuring full compliance with existing legislation on such matters, assess any damage caused by an oil spill, expeditiously process and grant approval requests made by an oil spiller for the use of any technology considered vital in ameliorating the effect of an oil spill, advise and guide the response efforts as to ensure the protection of highly sensitive areas, habitats and endangered or threatened wild life, and monitor the clean-up operations to ensure full rehabilitation of the affected areas. In addition, as lead agency, the NOSDRA co-operates with oil spillers to determine appropriate and expeditious response measures to prevent excessive damage to the environment and the communities. NOSDRA is also required to mobilize all internal and external human and financial resources that may be required to combat oil spills.[178]

10.4.2.2 Secondary Legislation and Guidelines

In addition to primary legislation on environment, African countries have released guidelines and other legislation that stipulate environmental standards for the

[177] See s 6, ibid.

[178] s 19(3), ibid.

conduct of extractive operations. In Nigeria for example, the *Environmental Guidelines for the Petroleum Industry in Nigeria ("EGASPIN")* specifically deals with the control of pollutants and pollution from the various aspects of petroleum operations, and regulates the environmental aspects of petroleum operations.[179] Its primary objective is to regulate environmental management practices in the production and discharge of produced formation waters, oily wastewaters, sludge and accidental spills of oil from oil and gas production installations. With regards to oil exploration activities, it sets out the procedure of the treatment and control of wastes connected with the oil production process. It also stipulates that to preserve, restore and maintain the chemical, physical and biological integrity of Nigeria's waters, oil and gas installation operators are to ensure that their levels of pollution control are in line with the Best Practicable Control Technology Currently Available. Other wastes such as oil sludge, solid wastes, etc are to be disposed of by sanitary land filling, encapsulation, solidification, chemical treatment and recycling amongst other treatments. It also establishes monitoring requirements for production operations.

Furthermore or, it prescribes flare distances, appropriate burn technology, allowable heat radiation, and noise levels during gas flaring amongst other things.[180] It also has provisions for monitoring gaseous point emissions, which include a mandatory frequency sampling of gaseous point emission on a weekly basis or using an approved continued emission monitoring system. Reports on gaseous point emissions are also to be submitted to the Director annually. The EGASPIN also sets out national air quality guidelines to be adopted in regulating air quality. It regulates pollution resulting from the treatment and processing of associated and non-associated gas into Liquefied Natural Gas.[181] Additionally, it prescribes the relative density of emitted smoke, permissible heat radiation, ground temperature, and operations of flares without visible smoke for gas processing and conversion operations.[182]

10.4.2.3 Environmental Impact Assessment

In addition to industry standards and guidelines in the environment, the conduct of an environmental impact assessment (EIA) has become recognized as a precondition for petroleum and mining operations in all surveyed African countries. EIA is a formal process by which a proposed activity with potentially significant environmental, social and economic costs is studied with a view to evaluating its impacts, examining alternative approaches and developing measures to prevent or mitigate

[179] The Environmental Guidelines and Standards for the Petroleum Industry in Nigeria (2002), paras 44–46, 54–59.

[180] Ibid, paras 3.8.8.1 (iii)–(ix), ibid.

[181] See pp. 90–91, ibid.

[182] P 95–96, para 5.6.2.2, *ibid.*

the negative impacts.[183] As Hunter et al. noted '[o]ne of the primary reasons for conducting EIAs is to inform the public of the proposed projects and to engage them in a meaningful dialogue about the potential benefits and environmental and social costs of a proposed activity'.[184] It entails an examination, analysis and assessment of planned activities with a view to ensuring environmentally sound and sustainable development. Summarily, an EIA is an information gathering process through which environmental impacts potentially resulting from a proposed project are identified and assessed early in the planning process, thereby enabling decision makers to make informed decisions on whether or not to grant approval or consent to that project.

An EIA is a preventive tool that enables the anticipation and minimization of the negative effects of development. It informs decision-makers of the environmental consequences of a project or decision; it serves to integrate environmental matters into other spheres of decision-making. Furthermore, it identifies steps that can be taken to avoid negative environmental impacts or reduce them to acceptable levels before they occur. EIA therefore represents a proactive, preventive approach to environmental management and protection. The principle of prevention stipulates the use of special techniques such as risk analysis and assessment of the potential effects of planned activities to reduce or eliminate environmental harm.[185] As such, when used in project design and implementation, EIA helps to shape development in a manner that prevents harm to the environment.

The need for EIAs arose from increased environmental awareness in the 1950s and 1960s, when it became evident that extractive and other development projects were producing undesirable consequences on the environment. Pollution and degradation of the human environment caused amongst other things by rapid population growth, industrialization, infrastructural development, agricultural development, technological progress and natural resource exploitation forced the international community and governments to realise the need for a mechanism to ensure that the environmental consequences of major projects and plans were examined before a formal authorization of their execution and implementation. Since the United States of America passed its first EIA related legislation in 1969, over 100 countries, including Nigeria, have followed suit.

EIA has emerged as a formal procedure through which decision makers can gather and assess the impacts of a proposed project on the environment before work on the project commences. It provides governments with information about the consequences of development programmes that they have under consideration, as well as helping to formulate proper development policy. In addition, it provides an avenue for public participation in the decision making process for developmental projects which may impact lives, environment and livelihood. In summary, an EIA

[183] See Hunter et al. (1987), also UNEP (1987).

[184] Ibid.

[185] See Olawuyi (2015a, b), pp. 262–263.

serves three primary purposes namely: the integration of environmental issues into planning and decision-making; anticipation and minimization of environmental damage; and public participation in decision-making and environmental conservation.

EIAs can be tailored to be country specific and meet any development situation. Some of the basic features of an EIA however include: screening to determine whether certain projects should be subject to the EIA; scoping to determine which impacts should be taken into account; impact analysis to evaluate the likely environmental impacts; mitigation and impact management to develop measures to avoid, reduce or compensate for negative environmental impacts; and reporting to record and track the results of the EIA for decision makers and the public; review of the quality of the EIA report to determine if it includes all the required information; decision making to approve or reject project proposals, or set conditions under which projects can proceed; and implementation and follow up to ascertain if the project is going as planned, to monitor the effects of the project, and to take actions to mitigate problems that may arise.[186]

In Nigeria for example, the *EIA Act* regulates the protection of the Nigerian environment.[187] The chief objective of an EIA is to establish and take into account the extent to which any activity to be authorized by any level of government may affect the environment.[188] The *EIA Act* prohibits the undertaking of, or embarking on projects, which may significantly affect the environment without prior consideration of their environmental effects.[189] Petroleum operation activities, such as oil and gas field development, are listed in the mandatory study list provided in the *EIA Act*.[190] For such activities, the Agency can either ensure that a mandatory study is conducted, and a report is prepared and submitted to it; or in the alternative, refer the project to the Council, who will then refer it to mediation or a review panel. The Council refers the project to mediation or review where it determines that it is likely to cause immitigable and significant adverse environmental effects, or public concerns on the environmental effects of the project warrant it.[191] Where the project is likely to result in unjustifiable, immitigable and significant adverse environmental effects, the Agency will not permit the project to be carried out.

---

[186] See Olawuyi (2015a, b), pp. 262–263.

[187] *The Environmental Impact Assessment Act of 1992*, Laws of the Federation of Nigeria 2004, c E12.

[188] See s 1, ibid.

[189] Ss 2(1)–(3), ibid.

[190] See the Schedule, ibid.

[191] Ibid, s 26.

10.5 Environmental Institutions

As earlier noted, environmental aspects of extractive operations are directly under the supervision and purview of the supervising Ministry or department (*i.e.* the ministry in charge of mining or petroleum resources). In Nigeria for example, the Ministry for Petroleum Resources is charged with the formulation and implementation of government policy and general management of the operations of the petroleum industry. The Ministry issues licenses and leases; prescribe rules, regulations and guidelines; co-ordinates policies; manages petroleum resources; and administers relevant laws and regulations. Within the Ministry, technical, policy and regulatory issues are handled by the Department of Petroleum Resources ("DPR"). Pursuant to Section 8 of the Petroleum Act (PA), the Minister of Petroleum Resources has broad oversight and supervisory powers over all petroleum operations carried out under subsisting licenses and leases in Nigeria. The Minister's oversight powers include: accessing at any time, oil exploration licence (OEL), oil prospecting licence (OPL) and oil mining lease (OML) areas, refineries and installations to enforce all applicable laws and regulations; ordering the arrest of any person found committing or to have committed an offence under the PA or other laws or regulations; summoning a licensee or lessee or any of its subcontractors to provide information on their respective operations and where necessary suspending the operations (such as in cases where the operations are not conducted in accordance with good oilfield practice); and fixing or controlling the prices of petroleum products in Nigeria (section 6 of the PA). The Minister also has powers to revoke OPLs and OMLs in certain instances and to make regulations on any matters requiring regulations pursuant to section 9 of the PA.

The DPR, the technical arm of the Ministry of Petroleum Resources, also regulates and ensures compliance with environmental standards in Nigeria's oil and gas sector.[192] In its regulatory capacity, the DPR is responsible for the enforcement of environmental guidelines, rules and regulations established by the Ministry, with respect to all activities in petroleum operations; upstream and downstream, as well as petrochemicals. The core functions of the DPR include: supervising all petroleum industry operations being carried out under licences and leases in the country in order to ensure compliance with the applicable laws and regulations; enforcing safety and environmental regulations; and ensuring that petroleum operations conform to national and international industry practices and standards.[193]

[192] See generally, DPR (2010).

[193] See DPR (2010).

10.5.1 Secondary Institutions

In addition to the overseeing ministry for extractive operations, there are at least over hundred distinct committees, departments, and agencies established across different levels of government across Africa with mandates to protect the environment, preserve cultural heritage and ensure sustainable utilization of natural resources for present and future generations.[194]

10.5.1.1 Department or Ministry of the Environment ("MOE")

All African countries have established ministries or departments responsible for implementing Government policy on environmental protection and sustainable development.[195] Ranging from the Federal Ministry of Environment in Nigeria established in 1999, to the Ministry of Environment and Natural Resources in Kenya, the Department of Environmental Affairs in South Africa established in 1994 (as the Department of Environmental Affairs and Tourism); and the Ministry for Environmental Affairs in Egypt established in 1997.[196] The broad mandates of MOEs include formulating and implementing environmental policies; preparing the necessary action plans for environmental protection; stipulating procedures for environmental impact assessment for projects; managing and protecting oceans and coastal resources, promoting conservation and the sustainable use of natural resources; promoting cooperation in environmental science and conservation technology; and cooperating with federal and state ministries, local government, statutory bodies and research agencies on matters relating to the protection of the environment and the conservation of natural resources.[197] Typically, headed by a

---

[194] There is a plenitude of examples across Africa. See for example Kenya's National Environment Council (NEC) established under the Section 4 (1) of Kenya's Environmental Management and Coordination Act no. 8 of 1999; also Kenya's National Environment Management Authority (NEMA) established under the same law. Also South Africa's Department of Environmental Affairs; the Nigerian National Environmental Standards and Regulations Enforcement Agency (NESREA) established under the NESREA Act (NESREA) Act; the Niger Delta Development Commission (NDDC)of Nigeria established under the NDDC (Establishment etc) (Amendment) Act of 2017, the Energy Commission of Nigeria, the National Oil Spill Detection and Response Agency (NOSDRA), National Water Resources Institute (NWRI) and the Nigerian Tourism Development Corporation (NTDC). Recently the Government of Nigeria established the Hydrocarbon Pollution Restoration Project (HYPREP) to investigate, evaluate and make recommendations on the clean-up of hydrocarbon impacted sites in the Niger Delta region of Nigeria.

[195] For a comprehensive list of all environment ministries in Africa, see The World Bank, Africa Environmental Agencies.

[196] See Kenya's Ministry of Environment and Natural Resources, About Us, see also South Africa's Department for Environmental Affairs, *Introduction and Background*, Nigeria's Federal Ministry of Environment, About Us.

[197] Ibid.

cabinet Minister or Secretary for Environment, MOEs supervise and enforce environmental protection measures on water quality, effluent limitations, air quality, atmospheric protection, ozone protection, noise control as well as the removal and control of hazardous substances. At provincial or state levels, ministries of environment also play similar roles and are supervised by Commissioners for environment.

However, since their respective establishments in the 1990s, African MOEs have been inconsistent in the design and implementation of environmental protection programs in the extractive sector. Due to differing priorities placed on environmental matters by different governments, environmental policies and projects are inconsistently implemented in accordance with the visions and priorities of the government in power. This lack of continuity and political commitment often result in gaps or the disappearance of programs and structures designed for embedding sustainable development in resource use. Similarly, since MOEs are supervised by cabinet Ministers they are accountable to the Government and therefore arguably lack the required level of institutional independence required to hold policy-makers to account for their actions. MOEs lack the status and independence needed to serve as "watchdogs" to monitor and report on the sustainability of government operations and policies. Equally intertwined with this is the lack of independent funding opportunities for MOEs to drive and implement environmental protection programs. The abilities of MOEs to launch and implement long-term environmental protection programs have been hampered by lack of consistent funding for sustainable development by successive government. In Nigeria for example, activities of the Nigerian MOE have often been hampered due to diversion of funds meant for the Ministry and also because the political will to fund the Ministry has not been maintained by successive governments.

10.5.1.2 Environment and Sustainable Development Agencies and Commissions

A number of African countries have also established special commissions and agencies mandated to monitor, enforce and report on environmental protection and the sustainability of government operations and policies. Examples include Nigeria's National Oil Spill Detection and Response Agency (NOSDRA), the National Environmental Standards Regulatory and Enforcement Agency (NESREA), National Parks Service (NPS), and the Niger Delta Development Commission (NDDC). There are similar environmental agencies at the state level, such as the Lagos State Environmental Protection Agency (LASEPA), Oyo State Environmental Protection Commission, and the Rivers State Sustainable Development Agency (RSSDA) amongst others.[198] All these agencies have the protection of the environment and cultural heri-

[198] See Ministry of Environment of Nigeria.

tage, for present and future generations, as part of their broad mandates. In Kenya, there are similar national agencies such as the National Environment Management Authority (NEMA), Energy Regulatory Commission (ERC), Kenya Water Towers Agency (KWTA), Kenya Wildlife Service (KWS), National Musuems of Kenya (NMK), Kenya Forest Service (KFS) and Kenya Forest Research Institute (KEFRI). In addition to ensuring the preservation and conservation of the environment and its resources, these agencies are established to ensure that development activities "do not compromise the capacity of the resource base to meet the needs of the present generation as well as those of future generations".[199] Similarly, the Egyptian Environmental Affairs Agency is mandated, amongst other things, to prepare studies on the state of the environment, formulate the national plan for the protection of the environment, and monitor the implementation of international and regional conventions related to the environment.[200]

All of the African environment agencies and commissions surveyed describe their work under four headings: advocacy—raising awareness of the concept of sustainable development and responding to Government policy initiatives; capacity building—establishing good working relationships with and between key parts of Government, and developing skills in departments; policy and advice—drawing on expert opinion to provide evidence-based advice to Government on sustainable development; and watchdog—monitoring performance against targets and reporting on them. Generally, environment agencies and commissions spearhead national implementation of norms of sustainable development and intergenerational justice envisioned by international treaty law. They also serve as watchdogs in scrutinizing and monitoring Government's progress on implementing the treaty obligations and national environmental plan and strategies. They also stipulate penalties for activities of the public that a detrimental to the environment.

Despite the plenitude of laws and institutions that mandate and oversee environmental standards and best practices in extractive operations, African countries continue to grapple with several complex environmental, social, and human rights challenges associated with extractive processes and operations. Considering the long term and sometimes irreversible impacts of pollution from mining and petroleum industries, African countries have to reassess how extant legal and institutional frameworks can be reformed to prioritize sustainable resource management in the design, approval, financing and implementation of extractive operations.

Depending on the socio-political structures and realities in respective African countries, it is possible that a specific environmental response and restoration institution could complement existing institutional arrangements. Another alternative is to grant environmental ministries and agencies direct and more specific oversight powers and functions over extractive operations. Whatever approach or model is adopted, what is clear is that the environmental impacts of mining and petroleum operations in Africa are complex, cross cutting, and therefore require urgent, sustained and committed action.

---

199 Section 3 Environmental Management and Coordination Act (EMCA), 1999.

200 Egyptian Environmental Affairs Agency.

10.6 Conclusion

The increased adoption of environmental legislation and guidelines across Africa mean that protecting the environment during extractive operations is no longer an option, but a necessity for ECs and business enterprises operating in the continent. Extractive operators can avoid costly liability, sanctions and penalties for environmental violations by keeping up-to-date with the meaning, nature and scope of environmental protection regulation and legislation in the extractive industry, and also by incorporating robust environmental due diligence into corporate risk management processes. There is a strong business case, in terms of cost, reputation and effectiveness for doing so.

Failure to effectively manage environmental and human rights issues associated with extractive projects carries significant financial, legal and reputational risks. Such risks may manifest themselves in community protests due to health and safety concerns, disruption of project construction and potential harm to employees due to such protests. There may be legal risks related to the court and regulatory actions to challenge projects, and reputational risks related to loss of life or injuries from actions of security forces in responding to protest situations. Failure to effectively address environmental and human rights risks can lead to significant costs in terms of the management time required to respond to crises, costly sus-pensions or closures of projects by supervisory bodies, and may affect a company's ability to receive project funding from carbon finance institutions or export credit agencies. Project proponents will be evaluated and judged by stakeholders (including financial institutions and supervisory institutions) based on the strength of the internal procedures, practices and approaches put in place to anticipate and manage human rights risks associated with the company's carbon investment and project. Furthermore, adopting sound internal human rights screening processes as part of corporate risk management frameworks can help business enterprises to build and maintain strong social license to operate and to foster relationships with communities, customers and other stakeholders based on mutual respect, acceptance and trust.

As for the role of lawyers representing companies and investors in extractive projects and investments, it is important to emphasize the significance of environmental and human rights impact assessments when providing counsel. The growing insistence of courts and tribunals on the rights to information, public participation and benefit-sharing for those persons potentially affected by proposed projects should lead lawyers to advise their clients of the wisdom of early compliance with environmental standards and requirements in extractive operations and projects. Nearly all cases of conflict over extractive projects arise from projects executed without proper consultations with the local communities. These communities are not generally opposed to such projects or investment, but they know and insist upon their rights, even if governments and supervisory bodies are unwilling or unable to protect them.[201]

---

[201] See Olawuyi (2016a, b).

The next chapter outlines general procedures and processes for integrating environmental and human rights due diligence into general corporate risk management processes, specifically in the design and implementation of extractive operations to avoid legal risks.

References

African Charter of Human and Peoples' Rights, 27 June 1981, Organization of African Unity Doc CAB/LEG/67/3 Rev 5, 21 ILM 58

Aghalino SO (2009) Gas flaring, environmental protection, pollution and abatement measures in Nigeria 1969–2001. J Sustain Dev Africa 11:4

Air Pollution-Protocols. UNECE. http://www.unece.org/env/lrtap/status/lrtap_s.html. Accessed 21 July 2017

Ajugwo A (2013) Negative effects of gas flaring; The Nigerian experience. J Environ Pollut Hum Health 1(1):6–8

Akcil A, Koldas S (2006) Acid Mine Drainage (AMD): causes, treatment and case studies. J Clean Prod 14(2):1139–1145

Anderson M (1996) Human rights approaches to environmental protection: an overview. In: Human rights approaches to environmental protection. Oxford University Press, Oxford

Apte SC, Kwong YTJ (2003) Deep sea tailings placement: critical review of environmental issues. CSIRO Australia and CANMET Canada 2003

Arbitration Protocol to the International Convention for the Prevention of Pollution from Ships 1973/78, 2 November 1973, 1340 UNTS 184

Ayoade M (2002) Disused offshore installations and pipelines: towards sustainable decommissioning. Kluwer Law International, Hague

Bailey H (1979) Semi-arid climates: their definition and distribution. In: Hall A, Cannel GH, Lawton HW (eds) Agriculture in semiarid environments. Springer, Berlin. Ecological studies No. 34

Bodansky D (2010) The art and craft of international environmental law. Harvard University Press, Cambridge

Boyd DR (2012) The right to a healthy environment: revitalizing Canada's constitution. UBC Press, Vancouver

Burges Salmon LLP (2014) Talking point--scope of environmental law. Lexis Nexis, New York

Case Concerning the Gabcikovo- Nagymaros Project (Hungry v Slovak Republic), 32 ILM (1993) 1293

Case of the Mayan Indigenous Communities from the Toledo District, Belize Report No. 40/04, Case 12.053, Merits, Mayan Indigenous Communities from the Toledo District, Belize, 12 Oct. 2004

Charter on Human and Peoples Right, adopted 27 June 1981, entered into force 21 Oct. 1986, O.A.U Doc. CAB/LEG/67/3 Rev. 5

Climate Justice Programme (2005) Gas flaring in Nigeria: a human rights, environmental and economic monstrosity. https://friendsoftheearth.uk/sites/default/files/downloads/gas_flaring_nigeria.pdf. Accessed 12 June 2018

Conference of the Parties to the United Nations Framework Convention on Climate Change. Paris Agreement. 2015. 21st Session, Paris. 30 November to 11 December 2015. https://unfccc.int/resource/docs/2015/cop21/eng/l09.pdf. Accessed 12 Oct 2017

Convention on Long-range Transboundary Air Pollution, 13 November 1979, 1302 UNTS 217, UNECE. http://www.unece.org/fileadmin/DAM/env/lrtap/full%20text/1979.CLRTAP.e.pdf. Accessed 25 June 2015

Cukrowska EM, Lusilao-Makiese J, Tutu H, Chimuka LK (2012) The impact of gold mining on mercury pollution in the Witwatersrand Basin, South Africa. In: Mapani B, Bohdan Kribek B (eds) Environmental and health impacts of mining in Africa. Czech Geological Survey, Prague, pp 15–18

Dappa-Biriye H, Briggs R, Idoniboye-Obu B, Fubara M (1992) The endangered environment of the Niger Delta: constraints and strategies. An NGO memorandum for the World conference of indigenous peoples on environment and development and the United Nations conference on environment and development, Rio de Janeiro

Davies TC (2012) Advances in mitigation and rehabilitation technology in major and abandoned mines in Sub-Saharan Africa. In: Mapani B, Bohdan Kribek B (eds) Environmental and health impacts of mining in Africa. Czech Geological Survey, Prague

Department of Environment, South Africa. Effects of Mining on the Environment and Human Health. https://www.environment.co.za/mining-2/effects-of-mining.html

Department of Petroleum Resources (DPR) (2010) About DPR. Historical background. http://www.dprnigeria.com/aboutus.html. Accessed 13 June 2018

Doha Amendments to the Kyoto Protocol. 2012 C.N.718. 2012.TREATIES-XXVII.7.c

Donnelly B, Bishop P (2007) Natural law and eco-centrism. J Environ Law 19(1):89

DPR (2010) Our evolution and history. Department of Petroleum Resources. http://www.dprnigeria.com/in_the_news.html. Accessed 13 June 2018

Earth and Mining Watch Canada (2012) Troubled waters: how mine waste dumping is poisoning our Oceans, Rivers and Lakes. https://miningwatch.ca/sites/default/files/Troubled-Waters_Full.pdf. Accessed 13 June 2018

Eboh M (2014) Nigeria loses N27bn to gas flaring in one month. Vanguard Newspaper. http://www.vanguardngr.com/2014/11/nigeria-loses-n27bn-gas-flaring-one-month/. Accessed 13 June 2018

Eckersley R (1992) Environmentalism and political theory: towards an ecocentric approach. SUNY Press, Albany

Egyptian Environmental Affairs Agency. http://www.eeaa.gov.eg/en-us/aboutus/aboutagency/mission.aspx. Accessed 12 June 2018

Elder P (1984) Legal rights for nature: the wrong answers to the right question. Osgoode Hall Law J 22:285

Elf Nigeria Ltd v Opere Sillo (1994) 6 N.W.L.R.

Elvidge C, Zhizhin M, Baugh K, Hsu F, Ghosh T (2016) Methods for global survey of natural gas flaring from visible infrared imaging radiometer suite data. Energies 9(1):14. https://doi.org/10.3390/en9010014

Emeseh E (2006) Limitations of law in promoting synergy between environment and development policies in developing countries: a case study of the petroleum industry in Nigeria. J Energy Nat Resour Law 24:574

Environmental Guidelines and Standards for the Petroleum Industry in Nigeria (2002) Lagos: Department of Petroleum Resources 1991

Environmental Impact Assessment Act of 1992, Laws of the Federation of Nigeria 2004, c E12

Environmental Rights Action (2005) Gas flaring in Nigeria: a human rights, environmental

European Environmental Bureau (1997) Environmental liability in Europe: concerning the need for a European Directive on environmental liability

European Renewable Energy Council (2007) Climate protection. In: Energy revolution – a sustainable World Energy Outlook. European Renewable Energy Council, Amsterdam

Federal Ministry of Environment of Nigeria. Agencies and Parastatals. http://environment.gov.ng/index.php/about-moe/agencies-parastatals. Accessed 25 Sept 2015

Feinberg J (1974) The rights of animals and future generations. In: Blackstone WA (ed) Philosophy and environmental crisis. University of Georgia Press, Georgia

Findikakis A, Law A (1998) Marine tailings disposal simulation. J Hydraul Eng 124(4):370–383

Framework Environment Law – Law No. 5/98, of 19 June 2015, Angola

French D (2017) The Sofia guiding statements on sustainable development principles in the decisions of international tribunals. In: Cordonier Segger MC, Weeramantry CG (eds) Sustainable development principles in the decisions of international courts and tribunals, 1992 – 2012. Routledge, Abingdon
Gardner J, Sainato P (2007) Mountaintop mining and sustainable development in Appalachia. Min Eng 59:48
Global Gas Flaring Reduction (GGFR) (2009) Preparing for potential opportunities. The News Flare 8. http://siteresources.worldbank.org/EXTGGFR/Resources/ggfr_newsletter_8.pdf?resourceurlname=ggfrnewsletter_8.pdf. Accessed 21 July 2015
Gorman D, Neilson J (1997) Decommissioning offshore structures. Springer, New York
Government of Algeria. 1982. Law no. 82-10 / 1982 on Fishing, Algeria
Government of Algeria. 1983. Law no. 83- 16-17 July 1983 on Water, Algeria
Government of Algeria. 1984. Law no. 84 -12/1984 on Forestry Order, Algeria
Government of Benin. 1990. Constitution of the Republic of Benin, 1990
Government of Cameroon. 1996. Constitution of the Republic of Cameroon, 1996
Government of Ethiopia. 1995. Constitution of the Federal Democratic Republic of Ethiopia, 1995
Government of Ghana. 1994. Environmental Protection Agency Act, 1994 of Ghana
Government of Kenya. 1999. Environmental Management and Coordination Act no. 8 of 1999, Kenya
Government of Kenya. 2010. Constitution of Kenya, 2010
Government of Mali. 1992. Constitution of the Republic of Mali, 1992
Government of Niger. 2010. Constitution of the Republic of Niger, 2010
Government of Nigeria. 1999. Constitution of the Federal Republic of Nigeria, 1999
Government of South Africa. 1996. Constitution of the Republic of South Africa, 1996
Government of the Republic of Burkina Faso. 1991. Constitution of the Republic of Burkina Faso, 1991
Government of Togo. 1992. Constitution of the Republic of Togo, 1992
Government of Zimbabwe. 2002. Environmental Management Act (EMA), Act No.13 of 2002, Zimbabwe
Greenpeace International (1994) Shell shocked: the environmental and social costs of living with Shell in Nigeria. http://archive.greenpeace.org/comms/ken/hell.html. Accessed 21 July 2015
Human Rights Watch (1999a) The price of oil: corporate responsibility and human rights violations in Nigeria's oil producing communities. Human Rights Watch, Washington, DC https://www.hrw.org/reports/1999/nigeria/nigeria0199.pdf. Accessed 12 June 2018
Human Rights Watch (1999b) The price of oil: corporate responsibility and human rights violations in Nigeria's oil producing communities. Human Rights Watch, New York http://www.hrw.org/reports/1999/nigeria/. Accessed 10 July 2015
Hunter D, Salzman J, Zaelke D (1987) International environmental law and policy. Foundation Press, New York, p 1998
Intergovernmental Panel on Climate Change (IPCC) (2007) The Fourth Assessment Report (AR4)
International Convention for the Prevention of Pollution from Ships 1973/78, 2 November 1973, 1340 UNTS 184
International Covenant of Economic and Social Rights (ICESR)
2002 International Law Association (ILA) New Delhi Declaration on Sustainable Development. 2002. International environmental agreements: politics, law and economics 2. UN Doc. A/CONF.199/8
Johnston D, Potter H, Jones C, Rolley S, Watson I, Pritchard J (2008) Abandoned mines and the water environment. UK Environment Agency, Bristol https://assets.publishing.service.gov.uk/government/uploads/system/uploads/attachment_data/file/291482/LIT_8879_df7d5c.pdf. Accessed 21 Jan 2018
Jonah Gbemre v. Shell Petroleum Development Company Nigeria Ltd and Others. No. FHC/B/CS/53/05, (Nov. 14, 2005) Federal High Court of Nigeria, http://www.climatelaw.org/cases/case-documents/nigeria/ni-shell-nov05-judgment.pdf. Accessed 23 June 2015

Kamminga M, Zia-Zarifi S (2000) Liability of multinational corporations under international law. Kluwer, Alphen aan den Rijn, pp 139–195

Kenya's Ministry of Environment and Natural Resources. About Us. http://www.environment.go.ke/?page_id=2. Accessed 21 Sept 2015

Knippertz P, Evans MJ, Field PR, Fink AH, Liousse C, Marsham JH (2015) The possible role of local air pollution in climate change in West Africa. Nat Clim Change 5:815–822

Kyoto Protocol to the United Nations Framework Convention on Climate Change, UN Doc FCCC/CP/1997/7/Add.1, December 10, UNFCCC

LacLanoux Arbitration (France v. Spain), 1957, 24 I.L.R. 101

Law 4/1994 for the Protection of the *Environment* of Egypt

Locke J. Two treatises of Government. http://www.gutenberg.org/files/7370/7370-h/7370-h.htm. Accessed 21 Jan 2018

Macdonald JM (1994) Artificial reef debate, habitat enhancement or waste disposal? Ocean Dev Int Law 25:87

Macias R (2010) Public participation in energy and natural resources development: a theory and criteria for evaluation. Canadian Institute of Resources Law Occasional Paper No 34. University of Calgary Digital Repository. https://dspace.ucalgary.ca/bitstream/1880/48390/1/CriteriaOP34w.pdf. Accessed 12 June 2018

Mapani B, Kribek B (2012) Environmental and health impacts of mining in Africa. Czech Geological Survey, Prague

McKinnon E (2002) The environmental effects of mining waste disposal at Lihir Gold mine, Papua New Guinea. J Rural Remote Environ Health 1:40–50

MINEO Consortium (2000) Review of potential environmental and social impact of mining. http://tupa.gtk.fi/raportti/arkisto/rs_2004_24.pdf. Accessed 21 Jan 2018

Mining Working Group at the United Nations (2014) A rights-based approach to resource extraction in the pursuit of sustainable development. Advocacy Brief. https://miningwg.com. Accessed 12 June 2018

Monica M (2012) Nigeria's penalty for gas flaring will not curb emissions, say campaigners. The Guardian. https://www.theguardian.com/environment/2012/may/31/nigeria-penalty-gas-flaring. Accessed 12 June 2018

Mutum R. Nigeria: ICPC returns N1 billion looted funds to environment ministry. allAfrica. http://allafrica.com/stories/201509160545.html. Accessed 12 June 2018

National Commission on the BP Deepwater Horizon Oil Spill & Offshore Drilling, Report To The President, Deepwater: The Gulf Oil Disaster And The Future Of Offshore Drilling (2011) https://www.gpo.gov/fdsys/pkg/gpo-oilcommission/pdf/gpo-oilcommission.pdf. Accessed 2 June 2018

National Environmental Management Act No. 107 of 1998, South Africa

National Environmental Management Laws Amendment Act, No. 25 of 2014, South Africa

National Environmental Standards and Regulations Enforcement Agency (NESREA) Act of 2007, Nigeria

National Oil Spill Detection and Response Agency (Establishment) Act 2006

Niger Delta Development Commission (NDDC) Act, Nigeria

Nigeria's Federal Ministry of Environment. About Us. http://environment.gov.ng/index.php/78-featured/72-welcome. Accessed 21 Sept 2015

Nkulu CBL, Asosa JN, Ngombe LK, Mwanza AM, Numbi OL, Ilunga BK, Kimpanga CM (2012) Pathways of exposure to cobalt in populations living in Katanga, D.R. Congo. In: Mapani B, Bohdan Kribek B (eds) Environmental and health impacts of mining in Africa

Nowak M (2000) Civil and political rights. In: Symonides J (ed) Human rights concepts and standards. Ashgate, Farnham

Nwadiaro v Shell Petroleum Development Company, Nigeria Ltd (1990) 5 N.W.L.R

Nwilo P, Badejo O (2006) Impacts and management of oil spill pollution along the Nigerian Coastal Areas. http://www.fig.net/pub/figpub/pub36/chapters/chapter_8.pdf. Accessed 10 July 2015

Ocholla GO (2013) Environmental issues and socio-economic problems emanating from salt mining in Kenya; A case study of Magarini district. Int J Humanit Soc Sci 3(3):213–223
Ogola J, Mitullah W, Omulo M (2002) Impact of gold mining on the environment and human health: a case study in the Migori Gold Belt, Kenya. Environ Geochem Health 24:141–158
Ojukwu-Ogba N (2009) Legislating development in Nigeria's oil-producing region: the N.D.D.C. Act seven years on. Afr J Int Comp Law 17:136–149
Olawuyi D (2012) Legal and sustainable development impacts of major oil spills. J Sustain Dev 9(1):1–15
Olawuyi D (2015a) Increasing relevance of right-based approaches to resource governance in Africa: shifting from regional aspiration to local realization. McGill Int J Sustain Dev Law Policy 11(2):113–158
Olawuyi D (2015b) Principles of Nigerian environmental law. Afe Babalola University Press, Ado Ekiti
Olawuyi D (2016a) Climate justice and corporate responsibility: taking human rights seriously in climate actions and projects. J Energy Nat Resour Law 34(1):27–44
Olawuyi D (2016b) The human rights based approach to carbon finance. Cambridge University Press, Cambridge
Oluduro O (2014) Oil exploitation and human rights violations in Nigeria's oil producing communities. Afrika Focus 25(2):160–166
Ossai E, Iniaghe G, Osakwe S, Agbaire P (1999) Pollution problems and environmental effects of chemicals. In: Ekechi F (ed) Reading in general studies: history and philosophy of science, Abraka General Studies Dept. Pub
Patin V (1999) The environmental impacts of the offshore oil and gas industry. EcoMonitor Publishing, New York
Paul J, Campbell G (2011) Investigating rare earth element mine development in EPA region 8 and potential environmental impacts. (908R11003). US Environmental Protection Agency, Washington, DC
Protocol on Provisions Concerning Reports on Incidents Involving Harmful Substances to the International Convention for the Prevention of Pollution from Ships 1973/78, 2 November 1973, 1340 UNTS 184
Redgwell C (1996) Life, The Universe and everything: a critique of anthropocentric rights. In: Human rights approaches to environmental protection. Oxford University Press, Oxford
Ritter L, Solomon K, Forget J, Stemeroff M, O'Leary C (1995) A review of selected persistent organic pollutants. International Programme on Chemical Safety (IPCS). PCS/95.39, vol 65, 66. World Health Organization, Geneva
Rosenthal E, Revkin A (2007) Science panel calls global warming 'Unequivocal' The New York Times
Saether K, Aguilera RV (2008) Corporate social responsibility in a comparative perspective. In: Crane A et al (eds) The Oxford handbook of corporate social responsibility. Oxford University Press, Oxford
Schrijver N (2017) Advancements in the principles of international law on sustainable development. In: Cordonier Segger MC, Weeramantry CG (eds) Sustainable development principles in the decisions of international courts and tribunals, 1992 – 2012. Routledge, Abingdon
Schuler D, Buchert M, Liu R, Stefanie G, Merz C (2011) Study on rare earths and their recycling. European Parliament/Institute for Applied Ecology, Öko-Institut e.V., Freiburg http://reinhard-buetikofer.eu/wp-content/uploads/2011/01/Rare-earths-study_Oeko-Institut_Jan-2011.pdf. Accessed 12 June 2018
Sensogut C (2007) Occupational noise in mines and its control – a case study. Polish J Environ Stud 16(6):6939–6942
Shell Petroleum Development Company Nigeria Ltd v Ambah (1991) 3 N.W.L.R
Shell Petroleum Development Company, Nigeria ltd v Farah (1995) 3 N.W.L.R
Shell Petroleum Development company, Nigeria Ltd v Isaiah (2001) 11 N.W.L.R
Shell Petroleum Development Company, Nigeria Ltd v Tiebo vii & Others (2005)

Sissoko K, van Keulen H, Verhagen J, Tekken V, Battaglini A (2010) Agriculture, livelihoods and climate change in the West African Sahel. Reg Environ Change 11(1):119–125

Social Development Integrated Centre (Social Action) (2009) Flames of hell: gas flaring in the Niger Delta. http://www.saction.org/home/saction_image/flames_of_hell.pdf. Accessed 12 Feb 2018

South Africa's Department for Environmental Affairs. Introduction and background. https://www.environment.gov.za/aboutus/department. Accessed 21 Sept 2015

Stevens W (2007) On the climate change beat, doubt gives way to certainty. The New York Times

Stockholm Convention on Persistent Organic Pollutants (adopted 22 May 2001, effective 17 May 2004) 2256 UNTS119

Stockholm Declaration on the Human Environment. 1972. UN Doc A/CONF48/14/ Rev.1

Sydney Morning Herald (2002) $1.3bn mine leaves a cyanide sea. https://www.smh.com.au/articles/2002/04/08/1017206313885.html. Accessed 12 June 2018

Tickell C (1968) Climatic change and World Affairs. Harvard University Press, Cambridge

Tiratsoo EN (1972) Natural gas: a study, 2nd edn. Scientific Press Ltd, Beaconsfield

Trail Smelter Arbitration (U.S.A. v. Canada), 1938–1941, 9 I.L.R. 315

UKOOA (1995) An assessment of the environmental impacts of decommissioning options for oil and gas installations in the UK North Sea. www.ukooa.co.uk/auris.html. Accessed 21 July 2015

UN Environment Programme (2011) Environmental assessment of Ogoniland. UN Environment Programme, Nairobi https://postconflict.unep.ch/publications/oea/unep_oea.pdf. Accessed 12 June 2018

UNDP (2006) Niger Delta human development report. United Nations Development Programme, Abuja

UNEP (1987) Goals and principles of environmental impact assessment. http://www.unep.org/regionalseas/publications/reports/RSRS/pdfs/rsrs122.pdf. Accessed 23 July 2017

UNEP. Environmental assessment of Ogoni Land. http://www.unep.org/nigeria/. Accessed 02 June 2017

United Nations (2010) Water, sanitation and hygiene in Nigeria. https://www.unicef.org/nigeria/NG_resources_wash2010.pdf. Accessed 23 June 2017

United Nations (2011) Guiding principles on business and human rights: implementing the United Nations "Protect, Respect and Remedy" Framework HR/PUB/11/04. https://www.ohchr.org/Documents/Publications/GuidingPrinciplesBusinessHR_EN.pdf. Accessed 21 May 2018

United Nations. The human rights based approach to development cooperation: towards a common understanding among UN Agencies. UN Practitioners' portal on human rights based approaches to programming. https://hrbaportal.org/the-human-rights-based-approach-to-development-cooperation-towards-a-common-understanding-among-un-agencies. Accessed 12 June 2018

United Nations Convention on the Law of the Sea, 1982, 21 ILM (1982), 1261 (adopted on 10 December 1982, and entered into force on November 16 1994)

United Nations Diplomatic Conferences United Nations Conference on The Law of The Sea 1958. UN Treaty. http://untreaty.un.org/cod/diplomaticconferences/lawofthesea-1958/lawofthesea-1958.html. Accessed 23 June 2015

United Nations Economic Commission for Europe (UNECE) (2018a) Member states and member representatives. https://www.unece.org/oes/nutshell/member_States_representatives.html. Accessed 21 May 2018

United Nations Economic Commission for Europe (UNECE) (2018b) The air convention and its protocols. http://www.unece.org/env/lrtap/status/lrtap_s.html. Accessed 21 May 2018

United Nations Environment-Chemicals: Persistent Organic Pollutants. UNEP https://www.unenvironment.org/explore-topics/chemicals-waste/what-we-do/persistent-organic-pollutants-pops. Accessed 11 June 2018

United Nations Framework Conference on Climate Change (Adopted 9 May 1992, effective 21 March 1994) 1771 UNTS 107

United Nations Framework Convention on Climate Change (2011) Outcome of the Durban climate change conference

United Nations World Commission on Environment and Development Report (the Brundtland Report) (1987) Our common future. Oxford University Press, Oxford

United States Environmental Protection Agency (2016) Terms of environment: glossary, abbreviations, and acronyms. http://infohouse.p2ric.org/ref/01/00402/eterms.html. Accessed 24 Jan 2018

Universal Declaration of Human Rights 1948

US Department of Interior, Bureau of Mines (1993) Potential for submarine tailings disposal to affect the availability of minerals from United States coastal areas

Vidal J (2011) Nigeria on alert as Shell announces worst oil spill in a decade. The Guardian. http://www.guardian.co.uk/environment/2011/dec/22/nigerian-shell-oil-spill. Accessed 23 May 2012

Water Resources Act of 1993, Laws of the Federation 2004, c W2

West and Central Africa Regional Framework Agreement on Air Pollution. 2009

World Bank (2009) Guidance on upstream flaring and venting policy and regulation. http://siteresources.worldbank.org/INTGGFR/Resources/fr_policy_regulations_guidance.pdf. Accessed 21 July 2017

World Bank. Africa Environmental Agencies. http://web.worldbank.org/archive/website01004/WEB/0__CO-75.HTM. Accessed 21 May 2018

World Health Organization (2016) What is the environment in the context of health? http://www.who.int/quantifying_ehimpacts/publications/preventingdisease2.pdf. Accessed 12 Aug 2018

Yanomami Case, Rep. No. 12/85, Case 7615 (Brazil), (1985)

Chapter 11
Business, Human Rights and Corruption Risks

Abstract This chapter provides an overview of business and legal risks that must be carefully considered when investing in African extractive markets. Such risks include resource nationalism; lopsided contractual provisions that confer unfair advantages on national oil or mining companies; arbitrary change in contract terms; delays in operational approvals; complex local content requirements; weak human rights culture; and high corruption potentials amongst others. Even though some of these risks are not peculiar to Africa, as this chapter will show, conventional means of addressing legal risks through contractual provisions, internal regulation, and consistent business practices alone are not adequate in African markets. The chapter discusses how investors, and their lawyers alike, can effectively deal with such risks, as early as possible, through extensive due diligence and diligent contract negotiations.

11.1 Introduction

All through the book, the unique opportunities across Africa, for international extractive companies (ECs) to spearhead extractive exploration and production and to acquire interests in fields with unexplored economic potentials, have been demonstrated and analyzed. Despite these investment opportunities however, entrants into extractive markets in Africa often face legal risks that pose monumental threats to the economic viability of extractive investments. This chapter provides an overview of the commonly encountered business and legal risk points in African extractive investments. It also discusses some of the more effective due diligence strategies that could be helpful to address legal risks that may arise from extractive investments in Africa.

Legal risks refer to the range of financial, reputational or investment loss; legal liability; and/or dispute settlement costs that a business enterprise may be exposed to as a result of a defective contractual arrangements or transaction.[1] In the African extractive context, such risks include resource nationalism; lopsided contractual

[1] See Olawuyi (2015); also McCormick (2010), pp. 1–3; also Connor et al. (2018).

D. S. Olawuyi, *Extractives Industry Law in Africa*,
https://doi.org/10.1007/978-3-319-97664-8_11

provisions that confer unfair advantages on national oil companies; arbitrary change in contract terms; delays in operational approvals; complex local content requirements; weak human rights culture; and high corruption potentials amongst others. Even though some of the legal risks discussed are not peculiar to Africa, as this chapter will show, conventional means of addressing legal risks through contractual provisions, internal regulation, and consistent business practices alone are not adequate in African markets. Much greater attention and legal due diligence is therefore necessary to effectively manage contract negotiation processes to prevent legal risks. Typically, legal due diligence involves investigating the facts, risks and opportunities of a transaction, entity or person before consummating a contractual relationship. Through detailed and careful due diligence, some of the legal risks commonly encountered in African markets may be comprehensively managed.

If not properly considered and managed, such legal risks may reduce the positive economic prospects and gains in a hitherto profitable investment enterprise and may frustrate a first time investor. While these legal risks have been perceived as disincentives to extractive investments in Africa, this book takes the view that these risks do not call for surrender. Rather, they could be efficiently managed at contract negotiation phases. For investors in Africa, and their professional advisers alike, an understanding of key risk points is very important for making investment decisions and implementing corporate and contractual structures to mitigate such risks to protect their investments.[2]

This chapter provides a summary of legal risks points that must be carefully considered when investing in African extractive markets. At the end of this chapter, you should be able to understand the nature and sources of legal risks in African extractive industries; and appreciate how investors, and their lawyers alike, can effectively deal with such risks, as early as possible, through extensive due diligence and diligent contract negotiations.

11.2 Distinctive Legal Risks in African Markets

An average investor going to do business in Africa would typically be concerned about how to obtain prospecting permits and licenses; incorporate a local subsidiary; seamlessly repatriate funds; and ensure compliance with international foreign corruption conventions amongst others. However, while these requirements are in themselves key to doing business in any African country, the enormous potentials for legal risks in many emerging African markets compel companies to, in addition to meeting entry requirements, adopt proactive strategies that eliminate or minimize other possible legal risks. The weaknesses of legal systems in emerging market legal systems compel a broad-based, proactive strategy to minimize legal risks, only a portion of which can be managed through traditional contractual documentation.

[2] Connor et al. (2018).

General problems include the diverse ownership regimes that apply across part of Africa. While this by itself is not a problem, the tendency to consider Africa as one homogenous region often creates problems for investors that fail to consider the unique legal requirements and applicable regime in each African country before going in to do business. Furthermore, specific problems arise due to poorly formulated laws that grant regulators broad discretion to negotiate lopsided contractual arrangements and to impose stringent participation requirements. There is also the high risk of corruption which could result in the violation of the diverse and increasingly more stringent anti-corruption requirements in many parts of the world. Generally, contractual provisions alone are not sufficient to address all of these legal risks. While is very important to include contractual terms that specify mitigate these risks, understanding the risks and taking elaborate steps to avoid them will go a long way in protecting the business interests in difficult emerging markets in Africa. This section explores some of the key legal risks in African markets.

11.2.1 Different Natural Resource Control and Ownership Regimes in African Countries

As already discussed in Chap. 4 of this book, natural resource ownership and control of extractive resources vary from one African country to the other, depending largely on the country's socio-political culture, historical and colonial background, the enduring legal regime, and strategic development framework. As such, when going into Africa, it is important to understand the specific ownership structure in the target country: who owns and controls extractive resources; is it the central or federal government, is it provincial or state government; or is there private ownership of resources. Answers to these questions are highly varied from one African country to the other and may at times fall outside the traditional ownership norms in other foreign jurisdictions.

Generally, in many dominial regimes, resource ownership is vested in the state, while the rights of participating entities are limited to the right to use, manage, control or enjoy benefits of resource production.[3] In dominial regimes there is no private ownership of subsurface natural resources.[4] In less dominial jurisdictions however, subsurface mineral, petroleum and natural gas ownership could be both freehold and leasehold (*i.e.* a majority owned by the government, while some are privately owned). In Canada for example, oil and gas resources are owned and

[3] See Clerk and Kohler (2005), p. 40. See also Omorogbe and Oniemola (2009), pp. 115–117.

[4] The dominial system can historically be traced to the Roman regalian system under which mineral resources came under the absolute ownership and control of the sovereign. It was regarded as *iura regalia*, a right enjoyed by the sovereign by virtue of its prerogatives and it provided the legal basis for collecting a royalty: a share in the production of minerals. The main justification for public ownership is that minerals should be considered public property to be conserved and managed for the welfare of all the citizens. See Graulau (2011), pp. 14–15.

controlled by the provinces, as constitutionally recognized in section 109 of the *Constitution Act, 1867* which grants each of the provinces ownership over all mines and minerals, including petroleum and natural gas, found within their respective borders.[5] Rights over hydrocarbon resources are granted by provincial governments in the form of Crown leases or licences. Apart from crown leases, there are also private ownership of resources in Alberta. Individuals who are the assigns or descendants of Alberta's original homesteaders are the registered owners of the petroleum and/or natural gas beneath approximately 6.0 million acres (4% of the Province's surface area). Freehold leases confer ownership over production reduced to possession by the lessee, while the freehold estate owner typically retains a royalty interest. The terms of these leases vary and are independently negotiated with the freehold estate owners, although standard form agreements drafted by industry organizations, most the Canadian Association of Petroleum Landmen (CAPL), are typically used.[6]

This distinction is unique and important because in many parts of Africa, the central government maintains radical title and ownership to all petroleum and natural gas within the country. As such, ownership of oil and gas by provincial or state government is not widespread. Secondly, freehold ownership is not widespread in African countries as all oil and gas resources are typically owned by the Federal Government. In such jurisdictions, investors have to deal directly with the government or national oil companies. A knowledge of this seemingly negligible fact may save the cost and time of dealing with wrong agencies and unscrupulous individuals who may lay claim to freehold ownership.

Furthermore, in some African countries, even after a resource exploration license is granted by the national oil or mining company, the grant must be formally ratified by parliament before it becomes effective. What this means is that investments cannot not start until a license is granted parliamentary approval. This requirement is not typical in many foreign jurisdictions. In Canada for instance, there is no legal requirement for the federal or provincial parliament to pass a vote on a crown lease granted by the responsible energy regulator. The required parliamentary approval in many African countries may take several months or in extreme cases years to come through. A knowledge of such seemingly minute requirements may save an investor from the embarrassment of making investment commitments or plans too early, before actual parliamentary approval is received.

It is also pertinent to understand the nature and scope of control and powers granted to local energy regulators and national oil or mining companies in the host country. For example, local enactments and constitutions in a number of African countries often confer excessive powers on national oil companies to unilaterally cancel or amend petroleum contracts and licences. Furthermore, ongoing volatility in the global economy means that governments increasingly look to oil companies

[5] RSC 1985, App II, No 26. Note that the Province of Alberta was granted petroleum and natural gas ownership rights pursuant to the Natural Resources Transfer Agreement, an agreement which was given constitutional status by the Constitution Act, 1930.

[6] Martin and Park (2010), p. 20.

to fund the shortfall in revenues and to regain control over natural resources through tax hikes, demand for greater state equity and indigenous participation, renegotiation of stability clauses in mining contracts and arbitrary imposition of additional levies and 'super profits'. For example, in countries such as Guinea, Tanzania, Democratic Republic of Congo and Sierra Leone, there have been cases of: increased government participation[7]; arbitrary confiscation and nationalisation of Oil Mining Licences (OMLs)[8]; unilateral modification of the oil license or contract[9]; forced contract renegotiation, retroactive change in legislation and licensing regime after exploration rights have been granted[10]; and breach of contract by national oil company. The resource nationalism trend in many parts of Africa often pose financial risks to investors that have committed resources to establishing oil operations.[11] Resource nationalism increases the risks IOCs face and lowers the expected returns on investments.

It is important to understand the ownership regime in a country, the long-term validity of a license or permit, and the rights that come with such permits. It should

[7] In September 2011, Guinea adopted a new mining code which vests the state a 15% stake in all new mining projects with the right to purchase a further 20%. This led to the exit of investors such as BHP Billiton from Guinea and a negative rating for the country as an investment destination. In the 2013 *Resource Nationalism Index*, Guinea was classified as being an 'extreme risk' country. Guinea was also ranked second to the last in investor protection in the World Economic Forum's 2012–2013 Annual Competitiveness Index.

[8] For example, In September 2012, Tanzania's Energy Minister announced a review of all existing PSAs to ensure that they are in the country's best interests. Additionally, the Tanzania Petroleum Development Corporation TPDC delayed a licensing round for nine deep-sea oil and gas blocks, originally set for September 2012, until a parliamentary vote on the new gas policy takes place. These measures have created some uncertainty for investors on the future direction of Tanzania's licensing regime.

[9] For example, Sierra Leone announced in March 2008 that the government will review all existing mining conventions; same in the Democratic Republic of Congo where the Government recently set up a revision commission which after reviewing 62 mining conventions approved the cancellation of 22 and the renegotiation of the remaining 40.

[10] On 7 December 2012, the 2012 Upstream Act was passed by the Ugandan Parliament, the Act introduces some far-reaching changes to the institutional and regulatory framework governing Uganda's upstream oil and gas industry. Similarly, in Mozambique, the government published in 2012, a revised draft of the Petroleum Law, which it is currently anticipated will become law by the end of 2013. Key changes proposed by the draft law include, for example, the introduction of a new type of concession agreement for the construction of offshore facilities and a requirement for government approval on any transfer of an interest in a concession agreement, whether directly or indirectly.

[11] Resource nationalism is when governments use nationalistic policies to assume greater control, influence or demand a larger share of revenues from companies engaged in the extraction and processing of the country's natural resources. Resource nationalism typically occurs when a State thinks that a foreign investor is getting too good a deal for their investment, especially when prices for the natural resource rise beyond the levels originally anticipated. In these cases, the State may seek to impose new terms or regulations on the investment or the foreign investors, beyond what was agreed during contract negotiation or licensing stages, to improve the position of the State. Although resource nationalism is a global concern, it is more prevalent in Africa due to frequent policy and government changes in many African countries. See Wilson (2015) and Pryke (2017).

also be an essential component of due diligence search to investigate and determine the ownership structure in specific countries; the nature of legislative power conferred on the national oil mining company; and the nature of investment protection and tenure security guaranteed to potential investors. During contract negotiations or licensing phases, it is also pertinent to ensure that stabilization, adaptation and foreign arbitration clauses are established as part of any agreements with the host state.

11.2.2 Effects of Regional Trade Harmonization Regimes

As discussed in Chap. 3, apart from domestic laws, regional harmonization treaties and regimes could also affect investments in a target country. Some of these regimes have supranational character and are directly applicable to investments in member states, even without domestication or implementation (for example Organisation for the Harmonisation of Business law in Africa (OHADA). It is therefore important to check whether the host country is a member or signatory to any regional trade regime and seek legal opinion on the key provisions that could impact an investment. As noted earlier in this book, the main African regimes to watch out for include:

In North Africa: Union of Arabic Maghreb (UMA) with Algeria, Morocco, Tunisia, Libya and Mauritania

West Africa: ECOWAS, WAEMU, OHADA[12];

[12] ECOWAS: Economic Community of West African Countries: is a regional group of fifteen countries (Benin, Burkina Faso, Cape Verde, Ivory Coast, Gambia, Ghana, Guinea, Guinea Bissau, Liberia, Mali, Niger, Nigeria, Senegal, Sierra Leone, Togo), its mission is to promote economic integration in "all fields of economic activity, particularly industry, transport, telecommunications, energy, agriculture, natural resources, commerce, monetary and financial questions, social and cultural matters". WAEMU: West African Economic and Monetary Union: 8 member States (Benin, Burkina Faso, Ivory Coast, Guinea Bissau, Mali, Niger, Senegal, Togo). Purpose: Strengthen competitiveness of the member States in an open and competitive market; create a common market between member States based on free circulation of persons, goods, services, funds; harmonisation of custom and exchange control regulations; coordination of national sector-based policy, including mining, telecommunication, energy and infrastructures. OHADA: Organisation for the Harmonisation of Business law in Africa; 16 Member States (Benin, Burkina-Faso, Cameroon, Central Republic of Africa, Côte d'Ivoire, Congo, Comoros, Democratic Republic of Congo, Gabon, Guinea, Guinea-Bissau, Equatorial Guinea, Mali, Niger, Senegal, Chad, Togo). There are 8 uniform acts all of which are directly applicable to investments in member states: General commercial law; Commercial companies; Securities; Simplified recovery procedure; Collective proceedings for wiping off debt; Arbitration; Accounting system; Road transportation of goods. Dispute Resolution is by the OHADA Common court of justice and arbitration which resolves disputes on matters which involve the uniform acts. Its decision is final and without appeal.

Central Africa: ECCAS, CEMAC, ECGL, OHADA; and[13]
East Africa: SADC, COMESA[14]

It is pertinent to ascertain the extents to which these regional trade groups or regimes may apply to the specific country of investment, and how these regimes may impose additional legal requirements such tax, reporting obligations, environmental liability, fair trade; fund repatriation, local participation amongst others; on foreign investments in that country.

11.2.3 Diverse Network of Applicable Laws and Regulations

As discussed in Chap. 4, different sources of law underpin and govern the extractive industry, most of which are scattered in different documents and legislative instruments. Ranging from international law to primary domestic legislation, secondary legislation and guidelines and reports by regulatory agencies. In addition to laws on ownership regime and structure, investors must also ensure a proper grasp of other network of laws that are applicable to extractives investments in the target country to avoid running into legal hurdles. It is important to understand laws on repatriation of funds; tax and custom regime; laws on the environment (gas flaring, reclamation, abandonment and clean ups); labour law; competition law; contract law; transportation law; and extractive specific laws such as energy law; mining law; electricity/water law and investment laws. It is particularly important to understand specific laws that stipulate requirements for extractive operations, to avoid any default.

Due to the fact that legal frameworks that are applicable to resource extraction are still evolving in many African countries, particularly countries such as Kenya, Tanzania and Ghana with recent oil discoveries, it is not uncommon to find that laws typically change very often. Apart from exploration laws, sector specific laws and regulation that impose requirements on environmental liability, gas flaring, remediation and cleanups must be carefully studied to ensure compliance. As studies in the last chapter, investors can avoid significant legal risks, fines and penalties by

[13] ECCAS: Economic Community of Central Africa States; ECGL: Economic Community of the Great Lakes Countries to promote economic integration and CEMAC: Economic and Monetary Community of Central Africa. 6 member States (Cameroon, Chad, Congo, Equatorial Guinea, Gabon, Republic of Central Africa) all seek to harmonise custom and exchange control regulations, national sector-based policy, including pharmacy, agriculture, airport and education.

[14] SADC: Southern African Development Community (to further socio-economic cooperation and integration as well as political and security cooperation among 15 southern African states: Angola, Botswana, Democratic Republic of Congo, Lesotho, Malawi, Mauritius, Mozambique, Namibia, Swaziland, Tanzania, Zambia, Zimbabwe, South Africa, Seychelles); COMESA: Common Market for Eastern and Southern Africa: establishes a common market and a free trade zone between nineteen member states, COMESA is also considering a common visa scheme to boost trade and tourism. There is a COMESA Common Investment Agreement (CCIA) Template applicable in COMESA areas.

understanding the full legal obligations and requirements under these laws as well. Failure to understand the unique legal provisions in a country may bring about liability and may result in incurring avoidable legal costs.

11.2.4 Increasing Local Content Requirements

Chapter 9 of this book has discussed the rise of local content laws and policies in Africa, that place obligations on foreign investors to ensure local participation in resource exploration and development.[15] The key aim of local content regimes is to create jobs for citizens, stimulate industrial development, increase local capability and technical know-how, create a competitive supplier base and ultimately further national economic development. They amongst other things mandate that: first consideration must be given to locals in employment and training; specifically only locals must be employed in junior and intermediate cadre positions[16]; community interests must be reflected, and the rights of locals to free prior informed consent on investment activities carried out on their lands must be respected and fulfilled.[17] There are also elaborate domestic laws that prohibit the importation of fabricated and welded products; and mandate retaining only the services of local financial institutions; mandate retaining a percentage of total revenue from operations in a local bank account; mandate retaining the services of only legal practitioners with local origin or a law firm(s) of legal practitioners within the country whose office is located in any part of the country; in most cases these laws establish a 50% threshold of local content for legal services relating to project management and consulting services; and establish mandatory corporate social responsibility requirements on investors amongst other things.

A breach of local content requirements is a criminal offence in many African countries. In Nigeria, for example, an operator, contractor or subcontractor who carries out any project contrary to the provisions of the Local Content Act, commits an offence and is liable upon conviction to a fine of 5% of the project sum for each project in which the offence is committed or cancellation of the project.

To avoid criminal liability and heavy fines, it is important for investors to seek legal advice on the local content requirements in the host country.

---

[15] See for example Section 14 of the Petroleum (Exploration and Production) Act 1991 (Act 2 of 1991) of Namibia.

[16] See for example Section 35 of the Nigerian Local Content Development Act, In Mozambique, Quota restrictions have been imposed on the employment of foreign citizens, since Decree No. 63/2011 was issued on 7 December 2011.

[17] The 2009 Land Act of South Sudan.

11.2.5 Business and Human Rights Risks

As discussed in Chap. 1, many resource rich African countries have weak human rights cultures and have, for many years, failed to respect, protect and fulfil the rights of local communities that may be impacted by extractive activities.[18] When investing in Africa, it makes a strong business sense, in terms of reputation, potential liability and risk exposure, for extractive companies to uphold strong human rights tenets in project design, implementation and material sourcing, irrespective of the weak human rights threshold in the host country.

Over the years, foreign extractive companies have been implicated in exacerbating human rights challenges in African countries, in the form of failing to address human rights impacts of their operations, or colluding with host governments to perpetuate land grabs, bribing local officials to cover up environmental pollution, tax evasion, procuring child labor, unfair pricing policies, anonymous company ownership, illegal or unethical sourcing of raw materials and natural resources especially from war or conflict zones, or making illegal payments to fast track licensing processes amongst others.[19] These human rights concerns associated with extractive operations in Africa have resulted in increased demand for, and international law recognition of, the obligations of transnational corporations and business enterprises to respect, protect and fulfil human rights in their operations and value chain. This is part of broader global efforts aimed at coordinating the systemic integration of human rights and sustainable development goals to ensure coherence, coordination, and harmonization.[20]

The Human Rights Based Approach (HRBA), as advocated by the United Nations, places emphasis on addressing and mitigating human rights impacts of resource development projects.[21] The aim of the HRBA is to ensure that existing resource regimes are reformed to include elements of: participation; accountability; equality and non-discrimination; access to information; and access to justice to ensure greater transparency and respect for human rights in extraction projects.[22]

---

[18] Many African countries score very weak in terms of human rights protection and fulfilment. See Human Rights Watch (2018).

[19] The Africa Progress Panel (2013), led by Kofi Annan, has highlighted how "tax evasion, illicit transfers of wealth and unfair pricing practices" deprive Africa of funds needed for development. According to the Report, Africa loses twice as much in illicit financial out flows as it receives in international aid. It calls on African governments "to set out a bold national agenda for strengthening transparency and accountability" in oil, gas and mining sectors.

[20] Mining Working Group at the United Nations (2014); see MacKay (2004); Petersmann (2002), p. 626.

[21] Office of the High Commissioner for Human Rights, *Claiming the Millennium Development Goals: A Human Rights Approach*, UN Human Rights, UN Doc HR/PUB/08/3 (2008) at 11–15, 37–39. [UNHCR, Claiming the Millennium Development Goals]

[22] According to the United Nations, other elements of good programming practices that are essential under a HRBA include: (i) People are recognized as key actors in their own development, rather than passive recipients of commodities and services; (ii) Participation is both a means and a goal; (iii) Strategies are empowering, not disempowering; (iv) Both outcomes and processes are

A more recent elucidation of the HRBA in international law is the extensive report of the United Nations Special Representative on the issue of human rights and transnational corporations and other business enterprises, John Gerald Ruggie, who was appointed in 2005 by the United Nations to study the issue of human rights and transnational corporations and other business enterprises.[23] After 6 years of research and extensive consultations, the Special Representative's final report—the *Guiding Principles on Business and Human Rights: Implementing the United Nations 'Protect, Respect and Remedy' Framework* (Ruggie Framework)—was unanimously adopted by the Human Rights Council providing, for the first time, a global standard for preventing and addressing the risk of adverse impacts on human rights linked to business activity. Though not legally binding, the Ruggie Framework provides normative guidance on how human rights requirements can be framed and integrated into resource governance. The Ruggie Framework documents the fact that if not properly planned, projects meant to advance economic goals could violate human rights. It encourages business enterprises to carry out 'human rights due diligence' or a 'human rights risk assessment' before executing projects, to demonstrate that human rights are respected.[24]

Furthermore, in 2009, the United Nations Development Group (UNDG) launched the UNDG Human Rights Mainstreaming Programme (UNDG-HRM), now known as the UNDG Human Rights Working Group (HRWG).[25] The UNDG-HRWG aims

monitored and evaluated; (v) analysis includes all stakeholders; (vi) Programmes focus on marginalized, disadvantaged, and excluded groups; (vii) The development process is locally owned; (viii) Programmes aim to reduce disparity; (ix) Both top-down and bottom-up approaches are used in synergy; (x) situation analysis is used to identity immediate, underlying, and basic causes of development problems; (xi) Measurable goals and targets are important in programming; (xii) strategic partnerships are developed and sustained; (xiii) programmes support accountability to all stakeholders. See UNHCR, Claiming the Millennium Development Goals, 11–15.

[23] United Nations, Press Release, SG/A/93 "Secretary-General Appoints John Ruggie of United States as Special Representative on the Issue of Human Rights and Transnational Corporations and Other Business Enterprises" (28 July 2005).

[24] Promotion and Protection of all Human Rights, Civil, Political, Economic, Social and Cultural Rights, Including the Right to development-Protect, Respect and Remedy: A Framework for Business and Human Right, HRCOR, 8th Sess, UN Doc A/HRC/8/5 (2008). See also Principle 17 of the Report of the Special Representative of the Secretary-General on the issue of human rights and transnational corporations and other business enterprises, John Ruggie, UNGA, 17th Sess UN Doc A/HRC/17/31 [2011 Ruggie Framework], stating that in order to identify, prevent, mitigate and account for how they address their adverse human rights impacts, business enterprises should carry out human rights due diligence. The process should include assessing actual and potential human rights impacts, integrating and acting upon the findings, tracking responses, and communicating how impacts are addressed.

[25] The UNDG was established in 1997 as a coalition of UN funds, programmes, specialized agencies, departments, and offices that play a role in development in over 150 countries. The group meets at least three times yearly to decide on issues related to country level coordination to achieve the internationally agreed development goals. The group's common objective is to deliver more coherent, effective and efficient programmes within the UN system. See UNDG, *About the UNDG*. The UNDG Human Rights Working Group (UNDG-HRWG) aims to facilitate the integration of human rights in the UN's development work: United Nations Development Group, "Human Rights Working Group".

to mainstream human rights norms and standards into the work and activities of all UN agencies and programmes.[26]

Similarly, the 2012 United Nations Conference on Sustainable Development (Rio+20 Conference) called on countries to recognize the interrelationship between sustainable development and human rights in mining and extraction activities.[27] A rights-based approach to sustainable development was recognized at the Rio+20 Conference as an approach that guarantees the achievement of the interlinked objectives of Rio+20: poverty eradication; transforming unsustainable consumption and production; and protecting natural resources, by ensuring rights to local autonomy and rights to participation in natural resource management.[28]

Efforts at the African Level The need for transnational corporations and business enterprises to consider, anticipate and prevent human rights implications of extractive projects has also received extensive consideration and mention at the regional level in Africa. Since 2006, African policy leaders have increasingly begun to introduce rights-based governance models and language into resource extraction policies and regimes, perhaps as an attempt to fill this gap in the African Charter. In 2009, the the African Commission on Human and Peoples' Rights (African Commission) established a Working Group on Extractive Industries, Environment and Human Rights Violations in Africa, to examine and address allegations of human rights violations committed in the extractive industries sector in Africa, in particular in mining, oil, gas, and timber extraction.[29] The general mandate of the Working Group is to monitor and report on how extractive activities affect the human rights and environment of the African peoples and to inform the African Commission on the possible liability of extractive companies for human and peoples' rights violations in extractive operations and projects.[30] The Working Group has collated a number of guiding principles and documents that could enable African countries to hold extractive companies accountable for breaching human rights provisions in the African Charter on Human and Peoples' Rights, when extractive operations or projects result in negative environmental and human rights impacts for local communities.[31] The activities of the Working Group has given prominence and visibility to the obligations of extractive companies operating in Africa to anticipate, measure and address the impacts of their activities on human rights and the environment.[32]

Furthermore, in 2012, the African Commission adopted a *Resolution on a Human Rights-Based Approach to Natural Resources Governance* (*2012 Resolution*) at its

---

[26] See UNDG (2011), p. 9.

[27] See *Final Report of the United Nations Conference on Sustainable Development* (UN Doc A/CONF.216/16) (New York: United Nations, 2012) at paras 9, 31, 227, 228.

[28] Ibid. See also United Nations Conference on Sustainable Development (2011).

[29] African Commission on Human and Peoples' Rights (2009).

[30] Ibid.

[31] See Manirakiza (2012, 2013, 2015).

[32] Azu (2016).

51st Ordinary Session held from April 18th to May 2nd, 2012 in Banjul, Gambia.[33] The *2012 Resolution* is arguably the most significant regional attempt so far by African governments to recognize, adopt, and mainstream human rights language into the development and use of natural resources. Amongst other things, it calls on governments to ensure that respect for human rights prevail in all matters of natural resources exploration, extraction, toxic waste management, development, and management. The overall aim of the resolution is to ensure the systemic integration of human rights into extractive projects and policies in Africa. The *2012 Resolution* hopes to harmonize, strengthen, and build on previous efforts by African governments to introduce human rights concepts into resource governance.

There are three key prior important regional efforts in Africa that serve as the foundation for the *2012 Resolution*. They are: the 2009 Economic Community of West African States (ECOWAS) *Directive on the Harmonization of Guiding Principles and Policies in the Mining Sector* (*ECOWAS Directive*),[34] the declarations of the Pan African Parliament,[35] and the *2009 African Mining Vision.*[36] The *2012 Resolution* hopes to bring together these previous efforts to ensure coherence, coordination, and consistency.

First, the 2009 *ECOWAS Directive*, which is applicable to fifteen West African countries that are members of the ECOWAS,[37] emphasizes the importance of stakeholder engagements and consultation in resource development. The *ECOWAS Directive* sets out guiding principles for harmonizing mining regulatory regimes across member states and is binding on ECOWAS members. It called on member states to respect the free, prior, and informed consents (FPIC) of local communities that face potential negative impacts from mining, hydrocarbon development, or natural resource projects more broadly. The *ECOWAS Directive* urges countries to ensure free, prior, and informed consent when communities will be affected by mineral or hydrocarbon projects. Specifically, it calls on companies to obtain the FPIC of local communities before exploration begins and prior to each subsequent phase of mining and post-mining operations.[38] The Directive emphasizes the importance of applying FPIC throughout the project lifecycle—from pre-mining, mining operations, closing, and post-closure periods. Furthermore, the directive defines "mineral" to include not only industrial minerals but also petroleum, thereby making the directive applicable to both the solid mineral and oil and gas sectors. The *ECOWAS Directive* therefore arguably remains the most significant FPIC policy document in

[33] African Commission on Human and Peoples' Rights (2012).

[34] Economic Community of West African States, Directive C/DIR.3/05/09, *On the Harmonization of Guiding Principles and Policies in the Mining Sector*.

[35] Sixth Ordinary Session of the Pan-African Parliament, Addis Ababa, Ethiopia, "Recommendations and Resolutions" [Ref: PAP(2)/RECOMS/(VI)] (January 16–20, 2012).

[36] African Union (2009).

[37] Namely: Benin, Burkina Faso, Cape Verde, Côte d'Ivoire, the Gambia, Ghana, Guinea, Guinea-Bissau, Liberia, Mali, Niger, Nigeria, Senegal, Sierra Leone, and Togo. See Economic Community of West African States. See The ECOWAS Commission at a Glance.

[38] See ibid. See also Mayer Brown LLP (2010).

Africa. It aims to improve transparency and stakeholder consultation and engagement in mineral policy formulation, implementation, and decision-making processes. It also establishes guidelines for countries to mitigate, to the extent possible, the negative impacts of resource development on the environment and the local communities, in line with international best practices.

Some of these human rights issues have been highlighted by the Pan-African Parliament, which has increasingly promoted regional efforts aimed at addressing human rights concerns of resource utilization in Africa. The Pan-African Parliament is the legislative body of the African Union. It has advisory and consultative powers to examine, discuss, or express an opinion on any matter, either on its own initiative or at the request of the Assembly or other policy organs, and make any recommendations it may deem fit relating to, inter alia, matters pertaining to respect of human rights.[39] In its Sixth Ordinary Session, held in Addis Ababa, Ethiopia in January 2012, the Pan-African Parliament expressed concern over the rise of large-scale land acquisitions and the impact of domestic and foreign investment on land, water, and related natural resources.[40]

In order to promote greater participation in decision making processes for resource extraction projects, the Pan-African Parliament called on states to ensure "effective consultations with local communities and various people affected by investment projects and [ensure] that any investment is approved through free, prior, and informed consent of affected communities."[41] The recommendations also call for enhanced land certification and registration systems that take into account pastoralist, women's, and communal rights in advance of investment. The Pan-African Parliament has therefore identified that human rights issues in resource utilization in Africa are exacerbated by pre-existing imbalances in power relations between: men and women, local governments and the governed, elders and young, rich and poor, indigenes and non-indigenes, educated and non-educated.[42] A significant demerit, however, is that the Pan-African Parliament has yet to attain full legislative powers, and as such can only advise and speak on the alarming impacts of the problem of exclusion, without effecting any real legislative or policy change.

Attempts to attain executive and binding policy shift in mainstreaming human rights protection into resource governance gained further momentum with the release of the *2009 African Mining Vision* (the "Vision Report").[43] The Vision Report was put together by the United Nations Economic Commission for Africa (ECA), in conjunction with the African Union (AU), the African Mining Partnership

---

[39] African Union (2009).

[40] Sixth Ordinary Session of the Pan-African Parliament, Addis Ababa, Ethiopia, "Recommendations and Resolutions" [Ref: PAP(2)/RECOMS/(VI)] (January 16–20, 2012). See Greenspan (2014), p. 14.

[41] Sixth Ordinary Session of the Pan-African Parliament, supra note 101 at para 6(d).

[42] Ibid at preamble (noting that the Pan African parliament is "fully alarmed by the negative impacts on human rights especially on women, including unequal access to land and disruption of access to water").

[43] African Union (2009).

(the intergovernmental forum of African ministers responsible for mining), the African Development Bank (AfDB), the United Nations Conference on Trade and Development (UNCTAD) and the United Nations Industrial Development Organization (UNIDO).[44] The Vision Report highlights human rights challenges that mining projects face, especially inadequate stakeholder engagement and consultation in resource extraction, together with the social and environmental side effects of resource extraction. It catalogues African initiatives aimed at developing a "new social contract for mining" that would balance local benefits with national poverty-alleviation efforts.[45] The Vision Report encourages African countries to develop new legal instruments to facilitate local community participation; multi-stakeholder partnerships of government, private sector, and local communities; and to promote "public participation to secure consent for government and industry actions. Importantly, the framework describes public participation not just as consultation, information sharing, and dispute resolution, but also as "participatory decision making."[46]

In December 2011, the African Union Conference of Ministers responsible for mineral resources adopted an Action Plan to implement the *African Mining Vision*: the *Addis Ababa Declaration on Sustainable Africa Extractive Industry—From Vision to Action* (*Addis Ababa Declaration*).[47] The *Addis Ababa Declaration* calls on African states to strengthen transparency, accountability, and access to information, improve public participation and provide capacity building programmes for local communities, civil society, and the legislature in order to provide effective oversight that will create a well-governed mining sector that is inclusive and appreciated by stakeholders. The *Declaration* calls on all African Union member states to "create a mineral sector that is environmentally friendly and is socially responsible through strengthened environmental, health, and safety standards, as well as through the implementation of environmental and social funds."[48]

Summarily, the *2012 Resolution* serves as an umbrella resolution aimed at harmonizing the cardinal objectives of the 2009 *ECOWAS Directive*, declarations of the Pan-African Parliament, and the *2009 African Mining Vision*, which are: to increase stakeholder engagement and accommodation in the design, planning, and implementation of resource extraction activities and projects. The *2012 Resolution* also expands its scope beyond West Africa and furthers the discussions beyond FPIC issues. It explicitly identifies other human rights challenges that threaten the sustainability of resource utilization in African countries, especially the lack of effective judicial remedies, fair compensation, and respect for the rights of indigenous peoples as well as the rights of women in project planning and implementation.

[44] Ibid at 1.

[45] Ibid at 12, 39.

[46] Ibid.

[47] African Union Commission (2011).

[48] African Union Commission (2011), pp. 32–35.

Generally, the *2012 Resolution* calls on African countries to carry out four important action plans aimed at integrating human rights principles and standards into national legal frameworks on resource development. First, it calls on African countries to strengthen the above discussed regional efforts, such as the 2009 ECOWAS Directive on Mining and the African Commission's Working Group on Extractive Industries and Human Rights, to promote natural resources legislation that respect human rights of all and require transparent, maximum and effective community participation in (a) decision-making about, (b) prioritisation and scale of, and (c) benefits from any development on their land or other resources or that affects them in any substantial way.[49]

Second, it calls on African countries are to "establish a clear legal framework for sustainable development as it impacts on natural resources, in particular water, that would make the realization of human rights a prerequisite for sustainability."[50] Thirdly, it calls on African countries: "set up independent monitoring and accountability mechanisms that ensure that human rights are justiciable and extractive industries and investors legally accountable in the country hosting their activities and in the country of legal domicile." This encourages African countries to review, monitor, and enforce compliance with human rights standards and obligations in the design and execution of resource development projects.[51] The fourth element advocated by the *2012 Resolution* is for countries to ensure independent social and human rights impact assessments (HRIA) that guarantee free, prior, and informed consent; effective remedies; fair compensation; women, indigenous and customary people's rights; environmental impact assessments; impact on community existence including livelihoods, local governance structures and culture, and ensuring public participation; protection of the individuals in the informal sector; and economic, cultural and social rights.[52] This threshold requires countries to flag the likely impact of a resource development project on fundamental human rights and to demonstrate or describe the efforts put in place to mitigate or avoid these results. This would include assessing actual and potential human rights impacts, integrating and acting upon the findings, tracking responses, and communicating how impacts are addressed.

The increased recognition of the linkages between business and human rights at international, regional and national levels show that the expectation of business enterprises to protect, respect and fulfil human rights in extractive investments and projects is no longer an option but a necessity. Business enterprises and organisations investing in extractive projects can avoid the backlash associated with such projects by incorporating human rights due diligence into corporate risk management processes. There is a strong business case, in terms of cost, reputation and effectiveness for doing so. Failure to effectively manage human rights issues associated with extractive projects carries significant financial, legal and reputational risks. Such risks may manifest themselves in community protests due to health and

[49] African Commission on Human and Peoples' Rights (2012).

[50] Ibid.

[51] See Hammer et al. (2011), pp. 22–25.

[52] African Commission on Human and Peoples' Rights (2012).

safety concerns, disruption of project construction and potential harm to employees due to such protests. There may be legal risks related to the court and regulatory actions to challenge projects, and reputational risks related to loss of life or injuries from actions of security forces in responding to protest situations. Failure to effectively address human rights risks can lead to significant costs in terms of the management time required to respond to crises, costly suspensions or closures of projects by supervisory bodies, and may affect a company's ability to receive project funding from carbon finance institutions or export credit agencies. Project proponents will be evaluated and judged by stakeholders (including financial institutions and supervisory institutions) based on the strength of the internal procedures, practices and approaches put in place to anticipate and manage human rights risks associated with the company's investment and project. Furthermore, adopting sound internal human rights screening processes as part of corporate risk management frameworks can help business enterprises to build and maintain strong social licence to operate and to foster relationships with communities, customers and other stakeholders based on mutual respect, acceptance and trust.

As for the role of lawyers representing companies and investors in extractive projects and investments, it is important to emphasise the significance of human rights due diligence when providing counsel. The growing insistence of courts and tribunals on the rights to information, public participation and benefit-sharing for those persons potentially affected by proposed projects should lead lawyers to advise their clients of the wisdom of early compliance with human rights requirements in climate investments and projects. Nearly all cases of conflict over extractive projects arise from projects executed without proper consultations with the local communities. These communities are not generally opposed to extractive projects or investment, but they know and insist upon their rights, even if governments and supervisory bodies are unwilling or unable to protect them.

11.2.6 Corruption Risks

Corruption is one of the key impediments to sustainable resource governance worldwide. According to the OECD Foreign Bribery Report, the cost of corruption equals more than 5% of global gross domestic product (GDP) with over US$ 1 trillion paid in bribes each year.[53] The OECD also estimates that one in five cases of transnational bribery occur in the extractive sector.[54] The problem of corruption in the extractive sector has particularly taken a systemic and endemic nature in African countries.[55] For example, estimates indicate that Nigeria has had at least $400

[53] OECD (2014).

[54] Ibid.

[55] See AU/UN Economic Commission for Africa (2014), Africa Progress Panel (2013), also Chikwanha (2016), pp. 2–4.

billion of its oil revenue stolen or misspent since independence in 1960.[56] Monumental corruption has also been identified in the Democratic Republic of Congo (DRC). In addition to holding the world's largest reserves of cobalt, the DRC produces more than half of the world's cobalt.[57] Despite these however, DRC remains one of the poorest countries on earth.[58] This is primarily because only 6% of the revenue generated from cobalt mining in the DRC go to the national treasury.[59] Other income from mining have been corruptly diverted and stolen. There is therefore an increasing global attention on the need for business enterprises, transnational corporations and other non state actors in the extractive sector to anticipate, address and eradicate all forms of corrupt practices when investing in Africa.

Article 16 of the United Nations Convention Against Corruption (UNAC) 2008, defines corrupt practices as the promise, offering or giving of an undue advantage to a foreign public official or an official of a public international organization in order to obtain or retain business or some other advantage in relation to the conduct of international business. It also encourages States to consider adopting legislation that would make it illegal for a foreign public official to solicit or accept such payments or undue advantage.[60] Similarly, Article 4 of the African Union Convention on Preventing and Combating Corruption, 2003 defines corruption to include the solicitation or acceptance directly or indirectly of any goods or monetary value or benefits such as gift, favour, promise or advantage, in exchange for the performance of an act, omission or public functions. This includes the diversion of property by public officials, trading in influence, illicit enrichment, money laundering and concealment of property amongst others.[61] The Convention creates "mechanisms required to prevent, detect, punish and eradicate corruption and related offences in the public and private sectors."[62] These definitions mirror the broad categorization of corruption in international and regional instruments as any form of dishonest or illicit payment made to a public or private official to secure advantage or favour in relation to the conduct of business.[63]

In Africa, corruption risks manifest in the form of demand for bribes and facilitation payments at various points in the extractive value chain: ranging from the license bidding and award processes for mining, oil and gas rights, to the extraction,

[56] Annan (2012).

[57] Ibid.

[58] DRC is ranked 176 out of 187 on the United Nations Human Development Index. See United Nations Development Programme (2017).

[59] Kazeem (2017).

[60] United Nations Convention Against Corruption (UNAC) UN Doc. A/58/422; (Adopted 31 October 2003, In force on 14 December 2005).

[61] African Union Convention on Preventing and Combating Corruption, July 11, 2003, 43 I.L.M. 5.

[62] Ibid.

[63] See also Inter-American Convention Against Corruption (1996); OECD Convention on Combating Bribery of Foreign Public Officials in International Business Transactions (1997); United Nations Convention against Transnational Organized Crime and its Protocols (2000); Council of Europe, Agreement Establishing the Group of States against Corruption (1999a); Council of Europe, Civil Law Convention on Corruption (1999b), Council of Europe, Criminal Law Convention on Corruption (1999c), Eur. T.S. No. 173.

transportation, logistics, operations, regulation and revenue collection phases.[64] Other prevalent instances of corruption in Africa's extractive sector include: concealment of extractive income or production quantity; giving or accepting gratification through an agent; fraudulently acquiring or receiving property; making false statements or returns; tax evasion; money laundering; customs fraud; smuggling and transfer mispricing; abuse of office; trading in influence, favoritism and extortion; and making illegal payments through a series of offshore transactions.[65] ECs operating in Africa now face significant regulatory fines, penalties and loss of license, both in Africa and at their home jurisdictions, whether willingly, negligently or inadvertently implicated in the web of instigating, facilitating or participating in corrupt transactions or conduct in Africa.[66]

ECs can no longer underestimate the business sense and value of integrating robust risk identification and management frameworks in their extractive operations in Africa to avoid being implicated in corruption incidents. Corporate governance and business ethics standards are becoming increasingly stringent worldwide. Recently, there have been increased attentions on corrupt practices in extractive investments globally, especially in African countries. These new developments create particularly significant risks in African markets where corruption and other forms of illegality have been historically systemic and rampant. For example, in a number of cases, officials of multi-national oil companies in Canada and the United States have been charged and jailed for facilitating bribery and corruption while doing business in Africa. For example, Kellogg, Brown and Root, (KBR), a former subsidiary of Halliburton, was charged together with its parent company for allegedly paying bribes totaling $180 million to Nigerian top government officials between 2002 and 2003.[67] The bribes were allegedly paid to secure contracts to build the Bonny Liquefied Natural gas (LNG) plant in Nigeria's oil rich Niger Delta region. Halliburton agreed to plea bargain with Nigeria to pay a combined $579 million as settlement following US Department of Justice settlements on US criminal and civil charges. Similarly, Paris-based Technip S.A., an engineering and construction firm agreed to pay a $240 million criminal penalty and enter into a deferred prosecution agreement for its alleged role in the Bonny LNG case. In another foreign corruption case, BAE Systems in 2010 agreed to pay up to $450m (£288m) in fines after it pleaded guilty to false statements and accounting practices, notably over deals with Saudi Arabia and Tanzania. BAE Systems was ordered to pay back £29.5 million to the people of Tanzania, as well as a £500,000 fine for bribing Tanzanian officials.[68]

---

[64] OECD (2016).

[65] Africa Progress Panel (2013) highlighting how "tax evasion, illicit transfers of wealth and unfair pricing practices" in the extractive sector deprive Africa of funds needed for development. According to the Report, Africa loses twice as much in illicit financial out flows as it receives in international aid. It calls on African governments "to set out a bold national agenda for strengthening transparency and accountability" in oil, gas and mining sectors. See also Annan (2012) and Chikwanha (2016).

[66] Olawuyi (2015); Ndikumana (2013), pp. 1–5.

[67] See CNN (2009); also Oil and Gas Nigeria.

[68] Whitehead (2011) and Olawuyi (2015).

These examples illustrate the quantum of liability, fines or loss a company could attract if it fails to carefully adopt stringent internal control measures and code of conduct for its officers to avoid falling into the net of corruption while investing in foreign extractive assets. These cases also highlight that it is not enough to simply ensure that a company's own personnel in emerging markets comply with the law. To avoid legal risks arising from corruption, companies must also evaluate the integrity and business practices of its local foreign partner and ensure that such partner complies with the laws applicable to the company, with its code of conduct and other internal policies.

The first step is to identify what practices could come under the umbrella of 'corrupt practices'. Without this knowledge, it is very easy for an EC to offer or accept what may be termed as "hospitality or entertainment gift" or "facilitative payment" which may be well fall under corrupt payments under international and local foreign corruption laws. Many corruption cases stem from: illegal payments to government officials; violations of legally mandated accounting and internal controls; use of consultants to negotiate and reduce tax assessments; offshore payments to officials made outside the local books and records; hospitality and gifts; failure to self-report crimes by a company; and when facilitation payments fall under bribery. All of these conducts fall within the broad categorizations of corruption that have been outlawed by domestic legislation in most African countries.[69]

[69] For example, despite the prevalence of corruption in Nigeria's extractive industry, it has one of the most comprehensive network of laws that criminalize all forms of corruption. See Nigeria's *Economic and Financial Crimes Commission Establishment Act 2004* which prohibits all economic and financial crimes in Nigeria; The *Independent Corrupt Practices and Other Related Offences Act 2000* which criminalises bribery, as well as attempted corruption, fraud, extortion and money laundering; *The Advance Fee Fraud and Other Fraud Offences Related Act 2006* which outlaws internet scam and currency fraud. It provides that it is an offence for anyone—by false pretence and with the intent to defraud—to obtain property from any other person for him or herself or for any other person; induce another person to deliver property to a third party; or invite or induce another person to visit Nigeria for any purpose connected with the commission of an offence. See also *The Money Laundering Prohibition Act 2012* which prohibits the laundering of the proceeds of a crime or any criminal or illegal activity, and provides for appropriate penalties for money laundering infringements; *The Code of Conduct for Public Officers, Fifth Schedule to the 1999 Constitution of the Federal Republic of Nigeria* 1999, which prohibits pubic officials from accepting bribes and undue gratification; *The Fiscal Responsibility Act 2010* which establishes the Fiscal Responsibility Commission, which has the power to compel any person or government institution to disclose information relating to public revenues and expenditure; and to recommend criminal fines and penalties for any corrupt transactions; *The Miscellaneous Offences Act of 1983* which criminalizes acts of forging and uttering negotiable instruments. Section 8 of the Act also prohibits any illegal importation, export or purchase of minerals without the permission of the appropriate national authority. Section 17 prohibits the illegal importation, export, sale or distribution of crude oil, petroleum or petroleum product in Nigeria, while s.18 criminalizes petroleum adulteration; *The Penal Code* (LFN 2004) which identifies bribery, corruption and advance fee fraud as punishable criminal offences in the northern part of Nigeria; and *The Criminal Code* (LFN 2004) which sets out offences and penalties relating to bribery, corruption and advance fee fraud in the southern part of Nigeria. More importantly, the *Nigerian Extractive Industries Transparency Initiative Act* contains provisions that mandate transparency and accountability in the reporting and financial disclosure of extractive companies.

Apart from domestic laws in Africa that criminalize corruption, the Canadian Corruption of Foreign Public Officials Act ("CFPOA"), the United States Foreign Corrupt Practices Act 1977 ("FCPA"), and the United Kingdom Bribery Act 2010 specifically criminalize and prohibit the bribery of foreign officials. Under the FCPA, it is illegal to give anything to a foreign official for the purpose of obtaining or retaining business. The Act recognizes an exception for "facilitating payments" meant to expedite or secure routine governmental action. A defense is also provided for payments that are legal under the relevant foreign law, or that would qualify as "reasonable and bona fide" payments related directly to "the promotion, demonstration, or explanation of products or services," or "the execution of performance of a contract with a foreign government or agency thereof". Corrupt payments are not the only basis for liability and legal risks. Under a recent decision by the U.S. Supreme Court, the violation of a foreign law—typically criminal tax evasion, money laundering, bribery or other financial crimes—can provide the basis for a prosecution under the FCPA which criminalizes fraud and money laundering. Furthermore, the FCPA also contains provisions requiring companies traded on an American stock exchange to maintain appropriate accounting standards.

Unlike the United States law, the Canadian CFPOA prohibits "facilitation or grease payments" made to speed up or facilitate routine transactions such as permits and licenses. In 2017, the Government of Canada announced that Canadian firms operating abroad will now be liable for corruption for paying "facilitation or grease payments" as they are now categorized as corrupt practices under the CFPOA.[70]

The CFPOA however provides a defense for payments that are permitted or required under the laws of a foreign state, or are incurred in good faith and related to the promotion, demonstration or explanation of a person's products and services or to the execution of a contract.

Evidently, facilitation or grease payments could easily be caught in the web of corruption in countries with more stringent corruption laws that outlaw such payments. This is why it is important to seek adequate legal advice before making such payments. To avoid legal risks, it is pertinent for companies to ascertain whether there are laws, regulations, or government codes of conduct that specifically restrict or prohibit the giving of gifts, facilitation payments or hospitality to government officials and/or public servants?; if so, are there specific currency limits either stated in such provisions or contained in associated guidance from regulators?; what are the possible penalties, if any, for those who violate such provisions or currency limits?. Furthermore, is there any obligation to self-report crimes by a company or by others when they become evident; what are the sanctions for failing to do so; and what is the actual practice?

It is also important for companies to implement adequate internal controls and compliance programs that can help reduce the risk of corruption. Whether acquiring an existing entity, entering a joint venture, or building an operation from scratch, due diligence may require more than simply evaluating management and controls

[70] Government of Canada (2017).

within the operation itself. Furthermore, it may be necessary to carefully evaluate and monitor relationships with vendors such as consultants and brokers, who may deal with governments on the company's behalf. Once companies have begun to do business in an African country, it is important to make sure that adequate internal controls are in place and are effective. This may require more than proper internal polices and control procedures. It may be necessary for management to set the tone and develop appropriate employee training programs. Once an investment has been made, ongoing monitoring and reviews of these programs may help to further reduce risk.

11.2.7 Lopsided or Unfair Contractual Provisions

When negotiating a joint venture or petroleum sharing contract with national authorities, it is important to watch out for unfair or lopsided contractual provisions which could: vest absolute rights on the host government to unilaterally change the terms of the contract; establish inequitable fiscal terms; or allow for insufficient participation of the State in the project. For example, there have been terms included in contracts which: impose obligations on licensee to supply free gas to the government during the term of the contract; (it is a standard term for petroleum sharing contracts (PSCs) to require an oil company to supply gas to satisfy domestic market requirements, but at the 'market price'); or establish a duty on the licensee to market all gas or oil produced including the government's share (typically, the licensee would only be entitled to market its own share of production).

It is important to ensure that the oil and gas contract is not lopsided or skewed in favour of the national oil company or host government. Appropriate attention is required to ensure that contracts are fairly negotiated to include inviolability mechanism which would neutralize the host State's powers to act arbitrarily; protect an investment from the application of new laws and regulations; establish fair and equitable contractual terms which will guarantee returns on investment; and ensure stability and security of tenure.

11.3 Mitigating Legal Risks Through Extensive Due Diligence

As with doing business in any region, mitigating legal risks involved in investments in Africa requires extensive legal due diligence. Through a meticulous and comprehensive due diligence review, an investor could ascertain the local laws, rights, obligations, roles of the different parties involved, and the nature of contractual risks and liabilities attached to a proposed investment. Such a review would also provide a background on the legal and political regime, investment climate and the general corruption profile of the host country. This information if compiled and reviewed by experienced legal experts, would help a company to take accurate investment decisions before going into a country. This section discusses five key due diligence steps that could enable an investor to take informed decisions about investing in an African market.

11.3.1 Survey and Review of Relevant Laws and Regulation

Effective due diligence starts with working with international law firms that have expertise in African law to undertake a detailed review of the legal requirements for doing business in the target country. Such a review would unpack the general legislative requirements for doing business in the country as well as requirements that are specific to hydrocarbon investments. By reviewing applicable legislation, regulations, institutions, approval structures, scope of exploration rights, pitfalls and risks and the net potentials of the investment activity, legal advisers are better placed to determine and to recommend whether to go ahead with the investment; reconsider the investment decision; to restructure the investment plans in order to mitigate or avoid the risks; or to go ahead with the investment albeit at a much-reduced purchase price.

The scope of such legal review must be as comprehensive as possible. A comprehensive review process will provide an investor with the appropriate information on whether to reconsider the investment decision, restructure the investment in order to mitigate legal risks or push to readjust the purchase price during negotiations. The review should also recommend the necessary legal compliance steps that must be taken in order for the business to proceed hitch-free. Is local incorporation necessary; does the country have a prescribed model form of contract; does the country require contracts to be written in local language; what are the regulations and the specific regulators for hydrocarbon activities; are there specific permits or licenses that are required for exploration activities to commence; is local participation of some sort required; what are the domestic gas/oil supply obligations; are there import and export restrictions or taxes applicable with respect to production materials; what are the procedures for fund repatriation; what rights would an investor acquire with respect to oil production; what are the requirements on abandonment and decommissioning; can government acquire interests in investments; and what level of protection is accorded to investments amongst other questions. Determining the answers to these questions with precision as early as possible and before contract execution are key to recording exploration success in any African country.

11.3.2 Careful Negotiation

An extractive investment requires stability in order for it to yield appropriate returns in the long term. Investors entering into extractive arrangements must protect themselves through careful contractual arrangements that prevent a change in the rules of the game, midway. This aside, the negotiation stage is equally a great opportunity to push for the inclusion of stability clauses and adaptation provisions that could protect the investor against the political and legal vicissitudes that are common place in many African countries, especially frequent changes in government or change in fiscal and policy regimes after the investor has invested capital in the project. These risks can also take the forms of governmental disruption in equity participation (through dilution or expropriation), forced renegotiation of contracts, forced contracting and procurement procedures, avoidance of agreed commitments, revision of regulation, or any political/cultural change which may impact priorities and disrupt your business plans.[71]

Legal arrangements to promote stability are underpinned by the regulatory taking doctrine.[72] The regulatory taking doctrine advocates for contractual safeguards to protect private investors from governments that unjustifiably expropriate property. Under the regulatory takings doctrine, a change of rules or regulation that undermines an investment's commercial viability may be deemed as a taking of property, and would require the host state to pay compensation.[73] While this theory has been mainstreamed into regional investment treaties and domestic property legislation in countries such as the United States and Canada, many African countries lack robust "regulatory takings" legislation that prevent government from arbitrary expropriation of private property without compensation.[74] Even when such provisions are available, they are often subjected to governmental discretions that create avenues for expropriation without due process or adequate compensation. To achieve stability, it is therefore important to negotiate contractual terms that prevent the state from changing the regulatory framework in a way that affects the economic

---

[71] See Irwin (1998).

[72] The U.S. Supreme Court first recognized regulatory takings in the 1922 case of *Pennsylvania Coal Co. v. Mahon*, 260 U.S. 393 (1922). Generally, Takings Clause was described in *Armstrong v. United States, 364 U.S. 40, 49 (*1960*)* as being "designed to bar Government from forcing some people alone to bear public burdens which, in all fairness and justice, should be borne by the public as a whole."

[73] Treanor (1995).

[74] For example, Article 1110 of the North American Free Trade Agreement (NAFTA) regarding expropriation and compensation, states that NAFTA parties may not "*directly or indirectly nationalize or expropriate* an investment of an investor of another Party in its territory or take a measure *tantamount to nationalization or expropriation* of such an investment. NAFTA makes exceptions in article 1105(1), where a government expropriates property for a public purpose, on a non-discriminatory basis, in accordance with due process of law, and on payment of compensation. See North American Free Trade Agreement, U.S.,-Can.-Mex., art. 1105, ¶ 1, Dec. 17, 1992, 32 I.L.M. 289 (1993).

equilibrium of the investment, and to compensate the investor if it does so.[75] It should also protect against unfair or lopsided arrangements that tend to tilt in favour of the host country. General economic balance in contractual relations can be achieved through stability provisions; intangibility or immutability clauses; and sound dispute resolution and arbitration clauses. Each of these drafting mechanisms are considered in what follows.

Stabilization Clause A stabilization clause is a contractual drafting technique used to mitigate legal and political risks that are typically common place in emerging markets; it mainly addresses the issue of changes in law in the host state during the life of the project.[76] A contract stabilization clause preserves the sanctity of the contract against the sovereign right to change the law. Generally, stability clauses can be classified under two categories, namely: freezing or equilibrium clauses—which preserves the contract and its legal and fiscal terms for a period of time. Under a freezing clause, the government agrees not to change the legislation applicable to the contract.[77] A freezing clause therefore "freezes" the law "as in force" as of the date of the conclusion of the contract, or stipulate that laws which may in the future impose higher rates or greater liability shall not apply to the investor.[78]

An economic equilibrium clause on the other hand allows new laws, but requires the economic terms of the agreement to be renegotiated to preserve the investors' original economic position despite any changes in law. In this case, the new law or regulation still applies, however, its impact on the investor may be mitigated for example through delayed implementation of the new law, indemnifying the foreign investor for all or a part of these associated loss or lowering royalty and tax rates to offset the costs.[79]

Although stabilization clauses insulate the contractual relationship from any material adverse change in law, they are still no full-proof guarantee against the state's exercise of sovereign authority. They can, however, provide a very strong ground for an aggrieved investor to seek and obtain a high amount of compensation for contractual breach as opposed to cases where such a clause is absent.

---

[75] Berger (2003); also, Maniruzzaman (2005), pp. 96–100.

[76] Comeaux and Kinsella (1994).

[77] A good example is a contract that provides that:

> In the event of any new provisions or changes or amendments to the law of Namibia which imposes on the Contractor new or higher taxes, duties, rights, payments or any other obligations with respect to those expressly applicable to the Licensee under this Contract, Parties agree to amend the Contract to adjust the profit sharing percentages accordingly, in order to maintain the economic equilibrium of this Contract.

See Practice Note, Understanding Stabilization Clauses in International Investment Agreements: Criticisms of Stabilization Clauses.

[78] See Alexander (2009), p. 244.

[79] Practice Note, Understanding Stabilization Clauses in International Investment Agreements: Examples of Economic Equilibrium Clauses.

Intangibility or Immutability Clause The second is the intangibility or immutability clause—in which the state agrees not to amend or terminate the contract without the mutual consent of the contracting parties. This provision normally insulates the contractual relationship from any governmental legislative or administrative measures that may have the effect of unilaterally amending or altering the contract and its fiscal regimes.[80] It essentially aims to freeze the economic, legal, financial and fiscal regimes in the petroleum contract at the time when the contract was concluded in order to neutralise the host government's powers to unilaterally change the terms of the contract midway.

In a nutshell, intangibility clauses provide that a host government cannot unilaterally nationalise a project or modify an investment contract, any changes require the consent of the investor.

Intangibility clauses provide the investor the confidence that the investment will not be subjected to unilateral contractual terms either due to a change of government which often happen at alarming rates in Africa; or through resource nationalism, policy somersault and reversal, that are common place in many African countries. The contract should also include practical tools through which such changes could be adequately managed when they are inevitable: for example, through renegotiation, compensation or arbitration.

11.3.3 Neutral Dispute Resolution Mechanisms

A third mitigation strategy is to include contractual terms that avoid the application of the legal system or dispute resolution mechanism of the host state. Clauses that specify that disputes arising from the contract will be resolved under the laws of a neutral country, most importantly a jurisdiction with well developed and tested laws on commercial transactions and contract interpretation help to neutralise any likelihood by the host state to influence legal and judicial processes in its favour when disputes arise. Neutral dispute resolution mechanisms provide parties with the best opportunity to resolve disputes before a disinterested and competent forum that is free from governmental influences.

Typically, except a contract expressly stipulates its preferred dispute resolution mechanism, disputes arising from the contract would be resolved through judicial apparatus of the host state.[81] This may be detrimental to the interest of the foreign

---

[80] For example, a Nigerian joint venture contract with an equilibrium stability clause provides that:

> The Parties agree that the commercial terms and conditions of this contract are based on the existing fiscal terms in accordance with the provisions of the Deep Offshore and Inland Basin Production Sharing Contract Act, 1999, if such fiscal terms are changed, the Parties agree, subject to Clause 27.3, to review the terms and conditions of this Contract affected by such changes to align such terms and conditions with the fiscal terms.

[81] See Article 3.1 of the *Rome I Regulation (EC) No 593/2008* which states that a contract shall be governed by the law chosen by parties. The choice shall be made expressly or clearly demonstrated

investor who is not well vast in the internal workings of the legal system in the host country, they are also issues with governmental influence on state authorities. As such, to escape the application of the dispute resolution system of the host country, it is important to provide arbitration clauses which will state that disputes would be resolved in a neutral venue and possibly in accordance with the UNCITRAL Arbitration Rules, adopted by the United Nations Commission on International Trade Law (UNCITRAL) in 1976. The UNCITRAL arbitration rules provide comprehensive sets of procedural rules upon which parties may agree for the conduct of arbitral proceedings arising out of their commercial relationship. The rules provide model arbitration clauses which cover all aspects of the arbitral process; they also set out the procedural rules regarding the appointment of arbitrators and the conduct of arbitral proceedings and establishing rules in relation to the form, effect and interpretation of the award.[82] Similarly, the London Court of International Arbitration (LCIA) Rules have also been frequently adopted in modern oil and gas contracts. The International Chamber of Commerce (ICC) Rules of Arbitration also provide arbitral processes that party may explore when negotiating petroleum agreements.[83] Generally, arbitration provisions provide neutral, competent, and transparent options for insulating commercial disputes from the influence and whims of the host state and from difficult judgment enforcement problems that may arise under local laws.[84]

11.3.4 International Insurance Coverage

An equally important mitigation strategy is to obtain international insurance coverage before making huge investments in potentially fragile and politically risky economies, such as African markets.[85] A substantial percentage of African countries

---

by the terms of the contract or circumstances of the case. By their choice, the parties can select the law applicable to the whole or to part only of the contract.

[82] UNCITRAL Arbitral Rules.

[83] See Article 21 of the ICC Rules: "The parties shall be free to agree upon the rules of law to be applied by the arbitral tribunal to the merits of the dispute. In the absence of any such agreement, the arbitral tribunal shall apply the rules of law which it determines to be appropriate".

[84] For example:

> any dispute arising out of or in connection with the agreement that is not amicably settled by the parties shall be finally referred to and resolved by arbitration in accordance with the London Court of International Arbitration (LCIA) rules, which rules are deemed to be incorporated by reference into the Agreement. The seat or legal place of arbitration shall be London, England. The Arbitral Tribunal shall consist of three (3) arbitrators. Each party shall be appointed by the LCIA Court. The Chairman of the Tribunal shall have at least ten (10) years' experience of shipping and energy trading matters.

See Energy Institute (2016).

[85] See Torrado (2005), p. 301.

have had one major political upheaval or conflict within the last decade. Ranging from the *Boko Haram* conundrum in Nigeria, political upheavals in Libya, Egypt, Algeria, Ivory Coast, Tunisia, Sudan, South Sudan, Zimbabwe and Madagascar; and wars and tensions in Democratic Republic of Congo, Liberia, Rwanda, Ethiopia and Somalia. These political challenges and instability weaken the investment climate in African markets and necessitate efforts by any investor to protect its investments against legal risks that may arise from political instability.

Extractive investors in Africa would therefore benefit from securing comprehensive international risk insurance for their investments. For example, the Multilateral Investments Guarantee Agency (MIGA) of the World Bank offers political risk insurance guarantees which help investors protect foreign direct investments against political and non-commercial risks in developing countries. Even though coverage is limited to 210 Million USD, it offers some form of protection against currency inconvertibility and transfer restriction; government expropriation; war, terrorism, and civil disturbance; breaches of contract; and dishonoring sovereign financial obligations. A risk insurance is often tailored to the prevailing risks in a country so as to transfer such political risks from the insured to the insurer. The World Bank also offers Partial Risk Guarantees (PRG) that covers lenders against the risk of Government non-performance of its contractual obligations to a specific project, for example the risk of non-payment by the host government, change in law/regulatory risk, expropriation risk, etc.

These international insurance programs could provide relative peace of mind to investors, knowing that they could be able to recover some of their investments back, in event of political upheavals.[86]

11.4 Conclusion

African countries undoubtedly present investors with unique opportunities to spearhead new discoveries and consolidate exploration and production in previously explored fields, especially marginal fields. Despite these potentials, extractive investments in Africa come with distinctive legal risks and challenges that must be carefully mitigated. A reoccurring dilemma for investors therefore is how to maximize investment opportunities in Africa while mitigating legal risks. This chapter has examined some such distinctive legal risks and has provided insights on how to they could be mitigated.

Summarily, as the saying goes, prevention is often better than cure. In developing investment plans for African markets, extensive due diligence, which includes a careful combination of surveying and understanding applicable legal regimes, negotiating protective contractual provisions, informed deal structuring, active management of political risk, and adequate insurance protection for political risks are highly

[86] What Lenders Need to Know About Political Risk Insurance: Conditions for Compensation; also see Issues to Consider When Submitting a Claim Under a Political Risk Insurance Policy.

important. By working with legal advisers that have experience in Africa, investors may be better prepared to avoid the pitfalls and challenges that may frustrate a worthwhile investment activity. Counsels should advise investors on the legal framework in the target country, and the requisite steps to negotiate and incorporate balanced and protectionist provisions such as stabilization, immutability and alternative dispute resolution clauses so as to minimize investment legal risks and/or its impacts, should it arise. Proper risk planning also includes obtaining risk insurance so as to ensure some form of cost recovery in the event of political upheavals.

Apart from risk planning, investors should also embrace opportunities to demonstrate the value of their investments to the host country by undertaking increased community development initiatives that provide direct social benefits, such as local employment, new infrastructure, schools, and improved health care delivery to the host country. While such initiatives may not directly lower all risks, they could increase public perceptions on the value that an investor brings to the local communities. They may also help an EC to retain its social license to operate in a country.

Furthermore, the tasks of reducing legal risks should not be construed as that of investors alone. Host governments must equally understand the importance of creating the enabling legal regime and conditions for extractive investments to thrive. The technical and financial investment flows from foreign extractive investments are undoubtedly significant sources of revenue and technology transfer needed to develop resource deposits across Africa. Africa must therefore remain competitive in terms of creating enabling investment framework, to ensure sustained flow of foreign extractive investment to Africa. With increased oil and gas discoveries in Mexico, Venezuela and Indonesia, investors will likely search for attractive investment locations. African policy leaders should therefore understand the need to evolve a more robust and balanced relationship, underpinned by mutual respect with investors, by reducing the identified legal risks and hurdles that stifle extractive investment opportunities in Africa.

References

Africa Progress Panel (2013) Equity in extractives: Stewarding Africa's natural resources for all. Africa Progress Panel, Geneva, Switzerland

African Commission on Human and Peoples' Rights (2009) Resolution 148(XLVI)09 *on the Establishment of a Working Group on Extractive Industries, Environment and Human Rights Violations in Africa*, adopted during the 46th Ordinary Session of the Commission, held in Banjul, The Gambia, from 11 to 25 November 2009. http://old.achpr.org/english/resolutions/resolution148_en.htm. Accessed 2 May 2018

African Commission on Human and Peoples' Rights (2012) Resolution on a Human Rights-Based Approach to Natural Resources Governance, African Commission on Human and Peoples' Rights, 51st Sess (2012)

African Union (2009) Africa Mining Vision. africaminingvision.org/amv_resources/AMV/Africa_Mining_Vision_English.pdf. Accessed 12 June 2018

African Union (2011) African Development Bank & United Nations Economic Commission for Africa, Building a Sustainable Future for Africa's Extractive Industry: From Vision to Action. Action Plan for Implementing the African Mining Vision

African Union/UN Economic Commission for Africa (2014) Conference of Ministers of Finance, Planning and Economic Development, Illicit financial flow report of the High-Level Panel on Illicit Financial Flows from Africa. www.uneca.org/publications/illicit-financial-flows. Accessed 2 May 2018

Alexander F (2009) Comment on articles on stabilization by Piero Bernardini, Lorenzo Cotula and AFM Maniruzzaman. J World Energy Law Bus 2:243–258

Annan K (2012) Momentum rises to lift Africa's resource curse. https://www.nytimes.com/2012/09/14/opinion/kofi-annan-momentum-rises-to-lift-africas-resource-curse.html4 September 2012. Accessed 2 May 2018

Armstrong v. United States, 364 U.S. 40, 49 (1960)

Azu M (2016) A review of the work of the African Commission's Working Group on Extractive Industries, Environment and Human Rights Violations in Africa. https://africlaw.com/2016/04/26/a-review-of-the-work-of-the-african-commissions-working-group-on-extractive-industries-environment-and-human-rights-violations-in-africa/. Accessed 2 May 2018

Berger K (2003) Renegotiation and adaptation of international investment contracts: the role of contract drafters and arbitrators. Vanderbilt J Transnatl Law 36:1347–1364

Chikwanha A (2016) Combating corruption in the extractive industry in Africa. https://issafrica.s3.amazonaws.com/site/uploads/iss-sida-1.pdf. Accessed 2 May 2018

Clerk A, Kohler P (2005) Property law, commentary and materials. Cambridge University Press, Cambridge

CNN (2009) KBR charged with bribing Nigerian officials for contracts. CNN. http://www.cnn.com/2009/US/02/06/KBR.bribery/index.html. Accessed 12 June 2018

Comeaux P, Kinsella S (1994) Reducing political risk in developing countries: bilateral investment treaties, stabilization clauses, and MIGA & OPIC investment insurance. N Y Law School J Int Comp Law 15

Connor R, Heffron RJ, Khan AA, Perkins E (2018) Legal strategies for the mitigation of risk for energy infrastructure projects. J Energy Nat Resour Law

Council of Europe (1999a) Agreement Establishing the Group of States against Corruption (GRECO), May 1, 1999. https://www.justice.gov/sites/default/files/criminal-fraud/legacy/2012/11/14/grecoagree.pdf. Accessed 12 June 2018

Council of Europe (1999b) Civil Law Convention on Corruption, Nov. 4, 1999, Eur. T.S. No. 174

Council of Europe (1999c) Criminal Law Convention on Corruption, January, 27, 1999, Eur. T.S. No. 173

Criminal Code Act, Cap C38, Laws of the Federation of Nigeria, 2004

Economic Community of West African States. The ECOWAS Commission at a Glance. afcac.org/en/index.php?option=com_content&view=article&id=57:economic-community-of-west-african-states-ecowas&catid=38:official-recs&Itemid=149. Accessed 12 June 2018

Economic Community of West African States, Directive C/DIR.3/05/09. On the Harmonization of Guiding Principles and Policies in the Mining Sector. http://documentation.ecowas.int/download/en/publications/Ecowas%20Directive%20and%20policies%20in%20the%20minning%20sector.pdf. Accessed 12 June 2018

Energy Institute (2016) Model general terms and conditions for Free on Board (FOB) sale and purchase of crude oil. https://publishing.energyinst.org/topics/hydrocarbon-management/hm-80-model-general-terms-and-conditions-for-free-on-board-fob-sale-and-purchase-of-crude-oil. Accessed 12 June 2018

Government of Canada (2017) Canada repeals facilitation payments exception in Corruption of Foreign Public Officials Act. https://www.canada.ca/en/globalaffairs/news/2017/10/canada_repeals_facilitationpaymentsexceptionincorruptionofforeig.html. Accessed 12 Aug 2018

Government of South Sudan, The 2009 Land Act of South Sudan

Graulau J (2011) Ownership of mines and taxation in Castilian laws, from the middle ages to the early modern period: the decisive influence of the sovereign in the history of mining. Contin Change 26
Greenspan E (2014) Free, prior, and informed consent in Africa: an emerging standard for extractive industry projects. Oxfam America. https://www.oxfamamerica.org/static/media/files/community-consent-in-africa-jan-2014-oxfam-americaAA.PDF. Accessed 12 June 2018
Hammer M et al (2011) Pathways to accountability II, the revised global accountability framework. One World Trust, London. https://acfid.asn.au/sites/site.acfid/files/resource_document/Pathways-to-Accountability-II.pdf. Accessed 12 June 2018
Human Rights Watch (2018) World report 2018: our annual review of human rights around the globe. https://www.hrw.org/world-report/2018. Accessed 12 Aug 2018
Irwin W (1998) Political risk: a realistic view toward assessment, quantification, and mitigation. In: Moran T (ed) Managing international political risk, pp 57–69
Issues to Consider When Submitting a Claim Under a Political Risk Insurance Policy. Practical Law. http://us.practicallaw.com/0-505-3583. Accessed 12 June 2018
Kazeem Y (2017) Two investigations into corruption in resource-rich DR Congo show why it stays so poor. Quartz Africa. https://qz.com/1038143/two-investigations-into-corruption-in-resource-rich-dr-congo-show-why-it-stays-so-poor/. Accessed 12 Apr 2018
MacKay F (2004) Indigenous peoples' right to free, prior and informed consent and the World Bank's extractive industries review. Sustain Dev Law Policy 4(2)
Manirakiza P (2012) Report of the Working Group on Extractive Industries, Environment and Human Rights Violations in Africa presented on the occasion of the 25th anniversary of the African Commission on Human and Peoples' Rights, commemorated as part of the 52nd Ordinary Session of the Commission, para 13, available at http://www.achpr.org/sessions/52nd/intersession-activity-reports/extractive-industries/. Accessed 28 Oct 2015
Manirakiza P (2013) Inter-session report submitted to the 54th ordinary session of the African Commission on Human and Peoples' Rights, paras 20–24, available at http://www.achpr.org/sessions/54th/intersession-activity-reports/pacifique-manirakiza/
Manirakiza P (2015) Inter-session report presented at the 56th ordinary session of the African Commission on Human and Peoples' Rights, para 13, available at http://www.achpr.org/files/sessions/56th/inter-act-reps/227/56os_intersession_report_comm_manirakiza_en.pdf. Accessed 19 Oct 2015
Maniruzzaman A (2005) Some reflections on stabilisation techniques in international petroleum, gas and mineral agreements. Int Energy Law Tax Rev 4:96–100
Martin AT, Park JJ (2010) Global petroleum industry model contracts revisited: higher, faster, stronger. J World Energy Law Bus 3
Mayer Brown (2010) Developments: recent legal developments in the mining sector of West Africa States. mayerbrown.com/files/Publication/a10390b1-79cb-4dee-b6bb-16bd5af12d96/Presentation/PublicationAttachment/653da675-bc14-439a-bc79-87abeb85ccbb/NEWSL_MINING_JAN10_BULLETIN_WEST_AFRICA.PDF. Accessed 12 June 2018
McCormick R (2010) Legal risk in the financial markets. Oxford University Press, Oxford
Mining Working Group at the United Nations (2014) A rights-based approach to resource extraction in the pursuit of sustainable development. Advocacy Brief
Miscellaneous Offences Act of 1983, Cap. M17 Laws of the Federation of Nigeria, 2004
Money Laundering Prohibition Act 2012, Laws of the Federation of Nigeria, 2004
Ndikumana L (2013) The private sector as culprit and victim of corruption in Africa. Massachusetts Amherst Working Paper Series No. 330
Nigeria (2004) Economic and Financial Crimes Commission Establishment Act 2004 Cap E1 Laws of the Federation of Nigeria, 2004
Nigeria (2006) Advance Fee Fraud and Other Fraud Offences Related Act 2006, Cap. A6, Laws of the Federation of Nigeria, 2004
Nigerian Extractive Industries Transparency Initiative Act, Cap N159 Laws of the Federation of Nigeria, 2004

Nigerian Local Content Development Act, Nigeria

North American Free Trade Agreement 1993

OECD (1997) OECD Convention on Combating Bribery of Foreign Public Officials in International Business Transactions, Dec. 17, 1997, S. Treaty Doc. No. 105-43

OECD (2016) Corruption in the extractive value chain: typology of risks, mitigation measures and incentives. OECD. http://www.oecd.org/dev/Corruption-in-the-extractive-value-chain.pdf. Accessed 2 May 2018

OECD Foreign Bribery Report (2014) An Analysis of the Crime of Bribery of Foreign Public Officials. https://doi.org/10.1787/9789264226616-en. Accessed 21 Apr 2018

Office of the High Commissioner for Human Rights (2008) *Claiming the Millennium Development Goals: A Human Rights Approach*, UN Human Rights, UN Doc HR/PUB/08/3

Olawuyi D (2015) Legal strategies and tools for mitigating legal risks associated with oil and gas investments in Africa. Opec Energy Rev 39(3):247–265

Omorogbe Y, Oniemola P (2009) Property rights in oil and gas under dominial regimes. In: McGarg A, Barton B, Badbrook A, Gooden L (eds) Property and the law in energy and natural resources. Oxford University Press, Oxford

Organization of American States (1996) Inter-American Convention Against Corruption, March 29, 1996, S. Treaty Doc. No. 105-39, 35 I.L.M 724

Pan-African Parliament (2011) Sixth Ordinary Session of the Pan-African Parliament, Addis Ababa, Ethiopia. 2011. Decision on the Implementation of the Previous Decisions of the Executive Council and the Assembly Doc. Ex.Cl/623(Xviii). https://au.int/sites/default/files/decisions/9645-assembly_en_30_31_january_2011_auc_assembly_african_union_sixteenth_ordinary_session.pdf. Accessed 12 June 2018

Pan-African Parliament (2018) https://au.int/en/organs/pap. 12 June 2018

Penal Code (Northern States) Federal Provisions Act (No. 25 of 1960). Laws of the Federation of Nigeria, 2004

Pennsylvania Coal Co. v. Mahon, 260 U.S. 393 (1922)

Petersmann E (2002) Time for a United Nations 'Global Compact' for integrating human rights into the law of worldwide organizations: lessons from European Integration. Eur J Int Law 13(3)

Petroleum (Exploration and Production) Act 1991 (Act 2 of 1991) of Namibia

Practice Note. Understanding Stabilization Clauses in International Investment Agreements: Criticisms of Stabilization Clauses. http://us.practicallaw.com/1-501-7863. Accessed 12 June 2018

Promotion and Protection of all Human Rights, Civil, Political, Economic, Social and Cultural Rights, Including the Right to Development, Protect, Respect and Remedy: A Framework for Business and Human Right, HRCOR, 8th Sess, UN Doc A/HRC/8/5 (2008)

Pryke S (2017) Explaining resource nationalism. Global Pol 8(4):474–482

Resolution on a Human Rights-Based Approach to Natural Resources Governance (2012) African Commission on Human and Peoples' Rights, 51st Session. achpr.org/sessions/51st/resolutions/224

Resolutions. Ref: PAP(2)/RECOMS/(VI). http://www1.uneca.org/Portals/lpi/Documents/Recommendation-Committee-on-agriculture.pdf. Accessed 12 June 2018

Rome I Regulation (EC) No 593/2008

Ruggie J (2011) United Nations Guiding Principles on Business and Human Rights: Implementing the United Nations 'Protect, Respect and Remedy' Framework. UN Document A/HRC/17/31

Torrado PM (2005) Political risk insurance and breach of contract coverage: how the intervention of domestic courts may prevent investors from claiming insurance. Pace Int Law Rev 17

Treanor WM (1995) The original understanding of the takings clause and the political process. Colum Law Rev 95:785–797

United Nations (2005) Press Release, SG/A/93 "Secretary-General Appoints John Ruggie of United States as Special Representative on the Issue of Human Rights and Transnational

Corporations and Other Business Enterprises". https://www.un.org/press/en/2005/sga934.doc.htm. 12 June 2018
United Nations (2012) Final Report of the United Nations Conference on Sustainable Development. 2012. UN Doc A/CONF.216/16. United Nations, New York
United Nations Commission on International Trade Law (1976) UNCITRAL Arbitral Rules. http://www.uncitral.org/uncitral/en/uncitral_texts/arbitration/1976Arbitration_rules.html. Accessed 12 June 2018
United Nations Conference on Sustainable Development (2011) The need for a rights-based approach to sustainable development. https://www.ohchr.org/Documents/Events/Rio20/Rio+20OHCHRPositions20.pdf. Accessed 12 June 2018
United Nations Convention Against Corruption (UNAC) (2005) UN Doc. A/58/422. Adopted 31 October 2003, In force on14 December 2005
United Nations Convention Against Transnational Organized Crime and its Protocols, December 13, 2000, S. Exec. Rep. No. 109-4, 40 ILM 335 (2001); UN Doc. A/55/383 at 25 (2000); UN Doc. A/RES/55/25 at 4 (2001)
United Nations Development Group (2011) UNDG Human Rights Mainstreaming Mechanism: Operational Plan 2011–2013. mptf.undp.org/document/download/7508. Accessed 12 June 2018
United Nations Development Group. About the UNDG. https://undg.org/about/. Accessed 12 June 2018
United Natons Development Programme (2017) Human Development Index and its components. http://hdr.undp.org/en/composite/HDI. Accessed 12 Aug 2018
What Lenders Need to Know About Political Risk Insurance: Conditions for Compensation. Practical Law. http://us.practicallaw.com/9-505-3828. Accessed 12 June 2018
Whitehead F (2011) BAE urged to pay £29.5m fine to Tanzania. The Guardian. http://www.theguardian.com/global-development/2011/jul/22/bae-tanzania-sfo-fine-delay. Accessed 12 June 2018
Wilson J (2015) Understanding resource nationalism: economic dynamics and political institutions. Contemp Polit 21(4):399–416
World Economic Forum (2012) Global Competitiveness Report. http://www3.weforum.org/docs/WEF_GlobalCompetitivenessReport_2012-13.pdf. Accessed 12 June 2018

The manufacturer's authorised representative in the EU is Springer Nature Customer Service Centre GmbH, Europaplatz 3, 69115 Heidelberg, Germany. If you have any concerns regarding our products, please contact ProductSafety@springernature.com

Printed and bound by CPI Group (UK) Ltd, Croydon, CR0 4YY
01/10/2026
02225834-0011